The Economics and Management of World Fisheries

Trond Bjørndal

Gordon R. Munro

OXFORD
UNIVERSITY PRESS

OXFORD
UNIVERSITY PRESS

Great Clarendon Street, Oxford, OX2 6DP,
United Kingdom

Oxford University Press is a department of the University of Oxford.
It furthers the University's objective of excellence in research, scholarship,
and education by publishing worldwide. Oxford is a registered trade mark of
Oxford University Press in the UK and in certain other countries

First Edition published in 2012

British Library Cataloguing in Publication Data
Data available

ISBN 978–0–19–957675–3

OXFORD
UNIVERSITY PRESS

Great Clarendon Street, Oxford, OX2 6DP,
United Kingdom

Oxford University Press is a department of the University of Oxford.
It furthers the University's objective of excellence in research, scholarship,
and education by publishing worldwide. Oxford is a registered trade mark of
Oxford University Press in the UK and in certain other countries

First Edition published in 2012

British Library Cataloguing in Publication Data
Data available

ISBN 978–0–19–957675–3

The Economics and Management of World Fisheries

Trond Bjørndal

Gordon R. Munro

OXFORD
UNIVERSITY PRESS

PREFACE

There has, over the past several decades, been increasing interest in, and concern over, the economic management of the world's capture fishery resources. The recent World Bank/FAO (2009) report, which maintains that the world is losing US$50 billion in resource rent per year because of inadequate management of these resources, provides but one example. With that interest and concern in mind, this book attempts to draw together the current economic theory of the management of fishery resources at both the national and the international level. In so doing the book examines, not only the theory itself, but also the empirical application of the theory, and the relevance of the theory to policy issues, affecting both developed and developing fishing states.

There are several underlying themes to the book. The first is the management difficulties created by the inherent common pool nature of the fishery resources. This is an old theme that goes back to the origins of modern fisheries economics. While the management of world capture fisheries has undergone a revolution since that time, the common pool nature of the resources continues to bedevil capture fisheries management. All is not gloom, however. While we can and do point to many capture fisheries management disasters, there are some important success stories as well. It is essential that we learn the lessons from the successes, as well as the disasters.

The second underlying theme is that the management of capture fishery resources should necessarily be seen as a problem of asset management through time. Along with the vast majority of natural resource/environmental economists, and international bodies such as the World Bank, we see capture fishery resources, along with all other natural resources, as natural capital assets. Hence, the economics of fisheries management is first and foremost an application of the economist's theories of capital and investment. Thus, the call from the World Bank, the OECD, and others for a massive rebuilding of world capture fishery resources is to be seen as a call for a massive natural resource investment programme.

The first consequences of taking this approach is that the underlying biology of the fishery resources must be recognized explicitly, hence the emphasis on *bioeconomics*. Every economic model of the fishery, we argue early on, has a biological model of the fishery as its foundation. Misspecify the underlying biological model, and the economic model of the fishery will, for policy purposes, be worthless at best.

The economist's theories of capital and investment are by definition, dynamic, in that they take time into account explicitly. The second consequence, therefore, is that we are continually being forced to recognize that fisheries management is subject to irreducible uncertainty through time. As the Great Recession has reminded us so painfully, those who believe that they can safely ignore uncertainty stand an excellent chance of being devoured by it.

A third theme is the strategic interaction among those exploiting the fishery resources, both at the individual level, and, in the case of international fisheries, at the state level. As a result, economists, in developing the economics of fisheries management, are being compelled more and more to blend the theories of capital and investment with the theory of strategic interaction, more popularly known as the theory of games.

The book is written primarily for those in the early stages of their postgraduate academic career. It is our hope, however, that the book, putting the mathematics to one side, will be accessible to a much broader audience.

It will be clear to the reader that we have drawn heavily upon the work of a large number of our colleagues. It is not possible, at this stage, for us to recognize them all individually. We do, however, want to recognize the assistance and the inspiration given to us by our colleague and friend, Professor Colin Clark, whose seminal book, *Mathematical Bioeconomics*, is now in its third edition. We see our book, not as a competitor, but rather as a complement, to that seminal work.

Since 2007, the authors have periodically mounted an intensive postgraduate course on fisheries economics at CEMARE, University of Portsmouth, based largely on material that we have attempted to cover in this book. We owe a great debt to the students in the course. Their comments and questions have forced us to sharpen our thinking and to do extensive re-drafting.

We must, finally, express our gratitude to Mr Christopher Martin, for his invaluable research assistance, and to Miss Megan Bailey for her skill in producing many of the figures in the book.

Trond Bjørndal and Gordon Munro

CONTENTS

LIST OF FIGURES

LIST OF TABLES

1 Introduction: Overview of World Capture Fisheries and the Foundations of Fisheries Economics

1.1 World capture fisheries: an overview

World fish production has expanded tremendously in the post-1950 period, as illustrated in Figure 1.1. Capture fisheries—marine and inland (fresh water)—stood at 18.7 million tonnes in 1950. In the following decades, catches increased from year to year, until the mid-1980s. Since then world catches have reached a plateau and levelled off at around 85–95 million tonnes per annum although there are substantial variations from year to year. Aquaculture, on the other hand, was of very limited importance in 1950. Production took off in the 1980s, with very impressive growth rates in the past two decades.

In 2010, the fisheries[1] of the world, including aquaculture, produced total harvests in the order of 145 million tonnes[2] per annum, having an estimated 'first sale' value of just under US$190 billion. Of these totals, aquaculture or fish farming, accounted for 58 million tones or 38 per cent of the total, and just over 50 per cent by value. Capture, or wild, fisheries accounted for 87 million tonnes (62 per cent of the total) and just under 50 per cent of value (FAO, 2010).

We can add, in passing, that over this time period there has been a shift in capture fisheries production from developed to developing fishing states. This is partly a consequence of the fact that the introduction of 200 mile Exclusive Economic Zones gave developing countries access to much larger fishery resources, as will be discussed below. In Table 1.1, the current top 12 capture fishery producers are listed, with annual catches for 2000–9. As can be seen, eight out of 12 are developing countries. In the past decade, China has every year been the largest producer in the world with an output of almost 15 million tonnes in 2009, representing almost 17 per cent of the world total. Peru is the second largest producer, however, with landings that exhibit substantial volatility from year to year.[3]

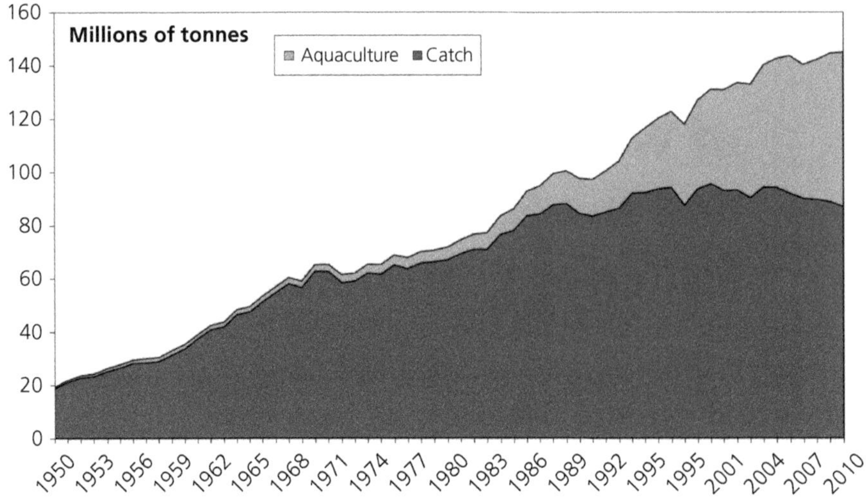

Figure 1.1. World fish production per year, capture and aquaculture (1950–2010)
Source: FAO statistics, 2011.

Although fish markets will not be analysed in this book, it is appropriate to point out that there is substantial international trade in fish. The export value of fish—capture and farmed—has shown tremendous increase in recent decades, from almost US8 billion in 1976 to US$98 billion in 2009 (nominal values; see Figure 1.2). What is also important is that developing countries represent more than 50 per cent of fish exports. This is due to the fact that, as a consequence of the introduction of 200 mile Exclusive Economic Zones developing countries gained property rights to a larger share of marine fish resources. In addition, developing countries are very important in terms of farmed fish production. In terms of fish consumption, it can be noted that three main markets—the EU, the USA, and Japan—combined represent slightly more than two thirds of world fish imports.

In 2010, world per capita food supply represented 17.3 kg, with 9.1 kg from capture fisheries and 8.2 kg from aquaculture.[4] Due to the substantial increase in aquaculture production, world per capita supply of fish has been maintained, or slightly increased, in recent decades, despite substantial population growth.

While aquaculture is unquestionably important, and indeed is likely in the future to actually surpass the production of capture fisheries, this book will confine itself, almost entirely, to the economics of capture fishery management. The current and ongoing importance of world capture fisheries is not in dispute. More to the point, so complex is the economics of capture fishery

Table 1.1. Top 12 producers (capture only) of fish and fishery products (tonnes)

Country	2000	2001	2002	2003	2004	2005	2006	2007	2008	2009
China	14,648,606	14,176,195	14,173,776	14,347,274	14,464,803	14,588,940	14,631,018	14,659,036	14,791,163	14,919,596
Peru	10,657,265	7,986,207	8,765,189	6,086,058	9,604,527	9,388,488	7,017,502	7,210,545	7,392,096	6,914,452
Indonesia	4,080,847	4,259,343	4,322,764	4,627,149	4,643,893	4,695,977	4,800,621	5,050,340	5,002,333	5,099,355
USA	4,717,638	4,944,336	4,937,305	4,938,956	4,959,826	4,892,967	4,852,284	4,767,596	4,349,853	4,222,052
India	3,666,427	3,777,092	3,736,603	3,712,149	3,391,009	3,691,362	3,844,837	3,859,293	4,099,227	4,053,241
Japan	5,072,718	4,713,773	4,373,840	4,680,360	4,330,029	4,312,113	4,318,136	4,277,682	4,323,590	3,847,017
Russian Federation	3,973,535	3,628,683	3,232,351	3,281,510	2,941,595	3,197,686	3,284,306	3,454,218	3,383,724	3,826,129
Chile	4,300,218	3,797,546	4,271,417	3,612,075	4,926,810	4,328,316	4,160,741	3,819,303	3,554,808	3,453,786
Burma	1,093,200	1,187,880	1,284,340	1,343,860	1,586,600	1,732,250	2,006,790	2,235,580	2,493,750	2,766,940
Philippines	1,896,149	1,948,823	2,029,914	2,165,876	2,211,301	2,269,721	2,319,031	2,499,680	2,561,237	2,602,454
Norway	2,699,365	2,686,942	2,740,475	2,548,803	2,524,377	2,392,594	2,256,405	2,380,425	2,431,371	2,524,437
Viet Nam	1,623,312	1,724,758	1,802,598	1,856,105	1,879,488	1,929,900	1,970,600	2,020,400	2,087,500	2,243,100
TOTAL	**93,543,412**	**90,775,621**	**90,968,368**	**88,262,381**	**92,444,505**	**92,190,236**	**89,886,690**	**90,073,134**	**89,587,030**	**88,918,060.10**

Source: FAO statistics, 2011.

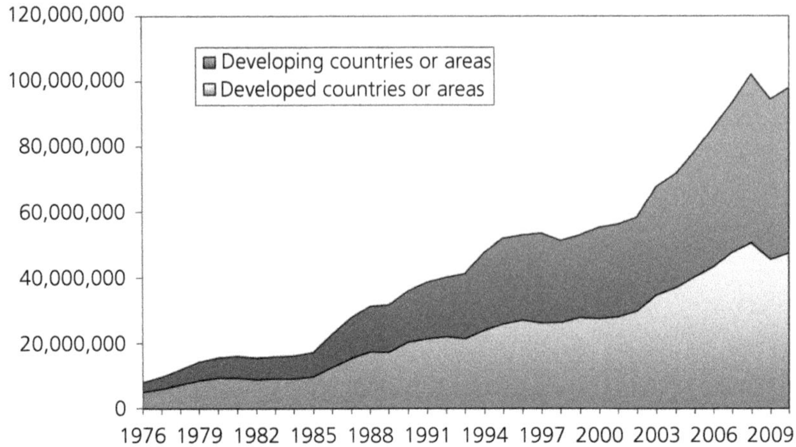

Figure 1.2. World fish trade in \$, export value, developing and developed countries (1976–2009)

Source: FAO statistics, 2011.

management that it is simply not possible to do justice to the economics of both capture fishery and aquaculture management in one book. The reader interested in the economics of aquaculture management is directed to the volume on this subject by Bjørndal and Asche (2011).

When we talk of world capture fisheries, we shall, for the most part, be talking of marine (ocean) capture fisheries. Inland capture fisheries account for just slightly more than 10 per cent of capture fisheries production (harvests) (FAO, 2010).[5] It has been estimated that marine capture fisheries, directly and indirectly, may account for the employment for over 120 million men and women, with most of these being in the developing world (World Bank, 2009).

The fact that capture fisheries production has reached a plateau is no great cause for concern. The capacity of the aquatic environment to support capture fishery resources is, after all, finite. What is of concern, however, is the growing evidence of excessive depletion of ocean capture fishery resources.

At a later point, we shall be introduced to the concept of maximum sustained yield—MSY—the maximum harvests that a fishery resource can support, year in and year out, on a sustainable basis. Using MSY as the appropriate criterion, the Food and Agriculture Organization of the UN (FAO) categorizes ocean capture fishery resources as: (i) underexploited to moderately exploited; (ii) fully exploited; and (iii) overexploited. A fishery resource being exploited at a level yielding the MSY is deemed to be *fully exploited*; a fishery resource driven down below the level that yields the MSY is deemed to be *overexploited*.

If we return to the first half of the 1950s, we find, according to the FAO, that no more than 5 per cent of the world's marine capture fishery resources were overexploited. By the mid-1970s, this figure had doubled, with 10 per cent falling into the overexploited category. According to the latest FAO data, over 30 per cent of world capture fishery resources are now overexploited. The percentage in the fully exploited category stands at 50 per cent (FAO, 2005, 2010). There is a lingering concern that a good many of the *fully exploited* resources are, in fact, candidates for future inclusion in the overexploited category.

When we look at the contribution that world capture fishery resources are making to the GDPs of the world, the results are even more depressing. A recent study by the FAO and World Bank, *The Sunken Billions* (World Bank, 2009), talks in terms of 'resource rent', a concept that we shall examine at a later point. For the time being, think of 'resource rent' as a measure of the net economic benefits forthcoming from a fishery to society.

The Sunken Billions estimates that, if world ocean capture fishery resources were managed optimally, they would yield 'resource rents' in the order of US $50 billion per annum. The study continues that these same resource are currently yielding no 'resource rents' whatsoever (World Bank, 2009). While there are capture fisheries that are yielding positive economic returns to society, there are others that are actually making a negative contribution to the GDPs of the states exploiting the resources. Recall that developing states, playing an increasingly important role in world fisheries, are states that simply cannot afford the luxury of non-producing natural resource assets.

How did this deplorable state of affairs arise? We shall argue that economics, and in particular the economic incentives confronting those who exploit the resources, lies at the heart of the problem, and further that economics lies at the heart of the solution.

To begin, capture fishery resources are, in the best of circumstances, difficult to manage. With few exceptions, capture fishery resources are mobile, and are not visible prior to capture. Furthermore, the relevant species interact, as competitors for food sources, or in predator–prey relationships, and are, in addition, subject to environmental shocks, all of which are not readily observable, let alone measurable or predictable. Finally, a given stock of a single species may be spread out spatially, and be better thought of as a set of interconnected stocks.[6]

The fact that capture fishery resources are mobile, and difficult to observe prior to capture has meant that, in the past, it has been very difficult, or more to the point costly, to establish effective property rights to these resources, be the property rights private or public in nature. As a consequence, the resources have, historically, been exploited on a 'common pool' (i.e., open to all) basis. The 'common pool' nature of the resources has, as we shall argue, created a set of economic incentives for fishermen that are highly perverse from society's point of view, leading inevitably to resource overexploitation and to economic waste. The 'common pool' scourge has come to affect both developed and developing fishing states, regardless of the nature of the political regimes of these states (Munro, 2010a). It will be seen that capture fishery management measures, both national and international, have been dominated by attempts to counter the baleful impact of the 'common pool' nature of the resources.

Prior to our undertaking to explore fisheries economics in depth, and to gain insight into the destructive nature of the 'common pool' nature of the resources in question, we have to deal with two preliminary tasks. The first is to examine, in rough outline, at least, the international legal framework within which world marine capture fisheries operate. We cannot make sense of the economics of capture fisheries management, unless we are cognisant of the international legal framework surrounding world capture fisheries. We shall see, for example, that the 'common pool' nature of the bulk of world marine capture fishery resources was, for centuries, actually enshrined in international law.

The second is to comment on the capital theoretic-biological foundations which underpin modern fisheries economics. It will be seen that, if we take a capital-theoretic approach to fisheries economics, as indeed we shall, we are compelled to enquire into the appropriate biological model for the resource(s) in question. In fact, we shall go so far as to assert that every economic model of the fishery, worthy of the name, has a biological model of the resource as its foundation. Mis-specify the underlying biological model, and the economic model of the fishery is certain to be of little or no value, or worse, may give rise to seriously misleading predictions.

With all of this in mind, let us turn to the first of our two preliminary tasks.

1.2 The international legal framework for world marine capture fisheries

Capture fishery resources contained within inland, or fresh, waters constitute the property of the state within which the waters lie. If the state shares the inland waters with another state, e.g., Lake Superior shared by Canada and the United States, then fishery resources crossing the boundary between the two states can be deemed to be jointly owned. Matters are not so simple when we come to marine capture fishery resources.

Prior to the end of the Second World War, international law recognized two broad types of marine areas, the coastal state (a state having significant marine coastline) territorial sea and the high seas (Orrego Vicuña, 1999).

Within the territorial sea, the coastal state had full sovereignty, subject only to right of innocent passage of foreign vessels. The coastal state's ownership of the resources (including fishery resources) contained therein was deemed to be absolute (McRae and Munro, 1989b). The territorial sea was, and is, however, very narrow. Traditionally it extended to no more than three nautical miles.[7] Today, the outer limit of the territorial sea is taken to be 12 nautical miles from shore (UN, 1982, Article 3). All marine waters, lying beyond the territorial seas, constituted the high seas.

Within the high seas, the principle of the freedom of the high seas prevailed, a principle first propounded in 1608 by the Dutch jurist Hugo Grotius. Under this principle, the resources of the high seas were to be seen as *res communis*, the property of all. Vessels of a given state should be allowed to navigate freely on the high seas, without interference from other states. The same applied to fishing on the high seas, in that it was an activity that was to remain free and open to all (Orrego Vicuña, 1999).

As Orrego Vicuña states, in his book on the international law of high seas fisheries, the principle of the freedom of the high seas rested upon two basic premises: (i) the impossibility of the high seas being subject to effective occupation; (ii) the inexhaustible nature of marine resources (Orrego Vicuña, 1999, p. 4). Both of these premises, as they applied to fisheries, became increasingly untenable over time. This, in turn, led to an ongoing diminution of freedom of the seas applied to fisheries, a diminution which has by no means ceased.

It may seem incredible, given what we have said about the overexploitation of world capture fishery resources, that marine capture fisheries could ever have been seen to be inexhaustible. One has to bear in mind, however, that, in marine fisheries, distance from shore does matter, and does so for obvious reasons. The farther one is out from shore, the more difficult and costly fishing becomes. In the time of Grotius, in the early seventeenth century, there was concern about the state of specific local inshore fishery resources. The great ocean fishery resources, on the other hand, did seem to be inexhaustible, because, given the then state of fishing technology, fishing far from shore was a costly, not to say dangerous, undertaking. The belief that the ocean fisheries were inexhaustible continued until late in the nineteenth century. In 1883, at the London exhibition on fisheries, one of Britain's leading scientists, Thomas Huxley, stated that 'probably all the great sea fisheries are inexhaustible—and any attempt to regulate these fisheries seems consequently—to be useless' (cited in Gordon, 1954, p. 126).

As Huxley spoke, however, fishing technology was changing and harvesting costs with them. The shift from sail to steam is the prime example. As harvesting costs fell the great sea-fishery resources lost their economic protection, and the realization gradually began to take hold that offshore fishery resources were not inexhaustible after all. It was noted by marine biologists that, after both world wars, fishery resources in the North Sea increased in abundance. The marine biologists, putting two and two together, surmised that the growth in the fishery resources was not unrelated to the sharp decline in fishing activities due to wartime conditions. They concluded that: (i) fishing did have an impact upon offshore fish stocks, and (ii) fish stocks could recover, if fishing was curtailed (National Research Council, 1999).

The growing realization that high seas fishery resources are exhaustible led gradually to curbs being placed upon the freedom to fish in the high seas, this being done initially through international agreements. An example is provided by the International Commission for the Northwest Atlantic Fisheries (ICNAF) established by Canada, the United States, and a few European countries in 1949. ICNAF attempted to encourage proactive management of high seas fisheries from the southern boundary of Greenland down to the US mid-Atlantic. ICNAF was originally a coordinator of management measures among members, with the regulations initially being confined to gear, but later being extended to harvest quotas.

The efficacy of fisheries management through international organizations became subject to question. Those who raised such questions (coastal states in particular) were to be given an alternative through the United Nations. Following the close of the Second World War, coastal states (beginning with the US) began asserting unilaterally claims of jurisdiction over marine resources off their shores, beyond the territorial sea. In response to what it saw as a disorderly attempt on the part of coastal states to extend their areas of marine jurisdiction, the United Nations convened an international Conference on the Law of the Sea in 1958. From a fisheries standpoint, the Conference resulted in coastal states being granted jurisdiction over shellfish resources out to the edge of the continental shelf,[8] but nothing more (Logan, 1974).

The 1958 conference was followed by the Second UN Conference on the Law of the Sea in 1960, which accomplished little or nothing, and then by the Third UN Conference on the Law of the Sea, 1973–82, which accomplished a great deal. From the perspective of fisheries, it can be said that the UN Conference of 1973–82 led to a revolution in the management of world marine capture fisheries.

The UN Conference of 1973–82 brought forth a convention, the UN Convention on the Law of the Sea (1982 UN Convention hereafter) (UN, 1982).[9] The key part of the 1982 UN Convention, as far as fisheries are concerned, is Part V, The Exclusive Economic Zone. Part V of the Convention grants coastal states the right to establish Exclusive Economic Zones (EEZs) out to 200 nautical miles from shore. Within the EEZ, the coastal state has 'sovereign rights for the purpose of exploring and exploiting, conserving and managing... the fishery resources contained therein' (UN, 1982, Article 56).

To all intents and purposes, the coastal state has property rights to the intra-EEZ fishery resources (McRae and Munro, 1989b; World Bank, 2009).[10]

If the coastal state has effective property rights to fishery resources within the EEZ, then the next, and obvious, question to raise is the significance of the EEZ regime to world capture fisheries. The answer is that this regime is highly significant. The EEZ regime has now become all but universal. In the mid-1970s, when the EEZ regime was clearly on the horizon, it was estimated that over 90 per cent of the world commercial marine capture fishery harvests, by volume, were taken within 200 nautical miles from shore (Alexander and Hodgson, 1975, p. 586). Hence, the advent of the EEZ regime could be seen as having effected a change in the status of a massive amount of renewable resource wealth from that of 'common pool' high seas resources, to that of coastal state property, which brought with it a corresponding radical reduction in high seas fisheries.

The New Law of the Sea, as it is commonly referred to, while having a dramatic impact on the status of world marine capture fishery resources, did bring with it its own set of resource management problems. First, it was hoped that the change in status of marine capture fishery resources from high seas 'common pool' resources to coastal state property would mitigate the 'common pool' problem, plaguing the management of such resources (Eckert, 1979). The problem may have been mitigated, but it was far from having been eliminated. As history has demonstrated painfully, it is one thing for a coastal state to be granted legal property rights to capture fishery resources within its EEZ. It is quite another for the coastal state to exercise these rights effectively, or to devolve these rights upon individuals, or groups, within the state. The 'common pool' problem is still very much with us.

Secondly, the mobility of capture fishery resources has meant that many coastal states find that fishery resources within the EEZ cross the EEZ boundary. The FAO uses the term 'shared' fish stocks to describe such resources. As we shall point out in chapter 7, there are two broad, non-mutually exclusive, categories relevant to coastal states: (a) transboundary stocks—fishery resources crossing the EEZ boundary into the EEZs of one or more coastal states; (b) straddling fish stocks—fishery resources crossing the EEZ boundary into the remaining high seas (Munro et al., 2004).[11] The two categories of 'shared' fish stocks account for just under one third of marine capture fishery harvests (Munro et al., 2004).

One might have thought that category (b), straddling stocks, would be of little interest, since the fishery resources beyond 200 miles appear, from our earlier comments, to be trifling. As we shall remark in chapter 7, this was indeed the view taken at the close of the Third UN Conference of the Law of the Sea in 1982. Then, for reasons to be examined in chapter 7, the management of these fish stocks achieved crisis status in the decade to follow, compelling the UN to convene another international conference in the early

1990s, popularly referred to as the UN Fish Stocks Conference, 1993–5. The Conference brought forth an agreement, popularly referred to as the UN Fish Stocks Agreement (1995 UNFSA) (UN, 1995).[12] The purpose of the 1995 UNFSA was, and is, to supplement and buttress the 1982 UN Convention. Under the 1995 UNFSA, coastal states and distant water fishing states (DWFSs)[13] exploiting straddling stocks are to come together to manage these resources through Regional Fisheries Management Organizations (Munro et al., 2004; UN, 1995). The Regional Fisheries Management Organization (RFMO) regime is, at the time of writing, at an early stage in its evolution. The economics of the RFMO management of straddling fish stocks will be a major topic for our consideration in chapter 7.

There is an ancillary issue raised by the New Law of the Sea, namely economic relations between coastal states and DWFSs. There is no really effective legal constraint on coastal states barring DWFSs from their EEZs (see n. 10). It may, however, be to the economic benefit of coastal states to grant such access. This is an issue that will be discussed in some detail, also in chapter 7.

To conclude, there are two complementary legal documents, both of which have come into force as international treaty law, that provide the 'rules of the game' for the international management of marine capture fishery resources. They are: the 1982 UN Convention on the Law of the Sea and the 1995 UN Fish Stocks Agreement (UN, 1982, 1995).

We now turn to the second of our preliminary tasks.

1.3 Capital theory, biology, and the foundations of modern fisheries economics

It is a fundamental proposition that the economics of the management of natural resources, be the resources renewable or non-renewable, is essentially an application of the economist's theory of capital and investment. This proposition was set forth with great force and clarity by Anthony Scott, over half a century ago, in his book: *Natural Resources: The Economics of Conservation* (Scott, 1955; 1983).[14]

Scott states that: 'in the pure theory of capital, [real] capital may be defined as the aggregate of things which are capable of yielding a stream of services extending into the future' (Scott, 1983, p. 14). Natural resources clearly fall within this definition, and differ from human-made or 'produced' capital, e.g., buildings, machinery, in that they come to us as endowments from nature.[15] As Scott emphasizes, the origin of such 'things' is irrelevant for their capacity to perform as capital (Scott, 1983, p. 15).

The idea that natural resources should be regarded as natural capital has now gained wide acceptance, well beyond the realm of academia. The World Bank report, *Where is the Wealth of Nations?* (World Bank, 2005), puts forth the principle, which can be traced back to Hicks (1946), that both the current national income, and the future prospects for development, of any nation rest upon that nation's portfolio of real capital assets. The portfolio is seen to consist of produced, natural and intangible (human and social) capital. Development is thus to be viewed as an exercise in real capital asset portfolio management through time (World Bank, 2005, pp. 1–5).

The World Bank report goes on to point out that natural capital is of particular importance in the real capital asset portfolios of developing nations. The net economic returns from the natural capital (loosely referred by the World Bank as resource rent) play two key roles in development (World Bank, 2005, p. 8):

- providing the basis of subsistence, particular in the poorest nations;
- providing a source of development finance, by furnishing the wherewithal for investments in other forms of capital, e.g., produced and human capital.

One of the natural capital assets to be managed in this real asset capital portfolio consists of capture fishery resources. These resources are renewable natural resources in that they are capable of growth. This means that, like other renewable natural resources such as forests, but unlike non-renewable resources, such as coal deposits, it is possible, within limits, to engage in positive investment in this form of natural capital by reducing harvesting sufficiently to allow the resources to grow.[16] Given the extensive depletion of capture fishery resources over the past several decades, there is now considerable discussion, on both a national and international basis, on the necessity of rebuilding hitherto depleted fishery stocks. Thus, for example, the World Bank/FAO study *The Sunken Billions* has a section entitled: 'Rebuilding Global Fish Capital', within which it is stated that 'there is enormous potential to rebuild global fish stocks and wealth' (World Bank, 2009, p. 50).[17] What, in fact, is being called for is a programme of positive investment in the natural capital that we call capture fishery resources.

The economics of fisheries management is then, at its heart, addressing a problem of (real) capital asset management, in that it is concerned with the management of a set of real capital assets, contained within society's overall portfolio of real capital assets, with the objective of maximizing the economic benefits, which these fishery assets are capable of providing to society, through time.

The initial economic models of the fishery that we shall examine are focused on single species fishery resources, or assets. We have already noted, however, that fisheries species interact with one another. We have also noted that many fish stocks must be thought of in spatial terms, as sub-stocks spread out over an area. This causes us no conceptual problems. It means rather that we cannot remain content with single resource asset models, but must go on to think in terms of *portfolios*, or perhaps one should say, *sub-portfolios*, of resource assets to be managed through time.

To this point, we have talked rather loosely about the theory of capital and investment. Ackley (1978) insists that one must make a distinction between the theory of capital and the theory of investment, even though the two theories are obviously closely related. We shall adopt the Ackley distinction.

The theory of capital, argues Ackley, is about stocks, and is concerned with determining the optimal stock of capital over time. The theory of investment, on the other hand, is about flows, and is concerned with the optimal rate at which the stock of capital is built up (or run down), as we approach the optimal stock level (Ackley, 1978).

Earlier, we referred to fish stock restoration programmes as investment programmes. Let us now take a specific example. Under its Sustainable Fisheries Act of 1996,[18] the United States is to undertake to rebuild depleted fish stocks under US jurisdiction to levels yielding the Maximum Sustainable Yield (MSY) levels. Once a stock has been determined to have been overexploited, the act mandates that the stock is to rebuilt within as short a time as possible, a period not to exceed 10 years (unless biology dictates a longer period) (Safina et al., 2005; Khwaja and Cox, 2010). The optimal stock level for this set of natural capital assets is the level yielding the MSY (theory of capital). The 10-year 'window' sets out the 'optimal' rate of investment in the natural capital (theory of investment).[19]

The rate of investment is not, in many cases, anything like the maximum. According to the scientists, many of the stocks could be rebuilt to the target levels within five years. Nonetheless, there is considerable controversy over the rate of investment, namely whether the investment rate, the approach to the target levels, is too rapid (Safina et al., 2005; Khwaja and Cox, 2010). Why this controversy should have emerged is a question that we shall explore in some detail in the chapters to come.

What sort of rule should one have in mind, in attempting to answer our capital theory question, namely the size of the optimal stock (biomass)? As noted, we shall, in the next few chapters, commence with the case of a single species, being managed by a social manager. In so doing, we must take into account, as Scott (1955) has emphasized, that investment in a particular natural resource is only one of many investment opportunities open to society. Recall our discussion of the World Bank (2005) report, *Where is the Wealth of Nations?* Since, at any given time, society's pool of saving is strictly finite, investment in the fishery natural resource will come at the expense of other investment opportunities.[20] In other words, there is an opportunity cost to be faced.

For the social asset manager, confronting a very broad asset portfolio, to have achieved an optimal portfolio mix, one condition to be satisfied is that all assets (of a common risk class) must be seen to be yielding a common rate of return. If this condition is not met, then some portfolio rebalancing is in order. We shall refer to this common rate of return as the social interest, or discount, rate. Thus, if the rate of return on a marginal investment in a fishery resource exceeds the social rate of discount (interest), then additional investment in the resource is clearly warranted. Hence, essentially our rule will be to invest (positively or negatively) in the fishery resource up to the point that the rate of return on the investment in this form of natural capital is equal to the social rate of discount. How does one measure the rate of return on the marginal natural capital investment? This is an issue to be discussed in detail in the following chapters.

Matters become more complex when there is species interaction that cannot be ignored, so that our single species models are no longer adequate. The problems raised thereby are difficult, but not insurmountable.

With regards to the question of the optimal rate of investment in a fishery resource(s), the question is basically how quickly one should approach the optimal stock target(s). As a general rule of thumb, the answer is: approach the target(s) with all possible speed, unless there are penalties associated with rapid rates of investment. If there are such penalties, then a more gradual approach, a slower rate of investment, is in order. Thus, for example, if positive investment in a fishery resource is called for, the maximum rate of investment in the resource will be achieved by declaring a harvest moratorium, i.e., closing the fishery down until the optimal stock level is achieved. If the declaration leads to havoc being wreaked in the affected fishing fleets and communities, then obviously there are penalties associated with the policy of most rapid rate of investment in the resource stock.

The following chapters will devote considerable time to deterministic economic models of the fishery, i.e., models in which it is assumed that the future is known with certainty. While such models are particularly useful for purposes of exposition, we cannot, and must not, rest content with such models. Asset portfolio management in the real world, be the assets real or financial, is always carried out under conditions of uncertainty. The problem of uncertainty in the management of capture fishery resource assets will be found to be exceptionally severe.

We next restate the assertion made earlier in the chapter, namely that every economic model of the fishery worthy of the name has a biological model of the fisheries as its foundation. Indeed, taking a capital theoretic approach to the economics of the fishery, forces us to examine the underlying biological model explicitly. We can also restate, without fear of contradiction, that, if the underlying biological model is mis-stated, we can easily come forth with an economic model of the fishery that is positively dangerous for policy purpose.

So close is the link between economics and biology in fisheries economics that we now talk in terms of *bioeconomics*.[21]

In the chapters to follow, we shall make extensive use of two biological models of the fishery, one developed by the American marine biologist, M. B. Schaefer, and the other developed by two British marine biologists, R. J. H. Beverton and S. J. Holt. It is, therefore, appropriate in introducing the biological underpinnings of fisheries economics models to turn to a survey paper on fisheries biology, written for non-biologists, by Schaefer and Beverton (1963). The paper is within sight of its 50th anniversary, but it remains very useful for our purposes.

In following Schaefer and Beverton, consider a fishery based upon a single species that is confined to a limited area, and which does not interact significantly with any other species. We measure the resource in terms of its biomass, the amount of fish expressed in terms of weight. The focus is on the 'fishable biomass', that part of the biomass subject to exploitation by fishing vessels, given the current state of fishing technology and regulations. A fishable biomass will grow as a consequence of:

(a) recruitment—entry of new fish to the biomass. Fish reproduce, and some fraction of the progeny survive to fishable age, and thus 'recruit' to the fishable biomass;
(b) growth of individual fish—individual fish in the fishable biomass experience, within limits, an increase in body weight as they grow and mature.

The growth of the fishable biomass will be kept in check by:

(c) natural mortality—fish die as a result of age, disease, environmental factors, or natural predators;
(d) fishing mortality—fish are removed by a particular class of predators known as fishermen.

If fishing mortality is absent, the fishable biomass will approach a natural equilibrium level at which: (a) + (b) are just offset by (c).

Denote the fishable biomass by x, measure in weight. The rate of growth of the biomass can be expressed as follows (Schaefer and Beverton, 1963):

$$\frac{\dot{X}}{X} = z(X) + g(X) - M(X) - f(E) + \mu \tag{1.1}$$

where $\dot{X} \equiv \frac{dX}{dt}$ is the time rate of change in the biomass, and where z, g, M, and f, respectively denote the rates of recruitment, growth of fish within the biomass, natural mortality, and fishing mortality. The term μ is a random variable with mean zero. z, g, and M all depend on biomass, while fishing mortality depends on 'fishing effort', denoted by E, which we, as economists, can interpret as the combined labour, produced capital, and ancillary services devoted to harvesting. It is commonly measured in terms of standardized fishing units (Clark, 1985), on a per period of time basis, e.g., fishing days.[22]

The biomass will tend towards equilibrium when (recalling that μ has a mean zero)

$$f(E) = z(x) + g(x) - M(x) \tag{1.2}$$

$$f(E)x = [z(x) + g(x) - M(x)]x \tag{1.2a}$$

that is to say, when the rate of fishing mortality is equal to the net natural rate of growth of the resource, or when harvesting is just skimming off the net natural growth of the resource in absolute terms.

Ideally, Schaefer and Beverton argue, marine biologists would like to be able to measure all of the rates of growth and mortality entering into equation (1.2). This has proven to be extremely difficult, with the consequence that the biologists have had to introduce various simplifications. This has resulted in two broad, general approaches being taken, commonly referred to as the Beverton-Holt approach, and the 'general production approach'.

In the Beverton-Holt model, a fish stock consists of a number of year classes, also known as age classes or cohorts. Each year class is described by parameters such as the number of fish, growth, and natural mortality. The model is 'updated' annually, when a new year class is recruited to the stock. As is easily understood, the model is very data intensive. The Beverton-Holt model has presented economists with intractable difficulties, when used as a foundation for analytical economic models of capture fisheries. This is so, even when recruitment is assumed to be constant. On the other hand, it has been used very effectively for the (biological) management of fisheries. The International Council for the Exploration of the Sea (ICES), a scientific organization that gives advice on the management of virtually all major fish stocks in the North Atlantic, uses the Beverton-Holt model as foundation for stock assessments.[23] The model is also used on an ad hoc basis for empirical economic models of capture fisheries, as will be seen in subsequent chapters.

The 'general production' models are lumped parameter models in the sense that no attempt is made to distinguish among the factors determining net natural growth. These models have served economists very well, as foundations for analytical economic models of capture fisheries. Let us take one example, the Schaefer model, which has been much used by fisheries economists. We shall, in preparation for the discussion of capture fisheries economic models in the following chapter, provide the 'bare bones' of the model. Those wanting an examination of the model in greater detail are encouraged to turn to Clark (2006, 2010).

The Schaefer model rests upon the logistic growth model, developed by P. H. Verhulst in the mid-1830s. The growth of a particular population is assumed to be limited by the finite nature of the environment in which it exists. The Schaefer model is as follows:

$$\frac{dx}{dt} = F(x, \bar{A}) \tag{1.3}$$

where x, as before, is the biomass, and $A = \bar{A}$ is the aquatic environment, assumed to be constant. Hence, we can rewrite eq. (1.3) simply as:

$$\frac{dx}{dt} = F(x) \tag{1.3a}$$

We next specify the functional form of the growth function, $F(x)$, as follows:

$$F(x) = rx\left[1 - \frac{x}{G}\right] \tag{1.4}$$

where G, a constant, is the carrying capacity of the environment, or natural equilibrium level of the resource, while r, a constant, is the 'intrinsic' (proportional) growth rate of the resource.[24] This is the famous logistic equation of population growth from the Verhulst model (Clark, 2006, 2010). The proportional growth rate $F(x)/x$ is a decreasing function of the size of x. A population model, in which the proportional growth rate is a decreasing function of x is described as a pure compensation model. We illustrate equation (1.4) in Figure 1.3.

Once harvesting is introduced, we have

$$\frac{dx}{dt} = F(x) - h(t) \tag{1.5}$$

where $h(t)$ is the harvest rate.

Note that, if for a given biomass level, $x^{\bullet}$: $F(x^{\bullet}) = h(t)$, then $dx/dt = 0$, which in turn implies that one could harvest at that rate indefinitely, with x

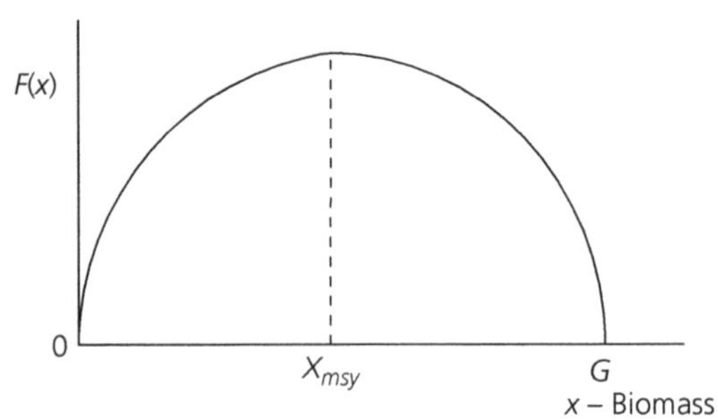

Figure 1.3. The Schaefer model introduced

remaining stable. Hence, $F(x^{\bullet})$ can be seen as the sustainable harvest, or yield, associated with the biomass level, $x^{\bullet}$, and the $F(x)$ curve, in Figure 1.3, can be seen to represent the sustainable harvest, or yield, for all possible levels of x. The maximum point on the $F(x)$ curve ($F'(x)=0$), represents the Maximum Sustainable Yield (harvest), with the corresponding biomass level: x_{MSY}.

The next, and final, step is to introduce the harvest production function:

$$h = qE^{\alpha}x^{\beta} \tag{1.6}$$

where E is, as before, fishing effort, the combined labour, produced capital, and ancillary services devoted to harvesting, while q, a constant, is the 'catchability coefficient', which can be seen as a measure of the state of fishing technology, and where α and β are constants. By assumption, in the Schaefer model, $\alpha = \beta = 1$. The implication of the assumption is that the resource is spread uniformly over the aquatic environment, and continues to be spread uniformly, when the resource is diminished.[25]

The 'bare bones' Schaefer model allows us to illustrate the importance of specifying the underlying biological model of the resource correctly, when constructing our economic model. Focus on what marine biologists refer to as the Catch Per Unit of Effort (CPUE); *h/E*. In the Schaefer model, we have:

$$\frac{h}{E} = qx \tag{1.7}$$

From the perspective of economists, the CPUE can be seen as the average product of fishing effort. In this model, the average product of *E* rises, or falls, in proportion to *x*. If the resource is steadily depleted, the average product of *E* will steadily fall, which, as we shall see, translates into steadily rising average cost of harvesting. Indeed, as *x* goes to zero, the average cost of harvesting goes to infinity. The implication is that, even if the fishery resource is subject to depletion, because of inadequate management, there is, *in the context of this model*, no risk of the resource being driven to extinction.

There are, however, many fisheries for which the Schaefer model is inapplicable. Return to equation (1.6): $h = qE^{\alpha}x^{\beta}$. In the Schaefer model, it is assumed that β and α are equal to one. There are fisheries in which β is not equal to one, but is rather much lower, even approaching zero in some cases (Bjørndal, 1988). In these fisheries, the average product of *E* will not fall in proportion to the decline in *x*. On the contrary, in some of these fisheries, the average product of *E* stays more or less constant up to the point that the resource is threatened with extinction.

If an economic model of a particular fishery rests upon the Schaefer model, when in fact, the average product of *E* will remain more or less unchanged as the resource is depleted, then consequences are stark and obvious. The policy implications of the economic model of the fishery will be dangerously misleading.

The foundations have now been laid. We are now in a position to commence our exploration of economic models of capture fisheries.

BIBLIOGRAPHY

For detailed descriptions and analyses of world capture fishery resources, one should in the first instance turn to the FAO. The starting point is: *The State of World Fisheries and Aquaculture.* The document is published by the FAO on a biennial basis. For those engaged in empirical research on fisheries, the FAO is a key source of data.

This chapter introduces the concept of bioeconomics—the fusion of biological and economic models. A primary source, which will be referred to many times in the chapters to follow, is Colin Clark's, *Mathematical Bioeconomics: The Optimal Management of Renewable Resources.* The book is now in its third edition (2010). The third edition differs significantly from the first (1976) and second (1990) editions. The wise reader will have copies of both the second and third editions on hand.

NOTES

1. For our purposes, we shall define a fishery as the industry, or occupation, devoted to the harvesting, processing and/or distribution of living aquatic resources. These resources consist, in turn, of: (1) fish: finfish/crustacea/molluscs; (2) mammals; (3) plants.
2. Excluding plants.
3. The main species in Peru is anchoveta which, with its short life cycle, exhibits great variation in size from year to year.
4. Source: FAO Food Outlook, June 2011.
5. It should be added that the economics of management of inland capture fisheries resource is virtually identical to that of marine capture fishery resources.
6. The term 'ecosystem management' has become popular in capture fisheries management. At a minimum, ecosystem management means that species interaction is recognized (National Research Council, 1999).
7. 1 nautical mile = 1.852 m.
8. The continental shelf is the relatively shallow seabed extending out from the continent, normally to a depth of no more than 20 metres (see Eckert, 1979).
9. The 1982 UN Convention achieved the status of international treaty law in 1994.
10. To quote *The Sunken Billions* report: 'most marine wild fish resources are considered to be the property of nations' (World Bank, 2009, p. 50).

    There is one apparent major limitation to the coastal state's property rights to these resources, which takes the form of the 'surplus principle', contained in Article 62 of the 1982 UN Convention. Article 62 states that 'where the coastal

state does not have capacity to harvest the entire allowable catch, it shall... give other states [distant water fishing states (see n.13)] access to the surplus of the allowable catch...' (UN 1982, Article 62). The 'surplus principle', it might be added in passing, has been a source of considerable confusion.

The 'surplus principle', impressive as it may sound, is empty. It is the coastal state that determines the total allowable catch and it is the coastal state that determines its own harvesting capacity. As Law of the Sea expert William Burke insists, if the coastal state declares there to be a surplus, it is because the coastal state finds it to be in its interest for there to be a surplus. To quote Burke (1983, p. 30): 'the short of the matter is that the treaty [1982 UN Convention] contains virtually no restriction on coastal state authority to forbid access to foreign fishing'.

If the coastal state does declare a surplus to exist, the 1982 UN Convention permits the coastal state to set a very broad range of terms and conditions on distant water fishing nations seeking access to the surplus. The coastal state is expected to give away nothing for free (UN, 1982, Article 62; Burke 1983, 1994).

11. The UN has an additional category, highly migratory stocks (tuna resources primarily), which cross the EEZ boundary into the remaining high seas. The UN uses the term 'straddling stocks' as a catchall term for all stocks crossing the EEZ boundary into the high seas, other than highly migratory stocks. The UN distinction arose primarily from political factors, and has essentially nothing to do with either biology or economics (see: Munro et al., 2004). We shall, therefore, merge the UN's two categories into one—straddling stocks.
12. The full title of the Agreement is: Agreement for the Implementation of the Provisions of the United Nations Convention on the Law of the Sea of 10 December 1982 Relating to the Conservation and Management of Straddling Fish Stocks and Highly Migratory Fish Stocks. The Agreement achieved the status of international treaty law in 2001.
13. A distant water fishing state can be defined as a fishing state, some of whose fishing fleets operate well beyond the state's EEZ. Prominent examples include Japan, Republic of Korea, and Spain.
14. Scott's book was subsequently revised and reissued. It is upon the reissued version that we shall draw (Scott, 1983).
15. Capture fishery resources clearly come to us as endowments from nature.
16. The World Bank refers to renewable natural resources, such a capture fishery resources, as 'truly a gift of nature' (World Bank, 2005, p. 7).
17. Another example is provided the OECD, which is completing a project on the Economics of Rebuilding Fisheries (OECD, 2010).
18. This act amended the Magnuson-Stevens Fishery and Conservation Act (MSA). The amendments to the MSA of 2007 are also relevant to the stock rebuilding issue (Khwaja and Cox, 2010).
19. Whether the perceived optimal stocks of 'natural' capital, and the perceived optimal rates of resource investment are optimal from an economic perspective, we shall not discuss at this point. We shall return to this American example in later chapters.

20. To quote Scott (1983, pp. vii–viii): 'conservation of resources [investment in resources] is not only analytically analogous to investment in capital goods [human made capital]; it is also in each planning period, an alternative—given an aggregate amount of saving available for investment in each period of time, increased conservation of [investment in] resources must mean also a reduced endowment of buildings and equipment for posterity'.
21. The first reference to bioeconomics we can find is Reinheimer's *Evolution by Co-operation: A Study in Bioeconomics* (1913). This term has been 'popularized' thanks to Colin W. Clark and the first edition of his book, *Mathematical Bioeconomics: The Optimal Management of Renewable Resources* (Clark, 1976).
22. For a practical example, see: Bjørndal and Brasão (2006).
23. See <www.ices.dk> for information about ICES.
24. The stock will grow at the rate of r when stock size is close to zero.
25. In our general discussion of biological models of the fishery, we have referred to the rate of fishing mortality, $f(E)$. Let us note that, in the Schaefer model, the rate of fishing mortality is:

$$f(E) = h/x = qE.$$

2 The Static Economic Model of the Fishery

2.1 Introduction

We come now to our first attempt to analyse the economics of capture fisheries management through the use of models. Before doing so, it is essential to bear in mind that any economic model we come to employ must always be seen as our servant, and never as our master. Clark (2010) makes this point with great clarity. Every model, be it in the social sciences or in the natural sciences, he states, does by necessity abstract from the world of reality. A model that merely duplicated reality would be all but useless in terms of providing insights. Model simplicity, however, comes with a cost. The model may be seriously flawed in some respects in that, while it provides a convincing explanation for some phenomena, it may prove to be very misleading in other ways. No model, he concludes, should ever be used uncritically, and further, 'every modeller—has an obligation to . . . publicize the underlying assumption and limitations of his or her model' (Clark, 2010, p. 3). We can add that, if an economic model comes forth with predictions that cannot be supported by sound economic intuition, we should go back and review the model with a very critical eye.

With this warning and admonition in mind, we turn to the article that marked the dawn of modern fisheries economics, H. Scott Gordon's seminal 1954 article, 'Economic Theory of a Common Property Resource: The Fishery' (Gordon, 1954). The article had, and continues to have, an immense impact upon policy-makers.[1] It merits our close attention.

Gordon's economic model of the fishery rests squarely upon the biological model of Schaefer, discussed in chapter 1. So squarely does Gordon's model rest upon this biological model that Gordon's model is commonly referred to in the literature as the *Gordon-Schaefer model*.

Gordon's economic model of the fishery is basically an application of undergraduate intermediate micro-economics. This gives the model the advantage of simplicity, but it has the disadvantage that the capital theoretic aspects of the problem are essentially suppressed. The suppression is due, not to an oversight on Gordon's part, but rather to the limited mathematical tools that were available in the mid-1950s. Gordon recognized fully that the ideal economic model of the fishery is a capital theoretic one.[2]

Prior to examining the model in detail, we have to undertake a preliminary step. The Gordon model is cast in terms of a relationship much used by biologists, namely that between sustained yield, or harvest, and fishing effort. Return to our discussion of the Schaefer model in chapter 1. There we saw that, if at a given level of x, $h(t) = F(x)$ (see: equation (1.5)), then it is said that the resource is being harvested on a *sustainable* basis. The resource management criterion employed by marine biologists for many decades was to stabilize the resource at the level where the sustainable yield, or harvest, is at a maximum—Maximum Sustainable Yield (MSY), x_{MSY}.

With respect to the Schaefer model, the growth function is (see: equation (1.4)) $F(x) = rx[1 - x/G]$. If harvesting is proceeding on a sustainable basis, then we have:

$$qEx = rx(1 - x/G) \tag{2.1}$$

from which we can derive an expression for x:

$$x = G\left(1 - \frac{q}{r}E\right) \tag{2.2}$$

Substitute for x in the harvest production function, $h = qEx$, and we have:

$$h_s = qGE - \frac{q^2}{r}GE^2 \tag{2.3}$$

where h_s denotes sustainable yield (harvest). We can then re-express equation (2.3) as:

$$h_s = uE - vE^2 \tag{2.3a}$$

where $u = qG$ and $v = q^2/r\,G$. The relationship between E and h_s is shown in Figure 2.1.

We now proceed to a detailed examination of Gordon's model.

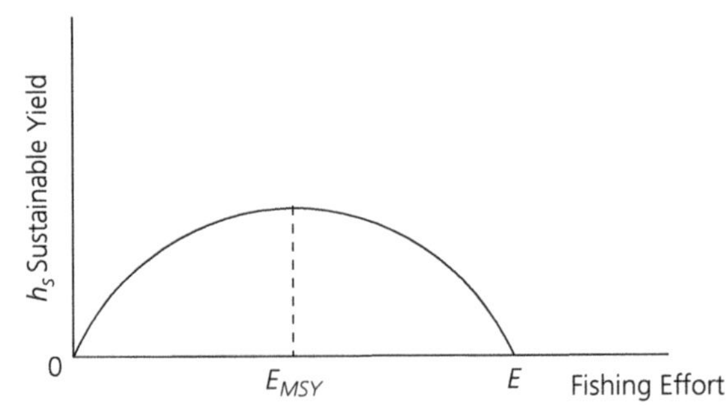

Figure 2.1. Sustainable yield (harvest) and fishing effort arising from the Schaefer model

2.2 **The Gordon-Schaefer model analysed**

Gordon (1954) transforms the sustainable yield (harvest) function into total revenue (with respect to *E*) function, and then introduces a fishing effort cost function. In so doing, he employs the following assumptions:

1. Both the demand for harvested fish and the combined supply of labour, produced capital, and ancillary services constituting fishing effort are perfectly elastic. Thus, the price of harvested fish, *p*, and the unit cost of fishing effort, denoted by *b*, are constants.
2. The price of harvested fish gives an accurate reflection of the marginal benefit to society of harvested fish (exclusive of fishing effort costs).
3. Private and social costs of fishing effort are identical, with private fishing effort costs reflecting the true social opportunity cost of fishing effort.

There are thus no tiresome second-best considerations to bother us. We are living in a first best world. It is next assumed that:

4. The fishing industry is perfectly competitive.
5. The fishery is characterized by what we shall term Pure Open Access. There are no regulations, national or international, governing the fishery. The fishery is open to exploitation by all. The fishery resource supporting the fishery is the ultimate 'common pool' resource.

Total revenue, with respect to fishing effort, TR_E, and the total cost of fishing effort, TC_E, are as follows:

$$TR_E = p\left[uE - vE^2\right] \tag{2.4}$$

$$TC_E = b \cdot E \tag{2.5}$$

Consider Figure 2.2. The total sustainable yield (harvest) is seen by Gordon as the total product of fishing effort (Gordon, 1954). We thus have the value of the average product of *E*:

$$VAP_E = \frac{TR_E}{E} = p[u - vE] \tag{2.6}$$

and the value of the marginal product of *E*:

$$VMP_E = \frac{d(TR_E)}{dE} = p[u - 2vE] \tag{2.7}$$

The average cost, and marginal cost, of fishing effort are simply:

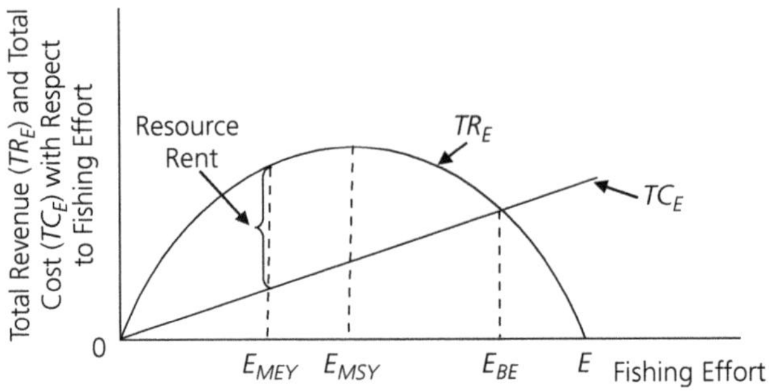

Figure 2.2. The Gordon-Schaefer model
MEY—Maximum Economic Yield
MSY—Maximum Sustainable Yield
BE—Bionomic Equilibrium

$$AC_E = MC_E = b \tag{2.8}$$

Gordon poses the following question. What is the optimal allocation of E to the fishery? The answer comes from the application of some elementary welfare economics. The optimal allocation of E to the fishery is given by the following equation, or rule:

$$VMP_E = MC_E \tag{2.9}$$

Recall that there are no second-best considerations and that MC_E is equal to the marginal opportunity cost of E.

The optimal level of E also happens to be the level of E that maximizes the total profit (sustainable) from the fishery. The total profit from the fishery TP_E is simply:

$$TP_E = TR_E - TC_E \tag{2.10}$$

The first order condition for a maximum is given by:

$$\frac{dTP_E}{dE} = p[u - 2vE] - b = 0 \tag{2.11}$$

which we can re-express as $VMP_E = MC_E$. The level of fishing effort, satisfying equations (2.9) and (2.11), is the level of fishing effort associated with Maximum Economic Yield (MEY), E_{MEY}, as opposed to Maximum Sustainable Yield (MSY), E_{MSY}.[3]

Gordon refers to the total profit from the fishery as 'resource rent'. If we adopt Joan Robinson's succinct definition of rent: '*... the essence of the* *conception of rent is the conception of a surplus earned by a particular factor of production over and above the minimum earnings necessary to induce it to do its work*' (Robinson, 1965, p. 102), then we can appropriately say that the total fishery profit can be described as rent accruing to the fishery resource. The fishery resource requires nothing 'to induce it to do its work'. Robinson goes on to say that the concept of rent is historically closely connected with the concept of the 'free gifts of nature' (ibid.).

Importantly, there is no consumer surplus; there is no producer surplus to complicate matters in Gordon's model. This is due to his assumptions about the constancy of p and b. Once these assumptions are relaxed, the net economic returns, or profits, from the fishery can no longer be described as constituting resource rent alone.

Now let us suppose that E is fact equal to E_{MEY}. Although the fish stock is in equilibrium, the fishery or fishing industry is not. Since there is no resource owner (sealord) to appropriate the resource rent, the resource rent will come into the hands of the fishing firms, and thus appear as true economic profit. The fishing industry, it will be recalled, is assumed to be perfectly competitive. According to the zero profit theorem, a perfectly competitive industry will expand (contract) to the point that true economic profit being earned by the firms is equal to zero. This would occur at the point where TR_E is equal to the total opportunity cost of E, i.e., at: $TR_E = TC_E$. Hence, in the Gordon model, the fishery would expand up to: $E = E_{BE}$, shown in Figure 2.2. At $E = E_{BE}$, the resource would then be in equilibrium, since harvesting is being undertaken on a sustained yield basis, and the perfectly competitive fishing industry would also be in equilibrium, since true economic profits equal zero. Gordon refers to this outcome as *Bionomic Equilibrium* (Gordon, 1954).

Now observe that, if Bionomic Equilibrium implies that $TR_E = TC_E$, then we can also say that at Bionomic Equilibrium:

$$VAP_E = AC_E = MC_E \tag{2.12}$$

Return to equations (2.6) and (2.7). It is obviously true that we shall always have: $VMP_E < VAP_E$, except in the uninteresting case in which $E = 0$. Thus, it follows that at Bionomic Equilibrium we will have: $VMP_E < MC_E$. The optimal allocation of E rule will be violated; there will be an overallocation of E to the fishery. Note as well, in passing, that at Bionomic Equilibrium, the resource rent will be fully dissipated.

In the Gordon-Schaefer model, we can be a bit more precise about the extent of overallocation of E at Bionomic Equilibrium. At Bionomic Equilibrium we have:

$$p[u - vE] - b = 0 \tag{2.13}$$

which implies that: $E_{BE} = pu - b/pv$, while at *MEY*, we have:

$$p[v - 2vE] - b = 0 \tag{2.14}$$

which, in turn, implies that:

$$E_{MEY} = \frac{pu - b}{2pv} \tag{2.15}$$

Thus, we conclude that:

$$E_{BE} = 2E_{MEY}$$

This says that the level of fishing effort at bioeconomic equilibrium is twice the fishing effort associated with maximum economic yield or, alternatively, that the level of fishing effort level associated with maximum economic yield is one half that associated with bioeconomic equilibrium.[4]

Every undergraduate student of economics is told that perfect competition leads to an optimal allocation of resources. Here we have a case in which perfect competition leads to a misallocation of resources. The cause lies in the 'common pool' nature of the resource, which leads to market failure. The price system sends out, from society's point of view, the wrong signals, creating a set of incentives among the fishermen that are perverse from society's point of view.[5]

Next let us observe that $E_{MEY} < E_{MSY}$ and that, in the context of Gordon's model, this would always be the case unless $MC_E = 0$, which it never is. This, in turn, leads to the conclusion that optimal economic management of fishery resources will lead to a more conservationist outcome than will the once revered MSY management criterion. We shall later see that this result is specific to the underlying Schaefer model and rests upon certain implicit assumptions in the Gordon model.

We turn now and re-cast the model in terms of revenues and costs with respect to sustainable harvests, rather than fishing effort. First, let us make the transition from fishing effort costs to harvesting costs. From equation (2.5) we have:

$$TC_E = b \cdot E$$

which we will now re-express as:

$$C(E) = b \cdot E$$

Moreover,

$$h = qEx$$

from which we get an expression for E:

$$E = \frac{h}{qx} \tag{2.16}$$

Thus we have total harvesting costs:

$$C(h, x) = \frac{bh}{qx} \tag{2.17}$$

while unit, or average harvesting costs $c(x)$ are:

$$c(x) = \left\{\frac{bh}{qx}\right\}/h = \frac{b}{qx} \tag{2.18}$$

Note that total harvesting costs are a function of x, as well as of h. We note that $c'(x) < 0$, i.e., unit harvesting costs are declining in stock size. Intuitively, as stock size increases so does catch per unit effort; consequently, unit harvesting cost must decrease.

Let us, in passing, be reminded that in chapter 1 it was claimed that the Schaefer model would lead to the conclusion that, under conditions of Pure Open Access, the resource could never be driven to extinction, because of the existence of what amounts to an economic 'brake' on stock depletion. The Gordon-Schaefer model elucidates and validates that claim. We need only note that:

$$\lim_{x \to 0} c(x) = \infty \tag{2.19}$$

We will in later chapters discuss the impact of this 'brake' on stock depletion.

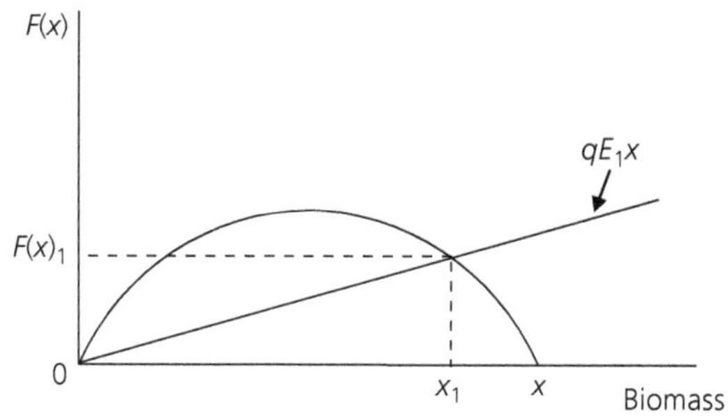

Figure 2.3. Schaefer model and sustainable yield

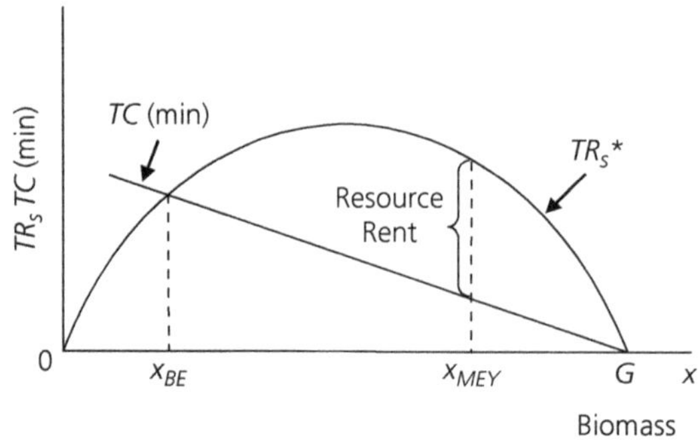

Figure 2.4a. The Gordon-Schaefer model, another view
$^*TR_s = p \bullet F(x)$

Now consider Figure 2.3, taken from the Schaefer model, which allows us to link the concept of sustainable yield (harvest) to both E and x simultaneously, and then allows us to move on to the concept of sustainable revenue in biomass (x) space.

In the figure $E = E_1$ is a given, fixed, level of fishing effort. With this level of fishing effort the resource will be in equilibrium $(dx/dt = 0)$ at $x = x_1$. We will have: $qE_1x_1 = F(x_1)$, and thus, $F(x_1)$ can be seen as the sustainable yield (harvest) associated with $E = E_1$ and with $x = x_1$. Hence, the $F(x)$ curve can, as we have claimed before, be seen as a sustainable yield curve representing the sustainable yields over the range $0 \leq x \leq G$. Now, multiply $F(x)$ by p and we have a sustainable revenue curve. Consider now Figure 2.4a. The total cost curve in Figure 2.4a needs some explanation. We refer to it as the minimum cost of harvesting the sustainable yield.

If harvesting is taking place on a sustainable yield basis $(h = F(x))$, then the total harvesting cost function becomes

$$C(F(x), x) = \frac{bF(x)}{qx} \tag{2.20}$$

From the Schaefer model, we have $F(x) = rx[1 - x/G]$. Hence, we can re-express equation (2.20) as:

$$C(F(x), x) = \frac{br}{q}\left[1 - \frac{x}{G}\right] \tag{2.20a}$$

$$C(F(x), x) = TC_{\min}$$

Thus, TC (min) is a linear, decreasing function of x.

Figure 2.4a makes clear the resource consequences of Pure Open Access. It is not just a matter of overallocation of labour and (produced) capital to the

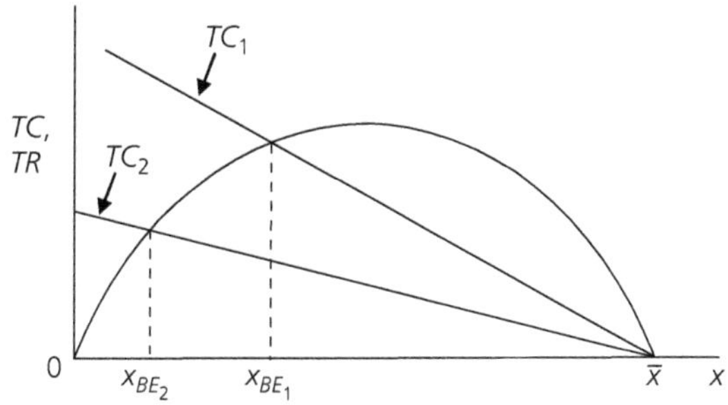

Figure 2.4b. The impact of technological improvements in harvesting technology, another view

fishery. There will be overexploitation—excessive disinvestment of natural capital—as well. If *MEY* represents optimality, then x_{MEY} is the optimal resource stock. Since $x_{BE} < x_{MEY}$, Pure Open Access, we conclude, leads to unequivocal resource overexploitation.

Figure 2.4a can be used to illustrate another, and very important, perverse aspect of 'common pool' fisheries, namely that technological improvements will make a bad situation worse. The parameter q, the catchability coefficient, can be viewed as a measure of the state of fishing technology, and hence of the productivity of labour and produced capital services constituting E. For the sake of simplicity, we treat E (often measured in terms of standardized fishing units per period of time) as being unchanged, and thus capture the entire technological change through q.[6]

In any event, assume that q doubles. In terms of our model (see: equation (2.20)), the doubling of q will, other things being equal, cause the minimum costs of harvesting the sustainable yield, for any given level of x, to be reduced by 50 per cent. Consider now Figure 2.4b.

The obvious effect of the technological improvement is to intensify the overexploitation of the resource. We note that Bionomic Equilibrium implies: $bF(x)/qx = pF(x)$. Solving for x we have:

$$x_{BE} = \frac{b}{pq} \tag{2.21}$$

It is clear that a doubling of q will, *ceteris paribus*, lead to a reduction of x_{BE} by 50 per cent. Similarly, a reduction in the cost of fishing effort, or an increase in the price of fish (p), will lead to reductions in x_{BE}.

As noted in equation (2.19), the unit harvesting cost goes to infinity as stock size approaches zero. Thus, x_{BE} will be positive. Having said that, economic forces tend to reduce x_{BE} over time through technological development and lower costs of some of the inputs used in the harvesting process. Similarly, subsidies that increase the price of output (p) or lower the cost of input (b) will also cause a reduction in x_{BE}. Thus, in many fisheries, the economic brake on resource depletion can over time be weak, in the sense that, although x_{BE} is positive, it will be small, perhaps dangerously so.[7]

As an aside, there is one important aspect of capture fishery resources that is not revealed by the Gordon-Schaefer model. This is the fact referred to in chapter 1, that many fishery resources are not concentrated at one point, but are rather spread out spatially, and are more accurately characterized as metapopulations, consisting of linked sub-populations (Sanchirico and Wilen, 1999).

Think now of a sink-source system, in which one or more sub-population(s) is (are) the source for other sub-populations, the sink sub-populations. Suppose further that the source sub-populations are well offshore while the sinks are near shore. An example is provided by the famous Northern Cod stock off Newfoundland and Labrador.[8]

The farther offshore the sub-populations are, the more costly it is to exploit them. Now consider a situation in which it is too costly to exploit the source sub-populations. Under Pure Open Access conditions, the sink sub-populations may well be overexploited, but the overall resource is protected from destruction by the unexploitable source sub-populations.

Through time, fishing technology steadily improves. The hitherto unexploitable source sub-populations become exploitable. Thus, due to improved technology the overall resource is at risk. Once again the Northern cod stock provides a striking example.[9]

Return now to Figure 2.2. The solution to the problem seems straightforward enough. Reduce fishing effort from E_{BE} to E_{MEY} and great benefits will promptly follow. In the example given, sustainable harvests, and thus revenues, will rise and costs will fall. Indeed some writers have argued that in cases such as that depicted in Figure 2.2, it is obvious that Bionomic Equilibrium is sub-optimal, because the same level of sustainable harvest could be taken with a much lower level of fishing effort.

Figure 2.4a, however, makes it clear that moving from Bionomic Equilibrium to *MEY* is not quite that simple. The resource has to be rebuilt from x_{BE} to x_{MEY}. In order to do this harvests must initially be reduced.

Observe the following. We have seen that, in terms of our model, $E_{MEY} = \frac{1}{2} E_{BE}$. Suppose that we commence with $qE_{BE}x_{BE} = h_0 = F(x_{BE})$. If, through government intervention, E were to be reduced from E_{BE} to E_{MEY}, the immediate consequence would be as follows:

$$q\frac{1}{2}E_{BE}x_{BE} = \frac{1}{2}h_0$$

Harvests would be reduced by 50 per cent!

Since harvests would be below $F(x)$, the resource would grow ($dx/dt > 0$). As the resource grew, harvests would gradually increase. This would take time. If the resource is slow growing (r is low), the time taken could be very long indeed.

What we are describing is, of course, a process of investment. Costs have to be incurred today in order to build up the stock of natural resource capital. The question then becomes whether the expected future benefits from rebuilding the resource make the current sacrifices worth bearing. These issues will be examined in detail in the chapter to follow.

2.3 **Regulated open access**

The case that we have discussed up to this point is that of Pure Open Access—a complete absence of property rights and government regulations. Many, if not most, national fisheries, at least, are subject to some form of government regulation, usually designed to protect the resource from overexploitation. Even if the conservation of the resource is successful, however, such government regulation can easily give rise to the 'common pool' problem in a second guise, which is now commonly referred to as Regulated Open Access (Wilen, 1985), or alternatively as the Class II Common Pool Problem (Munro and Scott, 1985). Gordon (1954) does not discuss this case in detail, but certainly hints at it.

We suppose that resource managers (government authorities of some form) intervene in the fishery to protect the resource from overexploitation, and indeed perhaps rebuild the resource. This is commonly done through gear restrictions and the imposition of harvest limits, as exemplified by total allowable catches—TACs. Where TACs are introduced, they are set on a season by season basis. When the TAC has been taken, the fishery is closed until the following season.

If nothing else is done, if fishing vessels are free to compete for shares of the harvest, the 'common pool' problem will emerge anew. The limited harvest now constitutes the 'common pool'.

Typically, in this sort of situation, the fleet becomes excessively large, while the season becomes steadily shorter. This brings in the question of 'non-malleable' vessel and human capital employed in the fishery, where the malleability of such capital refers to ease, and cost involved, in shifting the capital in and out of the fishery. We shall have to make considerable progress in developing the dynamic economic model of the fishery, before we can deal adequately with malleable vs. non-malleable vessel and human capital in the fishery. The consequence is that we shall have to wait until chapter 5 before analysing the Regulated Open Access issue thoroughly.

Nonetheless, we can readily obtain a general understanding of the Regulated Open Access issue by moving away only very slightly from our static analysis, through the use of a simple discrete time seasonal model. The model that we shall adopt is one employed by Munro and Scott (1985), which is based, in turn, on a model developed by Clark (1976).

The following assumptions are made:

(i) The resource managers have stabilized the resource by setting and enforcing a seasonal TAC, denoted by *H*.
(ii) During the fishing season, the resource experiences no natural growth. Net natural growth occurs off season, and is sufficient to restore the resource to its original level.
(iii) We denote the fishing season by *T*. The fishing season has a maximum length, $T = T_{max}$. The actual season length *T* will be: $0 \leq T \leq T_{max}$.
(iv) We denote the fleet by *B*. The fleet consists of identical vessels with identical crews. Fishing technology is constant through time.
(v) 'Crowding' externalities, in which vessels inadvertently disrupt or destroy one another's fishing gear, are non-existent.
(vi) All harvested fish is sold at the same (constant) price.
(vii) The fishing industry is fully competitive; there are no barriers to entry.

Denote the biomass at any time during the fishing season as $x(t)$, $0 \leq t \leq T$. Let $x(0) = X$. The harvest production function is: $h(t) = qEx$. During the fishing season, the fleet is fully employed so that we can re-write the harvest production function as: $h(t) = qBx$. Since there is no net natural growth during the fishing season, we have the following during the fishing season: $h(t) = -\ dx/dt$. From this it follows that at the end of the fishing season, we shall have: $x(\mathrm{T}) = X\,e^{-qBT}$

The actual harvest taken is equal to the TAC, which means both that the TAC harvesting constraint is binding and that the TAC is strictly enforced. We thus have:[10]

$$H = X(1 - e^{-qBT}) \tag{2.22}$$

Let us now solve for *T*, given *B* (or we could do the reverse, and solve for *B*, given T). We have:

$$Xe^{-qBT} = X - H \tag{2.22a}$$

Upon taking natural logs of both sides, we have:

$$\ln X - qBT = \ln\ (X - H) \tag{2.23}$$

$$T = \ln\left[\frac{X}{X - H}\right] \cdot \frac{1}{qB} \tag{2.24}$$

or:

$$T = \frac{N}{B} \tag{2.24a}$$

where N, a constant, is:

$$N = \frac{1}{q}\ln\left[\frac{X}{X - H}\right]$$

From equation (2.24a), it can be seen that the smaller B is, the longer T will be. The season T has a maximum length, $T = T_{max}$, which implies that there is a minimum fleet size, satisfying the following:

$$T_{max} = \frac{N}{B_{min}} \tag{2.25}$$

The actual fleet, B, is thus: $B \geq B_{min}$.

Now introduce seasonal cost functions, first for the individual vessel:

$$c(T) = c_0 + c_1 T \tag{2.26}$$

and for the fleet:

$$c(T)B = c_0 B + c_1 T B \tag{2.27}$$

where c_1 denotes salvageable costs per unit of time, and c_0 costs that are non-salvageable over the year. Salvageable costs are those which can be avoided by ceasing vessel operations, or which can be offset by turning to alternative activities. It is here that the malleability of vessel capital enters in. If $c_0 > 0$, this implies some degree of non-malleability in the fleet capital, in that not all of the capital can be easily and costlessly shifted out of the fishery.[11]

From equation (2.24a) we can express B as: $B = N/T$. This allows us to re-write equation (2.26) as:

$$c(T)B = c_0 B + c_1 N \tag{2.28}$$

If all costs were salvageable, i.e., $c_0 = 0$, then, given our other assumptions, it would not matter at all if the fleet approached infinite size, and H was taken in a second. All costs are, of course, not salvageable.

Now consider the consequences of $c_0 > 0$. To do so, we must introduce a fleet net revenue function. Suppose that all fish is sold at price p. We then have for the fleet total net revenue (NR)

$$NR = pH - [c_0 B + c_1 N] \tag{2.29}$$

Suppose that $B = B_{min}$ and that $NR >> 0$. Since there are no barriers to entry to the fishing industry, the industry is not in equilibrium, and will not be in equilibrium until the fishing firms are no more than covering their opportunity costs—the zero profit theorem—once again.

Thus the fleet will expand. The new vessels pouring into the fishery constitute redundant capital. The vessels cannot add to, or improve upon, harvest production. This matters nothing to the individual vessel owners, who are interested only in gaining a share of the valuable harvest.

The fleet will thus expand until $NR = 0$, which we can express as:

$$NR = pH - [c_1 N + c_0 B_{min} + c_0 (B - B_{min})] = 0 \tag{2.30}$$

Let us refer to $c_0(B - B_{min})$ as the redundancy deadweight loss in the fishery. It is to this that we attribute the complete dissipation of resource rent. As more and more vessels enter the fishery, the redundancy deadweight loss steadily increases, and continues to do so until NR is reduced to zero. At that point, there is no further incentives for vessels to enter the fishery—the zero profit theorem once again.

The model presented has many restrictive assumptions. What we find in the real world, and what will be discussed in greater detail in chapter 6, is that there are other avenues for rent dissipation. 'Crowding' is likely to appear, which will dissipate the rent. Technology is, in fact, seldom, if ever frozen. Rent can readily be dissipated through technological advances designed to enhance catching power. This is a method of enhancing fishing capacity, along with increasing the number of vessels.

The expansion of fleet capacity means (see: equation (2.24)) steadily reduced seasons. Thus, for example, the British Columbia Pacific halibut fishery, which has a T_{max} = 240–250 days per year, found by 1990 that: T = 6 days.[12] The shortened seasons can lead to numerous inefficiencies. Thus, for example, even if all of the harvested fish is going into the fresh market, the price is not independent of the nature of the fishing activity. Shortened seasons, and the resulting scramble, often lead to poor handling of the fish and inferior product. If the harvested fish is processed, then processing plants will face gluts, as a result of the short seasons, followed by famine, which provides a further avenue for rent dissipation.

Finally, the assumption that the resource managers can enforce the TACs absolutely is often not valid. Increases in fleet size often make monitoring very difficult, leading to chronic TAC 'overages', with obvious consequences for the conservation of the resources.

2.4 Implications, and limitations, of the static economic model of the fishery

The static economic model of the fishery can easily be extended by, for example, relaxing the assumptions of perfect elasticity of demand for harvested fish and perfect elasticity of supplies of labour, (produced) capital, and ancilliary services constituting fishing effort. We shall not do so, but shall rather deal with such relaxed assumptions in the context of the dynamic version of the economic model of the fishery to follow.

The Gordon-Schaefer model demonstrates in a forceful manner the economic consequences of Pure Open Access. From it one can readily infer the economic consequences of Regulated Open Access. This was to have, and continues to have, a major impact upon policy-makers. The 1989 Government of Australia document, *New Directions for Commonwealth Fisheries Management* (Australia, 1989) and the 1993 FAO document, *Marine Fisheries and the Law of the Sea: A Decade of Change* (FAO, 1992 are two examples of policy documents in which Figure 2.2 (in slightly modified form) appears, and in which extensive use is made of the Gordon-Schaefer analysis. A third, and much more recent, example is provided by the World Bank/FAO publication, *The Sunken Billions*, to which we referred to several times in chapter 1 (see World Bank, 2009, box 3.2).

Pure Open Access continues to exist in many high seas fisheries and also exists in national fisheries, where the fisheries regulations are weak and ineffective. The Gordon-Schaefer model can give us an explanation, in broad general terms, for *The Sunken Billions* study's conclusion that global resource rent in the world's capture fisheries is hovering around zero (World Bank, 2009). When we come to discuss international fisheries in chapter 7, we will be presented with examples of high seas fisheries that give tragic testimony to the predictive power of the Gordon-Schaefer model. With respect to Regulated Open Access, there are examples in abundance of regulated fisheries suffering from serious excess capacity in terms of produced and human capital. We shall discuss several examples in chapters 4 and 6.

While the Gordon-Schaefer model has had great policy influence, it has major drawbacks because of its static nature. First, it creates the illusion of certainty. A stock restoration programme, since it takes place over time, will yield results that can only be estimated. It is impossible to specify these results precisely. All management of capture fisheries inevitably takes place under a cloud of uncertainty.

In more general terms, the static model permits us to say nothing about the adjustment phase. The model appears to imply that, in commencing at Bionomic Equilibrium, it makes no difference, in determining optimal policy, whether the resource grows rapidly, such as say skipjack tuna, or whether it grows very slowly, such as orange roughy. Of course, it does matter, and indeed matters a great deal. This will be highlighted in some of the empirical case studies in chapters 4 and 8.

Restoration of a hitherto overexploited fishery resource is an undertaking in positive resource investment. As such, it must be analysed in a capital theoretic setting. Gordon did not do so in 1954, as we noted earlier, because of the inadequacy of available analytical tools. He did, however, clearly recognize the true nature of the problem as is indicated from the following quote taken from a conference paper, which he presented two years after the publication of his seminal article:

> The conservation problem is essentially one which requires a dynamic formulation... The economic justification of conservation is the same as that of any capital investment—by postponing utilization we hope to increase the quantity available for use at a future date. In the fishing industry we may allow our fish to grow and to reproduce so that the stock at a future date will be greater than it would be if we attempted to catch as much as possible at the present time. In theoretical terms this means that the optimum degree of exploitation of a fishery must be defined as a time function of some sort. That is to say, it is necessary to arrive at an optimum, which is a catch per unit of time, and one must reach this objective through consideration of the interaction between the rate of catch, the dynamics of fish populations, and the economic time-preference schedule of the community or the interest rate on invested capital. This is a very complicated problem and I suspect that we will have to look to the mathematical economists for assistance in clarifying it. (Gordon, 1956, cited in Munro and Scott, 1985, pp. 636–7)

Mathematical tools were subsequently to become available to make this 'very complicated problem' tractable. The basic ideas, arising from the dynamic, or capital theoretic, formulation of the economic model of the fishery can now readily be made accessible to undergraduates. With this in mind, we turn to our introduction to the dynamic economic model of the fishery.

BIBLIOGRAPHY

There are numerous presentations of the static economic model of the fishery, over and above Gordon's 1954 article, to which the reader can turn. Among the many are Anderson and Seijo (2010), Bjørndal and Munro (1998), Flaaten (2010), Munro and Scott (1985), and Whitmarsh (2011). An earlier, and very clear, exposition on the model, in which equations are kept to a minimum, is to be found in Christy and Scott (1965). For those not willing to work through the entire Gordon (1954) article, a good abridged version is to be found in Dorfman and Dorfman (1993), a book of readings on environmental economics. With respect to the issue of Regulated Open Access and the Class II Common Pool problem, the reader should, at this stage, turn to Wilen (1985), and to Munro and Scott (1985).

NOTES

1. The article also had an immense impact upon academics, outside of the community of fisheries economists. An example is provided by Robert Dorfman, who has a broad interest in environmental and resource economics. He published a widely used book of readings on the *Economics of the Environment* that went through several editions. In this book he states at one point that 'the economic theory of the use and misuse of common property [what we would today call a "common pool" resource] has nowhere been better explained than in H. Scott Gordon's analysis of the overuse of a typical common property resource, an ocean fishery'. Elsewhere, he refers to Gordon's '*careful* economic analysis' (Dorfman and Dorfman, 1972, p. 66; 1993, p. 76). Another example is provided by Baumol and Oates (1998, p. 28) who commend the rigour of the analysis in Gordon's 1954 article.

   The basic ideas contained within Gordon's article are not, in fact, original to Gordon, a fact that by no means detracts from the strength of Gordon's argument. The basic ideas can be found in an article by the Danish economist, Jens Warming, published in 1911 (Warming, 1911). The Warming article was published in Danish, and consequently gained very limited circulation. A translation of the Warming article, along with a commentary, was made available to the English-speaking world by Andersen (1983).
2. This point is discussed in detail towards the end of the chapter.
3. E_{MEY} is also different from the optimal effort level based on dynamic optimization, as we shall see in the next chapter.
4. The result that the E_{MEY} is half E_{BE} is model specific. It is, of course, true in general that E_{MEY} is less than E_{BE}.
5. Suppose that the fishery resource was not a 'common pool' one, but rather was one owned publicly by a coastal state. Suppose further that the coastal state exercised its property rights to the resource fully by absorbing the resource rent fully through taxation. Then perfect competition, and the market, would come into their own. We could look forward to an optimal allocation of resources. We shall analyse this possibility in detail in chapter 6, where we shall point out that that Gordon (1953) had put forth the public ownership and taxation scheme as a possible solution to the 'common pool' fisheries problem.
6. Admittedly this is somewhat artificial, but we justify the artificiality on the grounds of ease of exposition.
7. If we relax the assumption that the demand for harvested fish is perfectly elastic, then a further reason for the decline in x_{BE} may arise. If the price elasticity of demand for harvested fish is finite, then the depletion of the stock leading ultimately to a decline in harvest, will lead to an increase in the price of harvested fish, resulting in yet a further incentive to deplete the stock. An interesting example for North Sea herring is provided by Bjørndal and Conrad (1987b).

   We should also note that the Schaefer model assumes that there is no minimum viable stock size above zero. In many cases there are positive minimum viable stock sizes, in the sense that, if the resource is driven below that level, the resource on its own will go to extinction (Clark, 1990). In such cases, the effects of cost reductions, or price increases, in open access fisheries can be very dangerous.

8. Northern cod is a cod stock complex on the Grand Banks of Newfoundland. Historically it constituted the heart of the Newfoundland fishing industry, and provided the basis of a sustainable fishery for some 500 years. The cod spawn offshore from January to April. When the resource was abundant, a portion of the resource moved inshore during the summer months. Hence, the offshore portion of the resource was the source, and the inshore the sink (Munro, 1980).
9. Up until the late 1950s, the state of fishing technology was such that the Northern Cod fishery was largely an inshore fishery. Regardless of the extent of fishing inshore, the resource was safe. The source was protected by economics. In the late 1950s, advances in fishing technology made it commercially feasible to exploit the resource offshore during the spawning season—January to April—in spite of the harsh conditions. Even with fisheries regulations, first international and then Canadian, the resource eventually collapsed. In 1992, the government of Canada was forced to declare a 'temporary' harvest moratorium in the Northern Cod fishery (National Research Council, 1999). The 'temporary' harvest moratorium remains in place at the time of writing.
10. Intra-season, the resource, X, is decreasing exponentially due to harvesting. In the off season, the resource is restored to its original level due to growth.
11. There is an implicit assumption that human capital in the fishery is perfectly malleable.
12. Munro et al. (2009).

3 The Basic Dynamic Economic Model of the Fishery

The term 'dynamic' means in essence that time is taken into account explicitly—in contrast to the 'static', or timeless, models we encountered in chapter 2. When one discusses issues such as the restoration of hitherto overexploited fish stocks, time considerations become unavoidable. The rebuilding of a fishery resource is, as we have emphasized, a programme of investment in natural capital. Analysing such an investment programme in other than a dynamic framework is all but impossible.

The terms 'dynamic economic model of the fishery' and 'capital theoretic model of the fishery' are thus synonymous. Fishery resources are seen as stocks of natural capital assets, with the fisheries management problem being that of managing these assets in such a manner as to yield the maximum economic benefits to society through time.

With respect to any given fishery resource, the fisheries manager, as natural capital asset manager, is seen as being faced with a theory of capital question and a theory of investment question. Recall our discussion of theory of capital and the theory of investment in chapter 1. The questions are:

(i) What, from society's point of view, is the optimal stock of this form of natural capital, measure in biomass terms, through time? Resource managers will commonly talk in terms of 'target' biomass levels.
(ii) What, from society's point of view, is the optimal rate of investment (positive or negative) in this form of natural capital?

If we are already at the optimal stock level, then obviously the answer to question (ii) is simple—the optimal rate of investment is zero. If we are away from the optimal stock level—the normal case—then question (ii), as we discussed in chapter 1, is really about the speed at which the optimal, or target, biomass should be approached.

In the chapter, we shall commence, in sections 3.1 to 3.4, with the simplest version of the dynamic economic model of the fishery, one that is linear and autonomous, in the sense that the parameters are independent of time. We then move on in sections 3.5 to 3.7 to consider both the non-linear and non-autonomous versions of the model. Finally, we conclude by demonstrating in section 3.8 that the model discussed, although basic, can be used to analyse a complex policy problem arising from the developing world.

3.1 The basic linear autonomous dynamic economic model of the fishery

Our introduction to the dynamic economic model of the fishery takes the form of a dynamic version of the Gordon-Schaefer model discussed in the previous chapter. The model that we shall develop is a continuous time one. We do so for ease of exposition. At a later point, we shall digress and talk about discrete time versions of the basic model.

In any event, we focus upon a single species, allowing for no species interaction. Furthermore, we assume that the demand for fish and the supply of fishing effort (basically the combined services of labour and produced capital devoted to fishing and ancillary services such as fuel) are perfectly elastic. Hence, we can treat the price of harvested fish, p, and the unit cost of fishing effort, b, as constants. In addition, we assume that parameters, such as p and b, are invariant over time. The parameters are thus seen to be independent of time, or autonomous.

We make a further heroic assumption that the future is known to us fully. Our model is said to be deterministic. Many of these assumptions will be relaxed in the chapters to come.

Be that as it may, our dynamic economic model of the fishery, based upon the Schaefer model, is as follows. The first few equations will be found to be very familiar indeed from chapters 1 and 2. In the absence of harvesting we have:

$$dx/dt = F(x) \tag{3.1}$$

where x denotes the biomass.

$$F(x) = rx[1 - x/G] \tag{3.2}$$

where, it will be recalled, G denotes the equilibrium level of x in the absence of harvesting—often referred to as the carrying capacity of the environment—and r, a constant, denotes the intrinsic growth rate of the resource.

Upon introducing harvesting, we have:

$$dx/dt = F(x) - h(t) \tag{3.3}$$

$$h = qE^{\alpha}x^{\beta} \tag{3.4}$$

where, it will be recalled, that q, a constant, denotes the catchability coefficient, and that, by assumption, $\alpha = \beta = 1$.

We are reminded that the cost of fishing effort is given by

$$C(E) = b.E \tag{3.5}$$

with b denoting the unit cost of fishing effort.

The net rate of economic return (resource rent) from the fishery at any given point in time is given by:

$$\pi = [p - c(x)]h \tag{3.6}$$

where $c(x)$, the unit cost of fishing, is as defined in chapter 2. It will not be forgotten that: $c(x) = b/qx$ with $c'(x) < 0$.

We assume that the objective of the resource manager is to maximize the present value (PV) of the resource rent forthcoming from the fishery from $t = 0$ to $t = \infty$. We assume, as before, that p reflects the true marginal benefit (gross) to society from harvested fish and that b reflects the true marginal opportunity cost of the labour and produced capital services comprising fishing effort. Thus, there is no distinction between private and social resource rent.

We are presented with an optimal control problem, in which $x(t)$ is the state variable, the variable to be controlled through time, and $h(t)$ is the control variable. Let it be noted in passing that we could just as easily have let $E(t)$ serve as the control variable, with the choice being a matter of taste and convenience.

The resource manager's objective functional can be expressed as follows:

$$\begin{aligned} max\, PV &= \int_0^{\infty} e^{-\delta t}\pi(h(t), x(t))dt \\ &= \int_0^{\infty} e^{-\delta t}[p - c(x)]h(t)dt \end{aligned} \tag{3.7}$$

s.t.

$$\begin{gathered} dx/dt = F(x) - h(t) \\ x(0) = x_0 \\ 0 \le h(t) \le h_{max} \end{gathered}$$

For mathematical reasons, an upper bound has to be imposed upon $h(t)$. Finally, δ denotes the social rate of discount (interest), which we interpret as an opportunity cost, the rate of return open to the social manager on investment opportunities of comparable risk. Let it never be forgotten that the social manager is managing a broad portfolio of capital assets (World Bank, 2005), only some of which consist of capture fishery resources.[1]

Our optimal control model is linear. It is so in the sense that the objective functional, equation (3.7), is linear in the control variable h(t).

Before proceeding further, we must point out an important implicit assumption in the model. The produced capital (fleet and processing) and human capital employed in the fishery are perfectly malleable, with perfect malleability of capital being defined as it was in chapter 2. When this assumption, highly unrealistic for most fisheries, is relaxed, the analysis must be modified very substantially, and becomes much more complex. The issue of non-malleable fleet, processing and human capital will be touched upon later in the chapter, and will be addressed more fully in chapter 5.

We are now required to set up what is called the Hamiltonian and invoke the Maximum Principle. The Hamiltonian (H) can be expressed as follows:

$$H = e^{-\delta t}\{[p - c(x)]h(t) + \psi(t)(F(x) - h(t))\} \tag{3.8}$$

where $\psi(t)$ is the adjoint, or co-state, variable, which can be seen as the *shadow price* of the marginal addition to or investment in the stock of natural capital. The Hamiltonian is what Neher (1990) calls a 'performance indicator' for a fishery in the sense that it represents both the returns from harvesting and from investments in the stock. Fundamentally it reflects the ongoing trade-off, at the margin, between the return from current harvesting, and the return from investment in the stock of natural capital.

For ease of exposition, we find it convenient to re-express the Hamiltonian as:

$$H = e^{-\delta t}[p - c(x)]h(t) + \lambda(t)(F(x) - h(t)) \tag{3.8a}$$

where $\lambda(t) \equiv e^{-\delta t}\psi(t)$, i.e., the present value of the shadow price of fish.

The first condition laid down by the Maximum Principle is that at all times

$$\partial H/\partial h = e^{-\delta t}[p - c(x)] - \lambda(t) = 0 \tag{3.9}$$

Given the linearity of the model, we can easily be placed in a situation in which $\partial H/\partial h > 0$; $\partial H/\partial h < 0$. This possibility causes us to search out the so-called singular path along which: $\partial H/\partial h = 0$. Thus along the singular path, we have:

$$e^{-\delta t}(p - c(x)) = \lambda(t) \tag{3.10}$$

Equation (3.10) is easy to interpret in economic terms. It states at each point in time, the marginal cost of investing in the resource at time t—foregone current rent—is equal to the shadow price of marginal addition to the stock of natural capital, with both being discounted back to $t = 0$. Another way to put this is to say that, at the margin, the net value of one kg of fish harvested is equal to the value of a kg of fish left in the sea to reproduce and grow. We may think of the L.H.S. and the R.H.S. of (3.10) as being the 'supply price' and 'demand price' respectively of the 'natural' capital (both discounted back to $t = 0$) (see Clark and Munro, 1982).

The Maximum Principle also requires that:

$$\frac{d\lambda}{dt} = -\,\partial H/\partial x = e^{-\delta t}c'(x)h(t) - \lambda(t)F'(x) \tag{3.11}$$

Given equation (3.10), it can be said that, along the singular path we have:

$$\frac{d\lambda}{dt} = -\,\delta e^{-\delta t}(p - c(x))$$

Hence, by substitution we have

$$-\delta e^{-\delta t}(p - c(x)) = e^{-\delta t}c'(x)h(t) - e^{-\delta t}(p - c(x))F'(x) \tag{3.11a}$$

which reduces to:

$$F'(x) - \frac{c'(x)h(t)}{p - c(x)} = \delta, \tag{3.12}$$

where x is a steady state biomass level (see Clark and Munro, 1982 for details), which implies, in turn, that $h(t) = F(x)$. We thus re-state (3.12) as

$$F'(x^*) - \frac{c'(x^*)F(x^*)}{p - c(x^*)} = \delta \tag{3.13}$$

where x^* is the optimal biomass level. Equation (3.13) has come to be referred to in the literature as the *fundamental rule (equation) of renewable resource exploitation* (see, for example, Pearce and Turner, 1990, p. 256; Bjørndal et al., 2007, p. 2).

The L.H.S. of equation (3.13) is the rate of return on marginal investment in the resource, or the 'own rate of interest' of the resource. It has two components, with the first one being the 'marginal product' of the natural capital, which in turn reflects the impact of the marginal investment on sustainable harvests. The second component is termed the Marginal Stock Effect (MSE) (Clark and Munro, 1982). The MSE reflects the fact that (in the context of this model) a marginal investment in x has an economic benefit over and above that derived from sustainable harvests. The additional benefit comes through the impact of the resource investment upon harvesting costs. The larger the biomass x, the lower the harvesting costs, *ceteris paribus*.

Let us simplify further by re-writing equation (3.13) as:

$$F'(x^*)+\gamma(x^*)_{|h=F(x^*)}=\delta \tag{3.13a}$$

where $$\gamma(x^*)=\frac{-c'(x^*)F(x^*)}{p-c(x^*)}.$$

In any event, equation (3.13) (or 3.13a) is to be seen as an investment decision rule designed to answer our capital theory question pertaining to the optimal stock of this form of natural capital. The decision rule is in keeping with the concept of asset market equilibrium, which simply stated, is that asset markets will be in equilibrium, only when all assets of a common risk class are seen to be yielding the same rates of return (Solow, 1974). If this does not hold, then some portfolio rebalancing is in order (Gaudet, 2007).

Our investment decision rule (equations 3.13/3.13a) states: invest/disinvest in the resource biomass up to the point that the 'own rate of interest' on the marginal investment in the resource is equal to the social rate of discount, which can be seen as reflecting the common rate of return on relevant alternative investment opportunities open to society (Clark et al., 2010a).

It will, from time to time, be useful to re-express the fundamental equation as:

$$p-c(x^*)=\left[\frac{d}{dx^*}\{(p-c(x^*))F(x^*)\}\right]\frac{1}{\delta} \tag{3.14}$$

or as:

$$\delta(p-c(x^*))=\frac{d}{d(x^*)}\{(p-c(x^*))F(x^*)\} \tag{3.15}$$

The R.H.S. of equation (3.14) can be seen as the *PV* of incremental resource rent arising from the marginal investment in the stock. The L.H.S. can be seen, in turn, as the cost of the investment (the foregone current rent from the fishery), which the investment entails. Thus the investment rule expressed as (3.14) states: invest in the resource up to the point that the cost of the marginal investment is equal to the payoff from the marginal investment.

In equation (3.15), the R.H.S. denotes incremental sustainable resource rent arising from the marginal investment in the resource stock. The L.H.S. is a type of opportunity cost, the sustainable interest income that could have been earned, if instead of forgoing the marginal current resource rent, the rent had been realized and then put into an alternative investment offering a rate of return of δ.

One would expect that the 'own rate of interest' of the resource would be sensitive to whether the resource is slow growing or fast growing. The answer is: of course. Using the intrinsic growth rate, r, as a measure of whether the fishery resource is fast growing (e.g. skipjack tuna) or slow growing (e.g. orange roughy),[2] return to equations (3.13)/(3.13a) and note that $F'(x^*)$ can be re-expressed as: $r\{1 - 2x^*/G\}$. The impact of r upon the 'own rate of interest' of the resource then becomes transparent. This point will be illustrated in an empirical context in the case studies to be presented in chapter 4.

The fundamental equation, in its many guises, is an implicit equation for x^*. Given that the Schaefer model provides the foundation for our economic model, it is possible to solve the fundamental equation explicitly for $x = x^*$ (see Clark and Munro, 1982). We have

$$x^* = \frac{G}{4}\left\{1+\frac{b}{pqG}-\frac{\delta}{r}+\left[\left(1+\frac{b}{pqG}-\frac{\delta}{r}\right)^2+\frac{\lambda b}{pqG}\frac{\delta}{r}\right]^{1/2}\right\} \tag{3.16}$$

For other functional forms it may not be possible to derive an explicit solution for the optimal stock level.

We shall want to say much more about the fundamental equation, but before doing so, let us digress and answer the investment question, namely the optimal rate of investment (positive or negative) in the resource stock.

3.2 A digression on optimal rates of resource investment

Let it be supposed that the stock level, x^* has been determined, and that $x(0) \neq x^*$. At what rate should we plan to approach the target biomass level? We shall consider both the case in which $x(0) < x^*$ and the case in which $x(0) > x^*$.

Return now to equations (3.9) and (3.10). If $x(0) < x^*$, we shall have $\partial H/\partial h < 0$, which in turn implies that: $e^{-\delta t}[p - c(x)] < \lambda(t)$. The investment rule is clearly to undertake positive investment in the stock. This is because investments in the stock, represented by the shadow price, give a much higher return than the net revenues from harvesting. Since H is linear in $h(t)$, the inequality continues as $h(t)$ is reduced to zero. Thus, $h(t)$ should be set equal to zero, which upon returning to equation (3.3) implies invest (positively) in x at the maximum rate.

If $x(0) > x^*$, everything goes into reverse. The optimal investment policy is to set $h(t) = h_{max}$ so long as $x(t) > x^*$, which in turn implies that disinvestment in x should take place at the maximum rate.

The investment rule can be expressed as follows, in terms of the optimal rate of harvest—$h^*(t)$:

$$h^*(t) = \begin{cases} h_{max} & \text{if } x(t) > x^* \\ F(x^*) & \text{if } x(t) = x^* \\ 0 & \text{if } x(t) < x^* \end{cases} \tag{3.17}$$

The investment policy called for by equation (3.17) appears to be extreme, indeed draconian. It states that, if $x(t) < x^*$, an outright harvest moratorium should be declared, and should remain in place until $x(t) = x^*$. The instant that the biomass level x^* is achieved, $h(t)$ should jump from $h(t) = 0$ to $h(t) = F(x^*)$. These extreme resource investment policies are illustrated in Figure 3.1.

The extreme resource investment policies have often led to comments about the lack of realism of the dynamic economic model of the fishery. It must be realized, however, that these draconian, seemingly unrealistic, investment policies reflect our underlying assumptions. First, we have assumed that the price of harvested fish and the unit cost of fishing effort are independent of the harvest rate. Second, we have assumed, implicitly, that the relevant produced and human capital are perfectly malleable, in that they can easily and costlessly be shifted into, or out of the fishery.

As we stated in chapter 1, a general rule of thumb is that one should approach the target stock of capital (of any form) as rapidly as possible, *unless there are penalties associated with a rapid rate of investment/disinvestment.*

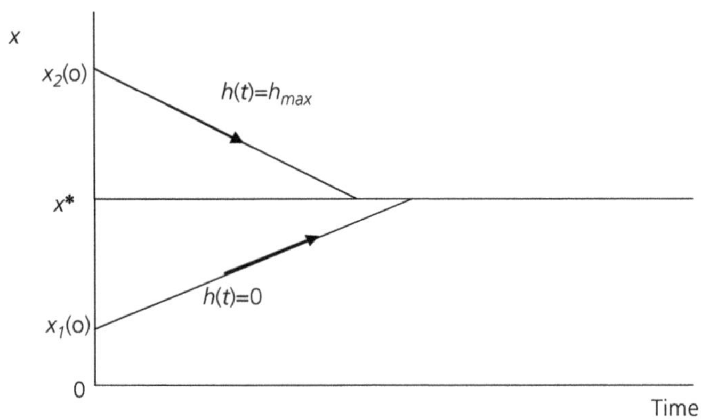

Figure 3.1. Optimal resource investment time paths: the linear model

Given our assumptions, no such penalties exist. If either of the two assumptions were to be relaxed, there would be penalties associated with a rapid investment/disinvestment policy. In the section to follow, we shall illustrate this fact by relaxing the assumption of constant prices for harvested fish and constant unit costs of fishing effort. In chapter 5, we shall show that a rapid investment/disinvestment policy would be decidedly non-optimal in the face of non-malleable produced and/or human capital.

3.3 The fundamental equation once again

With the investment policy digression complete, let us return to the fundamental equation, and first ask how it relates to our discussion of the static model in chapter 2. It will be recalled that, according to the static Gordon-Schaefer model, optimal fisheries management leads to achieving Maximum Economic Yield (*MEY*), i.e., maximizing sustainable resource rent. Hence, the target biomass x^* is: $x^* = x_{MEY}$.

Denoting sustainable resource rent as π_{sus}, we have: $\pi_{sus} = (p - c(x))F(x)$. Maximizing π_{sus} implies that we are at the point at which $d\pi_{sus}/dx = 0$. Now, return to equation (3.15) and re-order terms so that we have:

$$\frac{d\{(p - c(x^*))F(x^*)\}/dx^*}{p - c(x^*)} = \delta \tag{3.18}$$

The numerator of the expression on the L.H.S. of equation (3.18) is simply the first derivative of sustainable resource rent, $d\pi_{sus}/dx$, evaluated at $x = x^*$.

If, in fact $x^* = x_{MEY}$, then, by implication, that numerator is equal to zero. The equation can then hold, if and only if, $\delta = 0$. Hence, the static Gordon-Schaefer model implicitly assumes that $\delta = 0$. Whatever one's time preference may be, it is not reasonable to assume that the yield on alternative forms of investment is equal to zero.

In the discussion of the static Gordon-Schaefer model, considerable emphasis was given to the concept of Bionomic Equilibrium, the point to which the fishery would be driven, under conditions of Pure Open Access. In the aforementioned model, Bionomic Equilibrium is associated with the complete dissipation of the resource rent, which is to say that we are at a point at which $p - c(x) = 0$.

Could Bionomic Equilibrium be optimal? Could we ever find that $x^* = x_{BE}$? Return to equation (3.14), which we recall is:

$$p - c(x^*) = \left[\frac{d}{dx^*}\{(p - c(x^*))F(x^*)\}\right]\frac{1}{\delta}$$

The R.H.S. of equation (3.14) is just $d\pi_{sus}/dx$ divided by the discount rate, δ. Thus, as the discount rate goes to infinity, the R.H.S. will go to zero. Therefore, to maintain equality, we must have $p - c(x^*) = 0$, i.e., $x^* = x_{BE} = c/(pq)$, if and only if $\delta = \infty$. If $\delta = \infty$, then only today matters; economic returns coming from the fishery tomorrow, and thereafter, count for nothing.

It is all but inconceivable, even under the most dire circumstances, that the social rate of discount could approach infinity. If it is reasonable to assume that: $0 < \delta \ll \infty$, we can state unequivocally that Bionomic Equilibrium reflects excessive disinvestment, from society's point of view, of the fishery natural capital.

All of this gives further insight into the perverse fishermen incentives created by pure open access. Under pure open access, the fishermen are given the incentive to discount massively future economic returns from the resource, and to view the resource as a non-renewable resource that is to be mined with all possible speed.[3]

Next, we wish to examine the consequences of relaxing a key assumption underlying the Schaefer biological model. We do so by introducing the concept of *concentration profile* (Clark, 1985, 1990), which Clark (1990, p. 225) defines as the relationship between the density of the fish stock subject to exploitation by the fishermen, and the size of the stock (biomass). Recall our harvest production function: $h = qE^{\alpha}x^{\beta}$, and denote the concentration profile as $\rho(x)$. We can express $\rho(x)$ as:

$$\rho(x) = ax^{\beta} \tag{3.19}$$

where a is a constant (Clark, 1990, p. 227). We shall assume that $a = 1$.

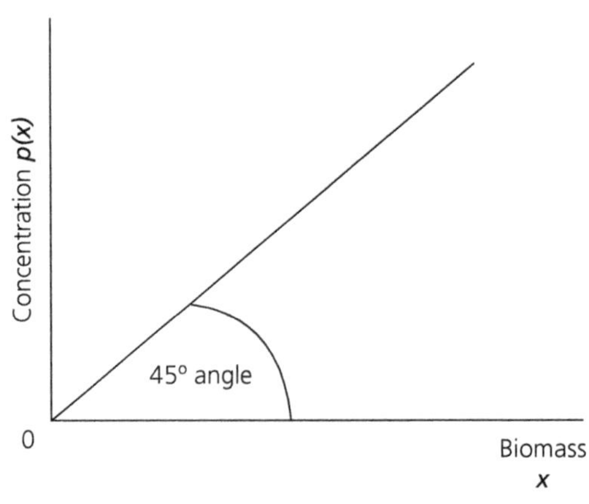

Figure 3.2. Concentration profile—diffusive fishery resources: the Schaefer model case

In the case of the Schaefer model, in which by assumption $\beta = 1$, there will be a strictly linear relationship between the concentration profile and the biomass size, and by implication, between the Catch Per Unit Effort (CPUE) and the biomass size. See Figure 3.2.

The Schaefer model assumes that the resource is spread uniformly and continues to be so, if the resource (stock) is reduced. This has profound economic implications. If the resource is reduced through exploitation, the CPUE will thus decline in proportion, which means that harvesting costs will steadily increase. This is what lies behind the Marginal Stock Effect.

We can go farther. The resource is protected from the threat of extinction, even under Pure Open Access. Let us be reminded that, given our other assumptions, we have $c(x) = b/qx^{\beta}$, where b, as before, is the unit cost of fishing effort. If $\beta = 1$, that is, as noted in chapter 2, $\lim_{x \to 0} c(x) = \infty$ (see equation 2.19). Thus, there is an economic brake on resource exploitation that may prevent stock extinction from occurring.[4]

As Clark (1990) and others have stressed, there are many fishery resources that do not exhibit this concentration profile. In particular, fishery resources that have strong schooling tendencies are not uniformly distributed, as the resource declines. Rather than $\beta = 1$, we have: $0 \leq \beta < 1$.[5] In extreme cases β will approach zero (Bjørndal, 1988). In such extreme cases $\rho(x)$ can take on the form illustrated in Figure 3.3 (Clark, 1990).

Consider now the cost consequences, by taking the extreme case of $\beta = 0$. We then have: $c(x) = b/qx^0$. If the resource is steadily depleted, unit harvesting costs remain constant up to the point that the biomass is at the verge of extinction (Bjørndal, 1988). The economic brake referred to earlier is completely non-functional. Such resources are, not surprisingly, highly vulnerable to severe overexploitation.

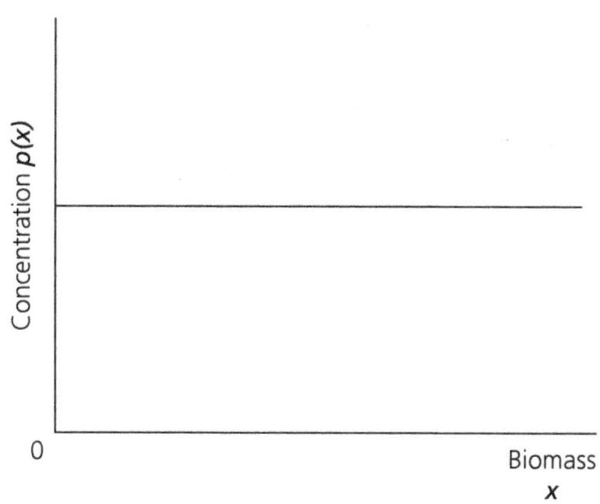

Figure 3.3. Concentration profile—schooling fishery resources: the extreme case

An example is provided by the Norwegian Spring Spawning Herring resource in the North Atlantic. Historically, the resource, which migrates between Norway and Iceland, has been one of the largest, and most valuable, fishery resources in the North Atlantic (Bjørndal, 2009). The resource is characterized by schooling characteristics (Bjørndal, 2009).

Prior to the 1970s, the herring fishery was a textbook example of a Pure Open Access one. It can be conjectured that, prior to the 1960s, some portions of the resource in the heart of the North Atlantic, far from shore, were protected economically from overexploitation. By the 1960s, technological development was such that all economic protection was gone. In the late 1960s–early 1970s the resource crashed.

Marine biologists often focus on that portion of the resource capable of reproduction—the spawning stock biomass (SSB). The International Council for the Exploration of the Sea (ICES) estimates that the minimum safe SSB level for the Norwegian Spring Spawning Herring resource is 2.5 million tonnes (Bjørndal, 2009). It is estimated that in 1973 the Norwegian Spring Spawning Herring SSB fell to 7,000 tonnes, in other words to less than three-tenths of 1 per cent of ICES declared safe minimum level. The resource had come within a hair's breadth of extinction (Bjørndal, 2009).

Let us note, in passing, that, if $\beta = 0$, and hence $c'(x) = 0$, then the fundamental equation reduces to (Bjørndal, 1988):

$$F'(x^*) = \delta \tag{3.20}$$

(given, of course, that $p > c$). This immediately raises the question of whether, in such circumstances, it could be optimal to drive the resource to extinction. In a much cited article, Clark (1973) demonstrates that this is a possible feasible outcome.

Suppose that our growth function is, as before, $F(x) = rx[1 - x/G]$, that p and b are constant and unchanging through time, that $\rho(x) = ax^0$, and that $p > c$. Assume, finally, that the only economic benefits forthcoming from the fishery are commercial benefits.

Next note that: $\lim_{x \to 0} F'(x) = r$. Then, given all of our assumptions, extinction will be socially optimal, if:

$$\delta_s > F'(0) = r \tag{3.21}$$

where, as before, δ_s denotes the social rate of discount.[6] The logic is that society would be better off by liquidating this form of natural capital and investing the proceeds elsewhere.

This result does not allow for the fact that parameters, such as p and b may vary through time. Perhaps the resource may be worth conserving at some point in the future?

Clark and Munro (1978) address this question, employing non-autonomous models, to be discussed at a later point in the chapter. They conclude that, if p and b vary through time, the requirements to be met, if extinction is to be optimal, become more stringent, but that the increased stringency provides no guarantee whatsoever of extinction always being non-optimal.[7]

A much more important consideration is the fact that, in many countries, extinction of fishery (and other living) resources is mandated against by law. One thinks, for example, of the Endangered Species Act in the United States, the Species at Risk Act in Canada, and international biodiversity treaties. How can we incorporate such considerations in the analysis? The answer is that the incorporation can be done with ease.

We invoke the concept of existence value. This is the idea that society enjoys benefits from the resource, over and above commercial benefits, by knowing that the resource is there and secure. Existence value can be measured in monetary terms through the use of such techniques as contingent valuation.

The inclusion of existence values in natural resource management models can be traced back to Vousden (1973), who incorporated existence value in a model on non-renewable resources. Now denote existence value, pertaining to the fishery resource as $\chi(x)$, with $\chi'(x) > 0$ and $\chi''(x) < 0$. This implies that the existence value increases with increases in the size of the stock, but at a decreasing rate. Assume, as is reasonable, that:

$$\lim_{x \to 0} \chi'(x) = \infty \tag{3.22}$$

This means that the marginal existence value becomes infinitely large if the stock is driven towards extinction.

Assume, as before, that p and b are independent of h and E respectively and that they are constant through time. The social manager's objective functional can now be expressed as follows (Clark et al., 2010a):

$$max\ PV = \int_0^\infty e^{-\delta t}\{(p - c(x))h(t) + \chi(x)\}dt \tag{3.23}$$

where δ is the social rate of discount, as before. Invoking the Maximum Principle, once again, it can be shown that the resource manager's fundamental equation now becomes:

$$F'(x) + \frac{\chi'(x^*) - c'(x^*)F(x^*)}{p - c(x^*)} = \delta \tag{3.24}$$

(see Clark et al., 2010a). Effectively, there is now an additional component of the Marginal Stock Effect.

Now let it be supposed that harvesting costs are completely insensitive to the size of the biomass $(\rho(x) = ax^0)$. The social manager's fundamental equation will reduce to:

$$F'(x^*) + \frac{\chi'(x^*)}{p - c} = \delta \tag{3.25}$$

(we assume that $p > c$).

Even if $\infty > \delta \gg F'(0) = r$, extinction is non-optimal. Given our assumption about $\chi'(x^*)$, the 'own rate of interest' of the resource (L.H.S. of 3.25) will be found to exceed δ well above $x = 0$. It would violate our asset market condition to liquidate x.

Following on from this, let us ask in general terms whether it is possible to find that $x^* < x_{MSY}$. The static Gordon-Schaefer model instructs us that x^* will always exceed x_{MSY} unless fishing effort costs are zero, which they never are. Some successors to H. Scott Gordon, employing dynamic economic models of the fishery, have insisted that the $x^* > x_{MSY}$ rule remains in force in a dynamic world.[8]

Our previous discussion makes it clear that the aforementioned successors to H. Scott Gordon were wrong. Return to equation (3.20), in which the Marginal Stock Effect (inclusive of existence value) is equal to zero. If $\delta > 0$, then clearly we have $x^* < x_{MSY}$. Herein lies the clue. Think of δ and the *MSE* pulling in opposite directions.

The larger δ, *ceteris paribus*, the smaller will be x^*; the larger the *MSE*, *ceteris paribus*, the larger will be x^*. In the static Gordon-Schaefer model, the *MSE* is positive *because* the underlying biological model is the Gordon-Schaefer model, while $\delta = 0$ by assumption (implicit). Now return to equation (3.13). If $MSE > 0$ and $\delta = 0$, the equation can hold only if $F'(x^*) < 0$, which in turn implies that $x^* > x_{MSY}$. If $MSE > 0$, and $0 < \delta < \infty$, then 'all bets are off'. All that we can say *a priori* is that: $x_{BE} < x^* < x_{MEY}$.

If $x^* < x_{MSY}$, then the logic is that it is not worth it to society to invest in the stock up to x_{MSY}. This can be seen more clearly by returning to equation (3.14), which shows the fundamental equation in terms of the cost of resource investment and the payoff from this investment at the margin. If $x^* < x_{MSY}$, then it follows that at some point below x_{MSY} it will be found that, if a further marginal investment in the resource is undertaken, the cost of the investment will exceed the payoff.[9]

For any given specific fishery, the question of whether x^* is above or below x_{msy} is ultimately an empirical one focused on the 'strength' of the *MSE* relative to the discount rate effect. This will be illustrated in two empirical case studies that will be presented in chapter 4.

All of this discussion is in the context of deterministic models. In a world of uncertainty, one might want to prescribe that x should never be allowed to fall below the estimated *MSY* level, for reasons of safety and security. This, however, is a different matter to be discussed at a later point.

3.4 Discrete time models

Up to this point, all of the models that we have discussed have been continuous time models, because, as we stated earlier, such models prove to be the most useful for expository purposes. As Clark (1990) writes, continuous time models are based upon the assumption that the response of the stock to external shocks, such as harvesting, is instantaneous. The models are unable to encompass delayed reactions, which are commonplace in the real world. Moreover, biological processes such as recruitment, growth, and migration are normally seasonal rather than continuous. In addition, for most fisheries harvesting is seasonal rather than continuous over time. Data are also usually available on a periodic basis, for example, annually.

For these reasons it is often not only necessary, but also appropriate to turn to discrete time models. This is particularly the case when employing empirical models, such as the ones that we shall examine in chapter 4. For these reasons, a digression on discrete time models of the fishery is in order.

We start with a discrete time version of a model of population dynamics:

$$x_{t+1} = x_t + (F(x_t) - h_t) \tag{3.26}$$

where x_t, h_t, and $F(x_t)$ are stock size, harvest, and natural growth, respectively, in period (year) t. The equation says that the stock size in year $t+1$ (x_{t+1}) is equal to stock size the previous year plus natural growth minus harvest.[10] Resource rent in year t is given by:

$$\pi_t = (p - c(x_t))h_t \tag{3.27}$$

where p is price of fish and $c(x_t) = b/qx_t$.

The net present value of resource rent is given by:

$$NPV = \sum_{t=1}^{\infty} \alpha^t (p - c(x))h_t \tag{3.28}$$

where $\alpha = 1/(1+i)$ is the discount factor and i is now the periodic (annual) discount rate. The objective of the resource manager is to maximize the *NPV* subject to the stock constraint.

A solution can be found by use of the method of Lagrange-multipliers (Clark, 1990):

$$L = \sum_{t=1}^{\infty} \{\alpha^t (p - c(x))h_t - \mu_t [x_{t+1} - x_t - (F(x_t) - h_t)]\} \tag{3.29}$$

where μ_t is the Lagrangian multiplier, or shadow price of the resource.

First order conditions are given by:

$$\frac{\partial L}{\partial h_t} = \alpha^t (p - c(x)) - \mu_t = 0 \tag{3.30}$$

$$\frac{\partial L}{\partial x_t} = -\alpha^t c'(x) h_t - \mu_{t-1} + \mu_t (1 + F'(x)) = 0 \tag{3.31}$$

$$\frac{\partial L}{\partial \mu_t} = x_{t+1} - x_t - (F(x_t) - h_t) = 0 \tag{3.32}$$

From equation (3.30), we have:

$$\mu_t = \alpha^t (p - c(x)) \tag{3.33}$$

It can be seen that this condition is analogous to equation (3.10). Next observe that μ_{t-1} is simply: $\mu_{t-1} = \alpha^{t-1}(p - c(x))$.

Substituting for μ_t and μ_{t-1} in equation (3.31), and noting that in the steady state we have $h_t = F(x_t)$, we obtain:

$$-\alpha^t c'(x) F(x) - \alpha^{t-1}(p - c(x)) + \alpha^t (p - c(x))(1 + F'(x)) = 0 \tag{3.34}$$

Then, after simplification, we have:[11]

$$F'(x^*) - \frac{c'(x^*)F(x^*)}{p - c(x^*)} = i \tag{3.35}$$

where x^* is the optimal stock level. We see that this is analogous to equation (3.13). Moreover, if we use the Schaefer function $F(x) = rx[1 - x/G]$, the optimal stock level is given by equation (3.16). In case the Marginal Stock Effect is zero, this reduces to:

$$F'(x^*) = i \tag{3.36}$$

We see this is analogous to equation (3.20). Given the Schaefer function, the equation can be solved for the following explicit expression for the optimal stock level x^*:

$$x^* = (i - r)G/2r.$$

This says that x^* is determined by the relationship between i, r, and G.

3.5 The basic non-linear dynamic economic model of the fishery

We now return to the continuous time dynamic economic model of the fishery, and relax one of our highly restrictive assumptions, namely that the demand for harvested fish and the supply of fishing effort are both perfectly price elastic. We retain the assumptions that all parameters are independent (autonomous) of time, and that produced and human capital in the fishery are perfectly malleable.

By allowing for the possibility of downward sloping demand curves and upward sloping supply curves of fishing effort, our model becomes non-linear in the control variable, $h(t)$. The main consequences, as we shall see, are twofold. First it will be found that the most rapid rate of investment/disinvestment in the resource, when $x \neq x^*$, is non-optimal. Rapid resource investment/disinvestment will lead to penalties. The second consequence is that we may be confronted with multiple equilibria.

Let us consider the following example. Suppose that the entire harvest from the fishery goes into the export market, and that the resource has no existence value ($\chi(x) = 0$). The net economic return from the resource consists of profits from sales in the export market. It will be further supposed that unit fishing effort costs are constant, but that the country as a whole faces a downward sloping demand for its exported fish, i.e., the price elasticity of the demand for its fish is finite. The social manager's objective functional can be expressed as follows:

$$PV = \int_0^\infty e^{-\delta t}[W(h) - c(x)h]dt \tag{3.37}$$

where $W(h) = p(h) \cdot h$ is the total (gross) revenue from harvested fish with price dependent on quantity, i.e., $p = p(h)$ with $p'(h) < 0$. As before δ is the social rate of discount.

It can be shown (see Clark and Munro, 1982) that the fundamental equation will now take the following form:

$$F'(x^*) - \frac{c'(x^*)F(x^*)}{W'(F(x^*)) - c(x^*)} = \delta \tag{3.38}$$

This looks exactly like the fundamental equation for the linear case, except for the denominator of the second term on the L.H.S. (the *MSE*). The term $W'(F(x^*))$ is, of course, simply the marginal revenue from harvesting the sustainable yield.

Suppose, for the moment, that a unique equilibrium biomass exists. Even in this relatively simple case, the optimal approach path (investment) programme will be quite different from what we encountered in the linear case.

It can be shown that the decision rule to be applied along the approach path is given by

$$F'(x) = \frac{c'(x)h}{W'(h) - c(x)} + \frac{\dot{\psi}}{\psi} = \delta \tag{3.39}$$

See Clark and Munro (1982) for details.

It will be recalled (see equation (3.8)) that $\psi(t)$ is the current value adjoint, or co-state, variable, which can be viewed as the shadow price of the natural capital in the form of the fishery resource. Along the approach path, $\psi(t)$ will be subject to continuous change, which means that capital gains (losses) are continually forthcoming. These capital gains/losses must be taken into account in the decision making process. When the equilibrium, optimal, stock, x^*, has been achieved, $\dot{\psi}$ will equal zero. Equation (3.39) will reduce to equation (3.38) (see Clark and Munro, 1982).

The optimal approach path is now asymptotic, meaning that investment/disinvestment in the resource should proceed at a gradual, rather than the most rapid, pace. This is pictured in Figure 3.4.

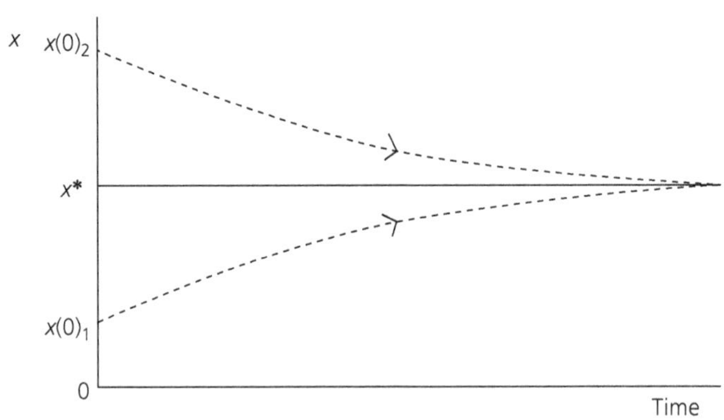

Figure 3.4. Optimal resource investment time paths: the non-linear model

Consider the logic of the implied resource investment programmes. Suppose that $x(0) > x^*$, so that a programme of resource disinvestment is required. Why not disinvest at the most rapid rate, by setting $h(t) = h_{max}$? The answer is simple. The temporary large harvests would 'flood the market', with a consequent possibly large fall in $p(h)$. One could well find that $W'(h)$ would become negative. Economic rationality dictates a gradual rate of disinvestment.

Consider the opposite case, as well, in which we have $x(0) < x^*$ and a programme of resource investment in required. If the rate of investment in the resource is maximized, $h(t) = 0$, then $W'(h)$ could go to astronomical heights. By maximizing the rate of investment, we would forgo the opportunity of harvesting and selling some fish at extremely high prices. Economic rationality now dictates a gradual rate of resource investment.

The second consequence of non-linear models is that we are faced with the distinct possibility of multiple equilibria. We can illustrate this possibility by taking the following straightforward example. Suppose, as before, that we are faced with a downward sloping demand curve for harvested fish, a constant unit price of fishing effort, but now suppose that harvesting costs are completely insensitive to the size of the biomass. The fundamental equation can now be expressed as:

$$[W'(F(x^*)) - c](F'(x^*) - \delta) = 0 \tag{3.40}$$

The biomass level at which $F'(x) = \delta$ is certainly an equilibrium biomass. However, there may be other biomass levels, at which $W'(F(x)) = c$. These other levels are also equilibrium biomass levels. Consider Figure 3.5. Suppose that $x(0) > x_3$. It would be nonsensical to reduce stock size below x_3. We

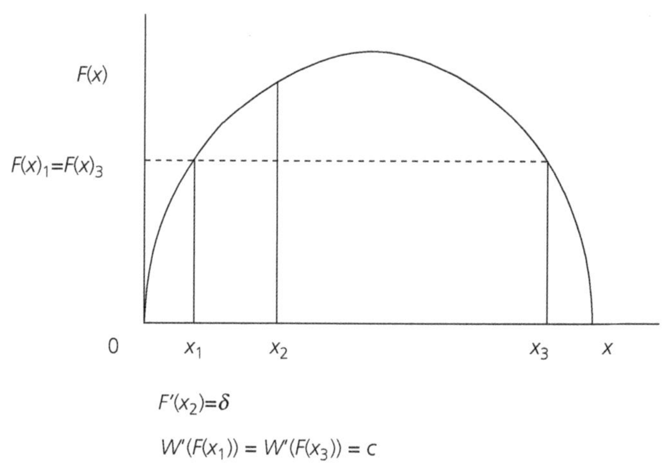

Figure 3.5. Multiple equilibria: the non-linear model

recognize that $W''(F(x^*)) < 0$. If there was a marginal reduction of x below x_3, then Figure 3.5 makes it clear that we would have: $W'(F(x)) < c$.

The example we have shown is straightforward. We have an unstable equilibrium, x_2, bounded by two stable equilibria. We would go to x_1 or x_3 depending upon our starting point. One could, however, conceivably encounter a case in which there were many more than three multiple equilibria. The determination of an *optimum optimorum*, in such circumstances, could prove to be daunting (Clark and Munro, 1982).

3.6 Dynamic economic models of the fishery and policy-making in the real world: the case of Australia

Before proceeding further, let us stop and ask if the dynamic economic models of the fishery we have developed thus far have had any perceptible impact upon policy-makers in the real world. The impact of the static economic models of the fishery on policy-makers has been noted by us many times.

We have already noted in chapter 1 that, in broad general terms, the dynamic economic model of the fishery has had an impact. The idea, or concept, of natural resources, including capture fishery resources, constituting natural capital has gained increasingly wide acceptance. We need only be reminded that the World Bank/FAO study *The Sunken Billions*, when discussing the depletion of capture fishery resources states that 'the depletion of a nation's fish stocks constitutes a loss of—the nations' stock of natural capital. Similarly, the depletion of global fish stocks constitutes a loss of global natural capital' (World Bank, 2009, p. 50).

More specifically, the dynamic economic model of the fishery, which we have discussed to this point, is being applied for policy purposes. The pioneering country, in this regard, is Australia. In Australia, all fisheries to be found from three miles from shore to the edge of the Australian Fishing Zone (AFZ)[12] are under the management jurisdiction of the federal government, with the actual day-to-day management being undertaken by the Australian Fisheries Management Authority (AFMA) (Gooday et al., 2009).

In 2005, the Australian Government released a document, *Commonwealth Harvest Strategy Policy*, which mandates that fish stocks are to be maintained at sustainable levels, while targeting maximum economic yield (*MEY*). In response to this policy directive, 'AFMA requires harvest strategies to maintain fish stocks, on average, at a target biomass equal to the stock size required to produce maximum economic yield' (Gooday et al., 2009, p. 52).

To date there is only one fishery to which this *MEY* policy has been fully applied, the Northern Prawn fishery (Gooday et al., 2009). The precedent has, however, been set, and other fisheries are certain to follow.

We have discussed the concept of *MEY* in the context of the static Gordon-Schaefer model. Is the Australian policy, then, no more than an application of the Gordon-Schaefer model? The answer is that the policy is much more. This fact is made very clear in the document produced by the Australian Bureau of Agricultural and Resource Economics (ABARE), *Development of Methods and Information to Support the Assessment of Economic Performance in Commonwealth Fisheries* (Kompas et al., 2009). The document provides the economic theory underlying the policy, and reports upon a fully developed bioeconomic model of the Northern Prawn fishery. The *MEY* discussed in the document is what might best be termed a 'dynamic' *MEY*, in that, by assumption, $\delta > 0$. The bioeconomic model for the Northern Prawn fishery assumes a social rate of discount of 3 per cent (Kompas et al., 2009, p. 38).

The discussion of the underlying theory in the ABARE document might best be characterized as a description (minus the equations) of the autonomous dynamic economic models of the fishery that we have analysed up to this point.

There is one aspect of the issue, which the ABARE document touches upon but briefly, namely that we do not live in an autonomous (let alone deterministic) world. Demand for fish, fishing effort supply, and environmental conditions are bound to shift continuously through time.

The Northern Prawn fishery bioeconomic model uses a 50 year time horizon. Surely, if one were to set up the model properly, it would be necessary to make projections, for example, of the demand conditions for Northern Prawn harvest over the 50 year time horizon. If the answer is yes, then the information requirements become overwhelming. One would have reason to doubt whether, in fact, the dynamic economic models of the fishery did have any real policy relevance. The answer to this critical question is to be found within the context of dynamic *non-autonomous* economic models of the fishery, to which we now turn.

3.7 The non-autonomous linear economic model

We restore our linearity assumptions, but now allow for the possibility that some (or all) of the parameters are functions of time. Thus, our model is now, by definition, non-autonomous.

We commence by assuming that the price of harvested fish, p, and the unit cost of fishing effort, b, are functions of time, while all other parameters remain independent of time. We thus have $p = p(t)$; $b = b(t)$. The fact that $b(t)$ shifts continuously through time automatically means that the unit cost of harvesting will also shift through time. We shall express unit harvesting costs as: $c(x, t)$. We deal with the shifts in $c(x, t)$ through time by introducing a variable coefficient, $\phi(t) \geq 0$. We thus have:

$$c(x, t) = \phi(t)c(x(t)) \tag{3.41}$$

(Clark and Munro, 1982).

The objective functional now becomes:

$$max\ PV = \int_0^{\infty} e^{-\delta t}[p(t) - \phi(t)c(x(t))]h(t)dt \tag{3.42}$$

We are now in a position to address the key question. In order to determine $x^*(t)$, is it necessary to have prior information on the time path of $p(t)$ and $\phi(t)c(x(t))$ from $t = 0$ to $t = \infty$? The great twentieth-century British economist, John Maynard Keynes, set forth an optimal investment rule, which implies that the answer to our question is yes. The Keynesian rule states that a marginal investment is justified if and only if the *PV* of the economic returns from the investment through time are not less than the cost of the investment (Keynes, 1936).

By the 1950s, economists began to suspect that the Keynesian optimal investment rule was in need of serious modification. The suspicions proved to be well founded.

The modifications were developed most fully by Arrow (1964, 1968). If the marginal investment is reversible, it is not necessary to be able to forecast (in our case) future price and cost shifts, except for the immediate future. The correct optimal investment decision rule is *myopic* (Arrow, 1964, p. 27). To return to our Northern Prawn fishery case, it is *not* necessary to make cost and price projections far out into the future. To quote Arrow, who offers a wonderful example of understatement, the myopic decision rule 'implies a considerable economy of information needs' (Arrow, 1964, p. 27).

If the marginal investment is not readily reversible, then, of course, the myopic investment decision rule has to be modified. It will still remain true, however, that it will not be necessary to make price, cost, and other projections far into the future, except in extreme cases. It is difficult to overemphasize the importance, for practical policy purposes, of the myopic optimal investment decision rule.[13]

Now return to the objective functional as expressed in equation (3.42). In setting about to maximize the PV of the economic rent from the fishery, we set up the Hamiltonian, invoke the Maximum Principle, and proceed as before. The calculations can be safely left to the reader. In any event, the fundamental equation now takes the following form (Clark and Munro, 1982):

$$F'(x^*) - \frac{\phi(t)c'(x^*)F(x^*)}{p(t) - \phi(t)c(x^*)} + \frac{\dot{p}(t) - \dot{\phi}(t)c(x^*)}{p(t) - \phi(t)c(x^*)} = \delta \tag{3.43}$$

Once again, the investment rule is myopic, in that only predictions of the immediate future are required, subject to the following condition. The shift in $(p(t) - (\phi(t)c(x^*))$ must not be so great as to cause the constraints on the control variable, $h(t)$, to become binding (Clark and Munro, 1982). Recall that $0 \leq h(t) \leq h_{max}$.

Next consider the third term on the L.H.S. of (3.43). If the term is positive, it indicates an increase in the *PV* of the resource rent stream in the immediate future. In effect, it means that the owner of the resource (public or private) can look forward to a capital gain. Conversely, if the term is negative, the resource owner can look forward to a capital loss in the immediate future. Let us, therefore, refer to this term as the Marginal Capital Gain Effect (*MCGE*), remembering that capital 'gains' can be both positive and negative.

To illustrate, let us focus on the impact of shifts in $p(t)$ through time, assuming for the time being, that all other parameters are independent of time. Return to equation (3.43), assume that $c'(x) \neq 0$, and consider the second term on the L.H.S., the Marginal Stock Effect (*MSE*). An increase in $p(t)$ will weaken the *MSE*, which in, of, and by itself, leads to a reduced $x^*(t)$. The increase in $p(t)$, everything else being equal, will increase the cost of a marginal resource investment, the forgone current resource rent entailed by the investment.

Now consider the third term on the L.H.S. of equation (3.43), the *MCGE*. This indicates that the *anticipation* of the increase in $p(t)$ tomorrow, will, in, of, and by itself, lead to an increase in $x^*(t)$ today. The rationale is straightforward. If you expect the fishery resource stock to be more valuable tomorrow, increase your holding of the fishery capital today. We can picture the singular path, the time path of $x^*(t)$, as illustrated in Figure 3.6.

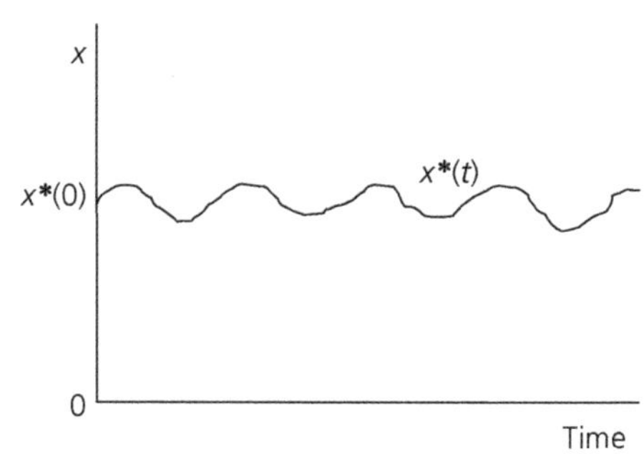

Figure 3.6. Non-autonomous models: the singular path, when myopic decision rules prevail

As a next step, we acknowledge the fact that the myopic decision rule cannot always be followed. If the parameter shift is so great that the constraints on the control variable become binding, we can be confronted with what Arrow (1968) terms a 'blocked interval' in which we are temporarily forced off the Singular Path. We continue with the example of shifts in $p(t)$, with all other parameters assumed to be independent of time.

Let it be supposed that $p(t)$ is constant up to $t = T$, at which point there is a large discrete increase in $p(t)$, $\dot{p}(T) = +\infty$. From there on $p(t)$ remains constant through time. This set of circumstances will lead to a 'spike' in the Singular Path at $t = T$, as illustrated in Figure 3.7.

Ideally one would like to stabilize the resource at $x = x_1^*$ until $t = T$, and then at that instant increase x to its carrying capacity, G, and then in the next instant reduce the resource to $x = x_2^*$. This is impossible. The constraints on the control variable become binding: one cannot reduce h below 0, or increase it above h_{max}. The consequence is that we are forced off the Singular Path and are thus confronted with a 'blocked interval', which in Figure 3.7 extends from $t = t_1$ to $t = t_2$.

The determination of $t = t_1$, the time at which we set $h(t) = 0$ (once t_1 is determined, t_2 is determined as well) involves resolving a trade-off problem. The later we wait to set $h = 0$, the less will be the sacrifice of forgone harvests, but the smaller will be capital gains to be enjoyed from the discrete increase in $p(t)$ at $t = T$. Clark and Munro (1982) provide the solution and come forth with a complex decision rule equation that we shall not repeat here. The interested reader is directed to Clark and Munro (1982, equation 26).

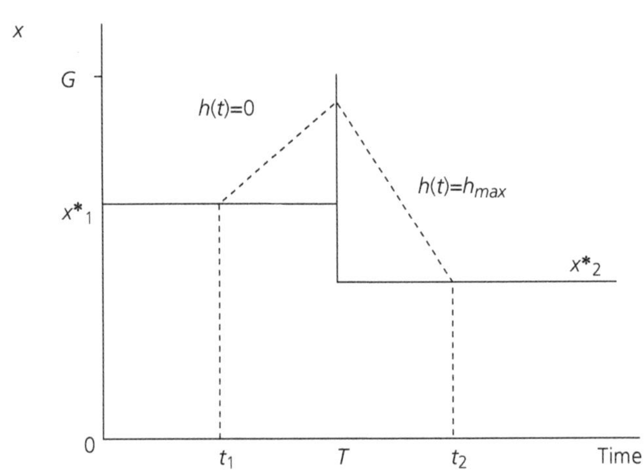

Figure 3.7. The singular path 'spike' and blocked intervals

If we are faced with blocked intervals, then clearly myopic decision rules will no longer do. In our example, the change in $p(t)$ has to be anticipated well in advance. The information demands could become significant.

Finally, let us note that the non-autonomous model can readily accommodate environmental fluctuations, although with less dramatic results. Let us now suppose that the carrying capacity, *G*, is a function of time: $G = G(t)$, and that we have:

$$F(x, t) = rx\left[1 - \frac{x}{G(t)}\right] \tag{3.44}$$

Assume, for the sake of simplicity, that the price of harvested fish, p, and the unit cost of fishing effort, *b*, are both constant through time.

In applying the maximum principle, it will be found that $G(t)$ is not differentiated with respect to *t*. The fundamental equation can be expressed as follows. There is no equivalent to the *MCGE* to be found in the equation:[14]

$$\left[\frac{\partial F(x, t)}{\partial x} - \frac{[\partial c(x, t)/\partial x]\cdot F(x, t)}{p - c(x, t)}\right]_{x = x^*(t)} = \delta \tag{3.45}$$

The optimal biomass is a function of time, i.e., $x^* = x^*(t)$, and, as before, the investment rule is *myopic*. One is not confronted with the impossible task of projecting environmental fluctuations into the far distant future.[15]

3.8 Myopic decision rules, blocked intervals, and a developing fishing state resource management problem

The simple linear non-autonomous economic model of the fishery has been shown to be useful in addressing some broad long-range fisheries management issues. The model can be used for much more than this, however, and can be used to gain insight into specific, complex medium to long-term policy issues.

We take as an example a specific policy issue that came to light in a 1970s multi-disciplinary study of fisheries management problems in the Southeast Asian fishing state of Malaysia. The issue continues to be entirely relevant in the twenty-first century, and is relevant to both developed and developing fishing states. Having said this, the issue is of particular interest to developing fishing states.

The 1970s study in question focused on the inshore fisheries of the neighbouring Malaysian states of Penang and Kedah. One of the authors (Munro) was the lead economist on the study project (Munro and Chee, 1978).

It was observed in the inshore fisheries of the two Malaysian states that, while produced capital was reasonably mobile between the fisheries and the rest of the Malaysian economy, human capital was not. There were many barriers, both economic and social, that impeded the mobility of labour from the fisheries to the rest of the economy. The government of Malaysia recognized the problem and planned to undertake measures to dismantle the barriers to labour mobility (Munro and Clark, 1979, p. 15). The fact that labour was blocked from leaving the fisheries meant that the opportunity cost of labour was low, perhaps approaching zero, and was almost certainly below the private cost of labour. If one is looking at the management of fisheries from the social perspective, then it is appropriate in such circumstances to 'shadow price' the labour at its true opportunity cost (Little and Mirrlees, 1968). It has been observed that this is a situation that can be found in fisheries in many parts of the world[16] (Copes, 1972; Crutchfield, 1975).

Noting that the inshore fisheries in the two Malaysian states were close to being Pure Open Access ones, the following questions are raised in the study. First, given that the social cost of labour in the fishing industries was almost certainly below the private cost, is it not possible that, for any one of these fisheries, the socially optimal biomass level, to begin with at least, $x^*(0)$, is, in fact, actually *below* the private Bioeconomic Equilibrium level, x_{BE}? In other words, is it not possible that Pure Open Access is leading to *underexploitation* of the fishery resources (Munro and Chee 1978; Munro and Clark, 1979)?

If, in fact, we have $x^*(0) < x_{BE}$, the appropriate policy would be clear. The private sector fishermen, the vessel owners, should be subsidized to the point that the resource is stabilized at $x(t) = x^*(0)$. This then leads to a second question.

The Government of Malaysia stated that it was its intention to break down the barriers to labour mobility. If the authorities were to succeed, the opportunity cost of labour would increase. It would be reasonable to suppose that the 'shadow price' and the private cost of fisheries labour would eventually coincide. The second question then is what would the appropriate policy be, if the authorities were to succeed in the plans to dismantle the barriers to labour mobility?[17] One had to assume that the dismantling of the barriers, and concomitant rise in the 'shadow price' of labour in the fisheries, would occur gradually over time.

Once again, the answer to the second question appears to be straightforward. In such circumstances, one should initially subsidize the private fishing industry, and then, as the 'shadow price' of labour in the fisheries begins to rise, one should steadily reduce the subsidy, until it becomes negative (i.e., a tax). Note that we have no option, other than to bring a non-autonomous economic model of the fishery to bear, in analysing this apparently straightforward policy issue.

The one difficulty with this straightforward optimal resource management policy is that it flies in the face of political reality. Throughout the world there are countless examples of 'temporary' subsidies (and taxes) that become permanent. In the case of subsidies, it commonly becomes politically very difficult to eliminate the subsidies once the economic justification for the subsidies has vanished. This caused the authors of the Malaysian study to raise a third question. What is the optimal resource management policy in the circumstances described thus far, if the authorities know full well that, upon introducing a fisheries subsidy, it will be very difficult to eliminate the subsidy in the future? The social benefits arising from the subsidy in the near to medium term could be followed by a permanent burden to society in the long term (Munro and Clark, 1979).

The analytical problem involves a solution that, as we shall see, involves myopic decision rules, and blocked intervals of permanent length.[18]

Munro and Clark (1979) proceed as follows. They address the problem by considering two polar extremes: (i) the case in which the authorities can effortlessly increase or decrease the fishing subsidy at will; (ii) the case in which, once a fishing subsidy is introduced, it can never be reduced.[19]

Munro and Clark (1979) consider both the case in which the economy outside of the fishing sector under consideration is enjoying full employment, and the case in which the outside economy is experiencing unemployment. We shall confine ourselves to the first case—full employment in the economy outside of the fishery.

Various simplifying assumptions are introduced. The underlying economic model of the fishery is the linear, Schaefer based model that we have now discussed extensively.

It is assumed further that produced capital is perfectly malleable with respect to the fisheries, while, initially, human capital is perfectly non-malleable—there are insurmountable barriers to labour mobility between the fishery and the rest of the economy. We commence with the fishery in Bionomic Equilibrium, as perceived by the fishing industry. The fishing industry will, it is assumed, always adjust to the point where the fishery is perceived by the industry to be in Bioeconomic Equilibrium, i.e., up to the point that the economic rent from the fishery, as seen by the private fishing industry, is equal to zero. The private industry's perception of the economic rent will obviously be affected by the subsidy, which can be negative as well as positive.[20]

It is assumed that unit private harvesting costs, $c_p(x)$, and the price of harvested fish, p, are independent of time. It is further assumed, if the barriers to labour mobility are fully dismantled—human capital is perfectly mobile—that $c_s(x) = c_p(x)$, where $c_s(x)$ denotes unit social harvesting costs. Finally, it is assumed that $\delta > 0$, where, as before, δ is the social rate of discount.

We now consider case (i)—there are no constraints on fisheries subsidy policy; the authorities can increase or decrease the subsidies at will. It will be supposed that the authorities' efforts to dismantle the barriers to labour mobility commence at $t = T_u \geq 0$, and that these efforts continue until complete labour mobility is achieved.

The shifts in the social cost of harvesting, arising from the rising social costs of fishing labour, after $t = T_u$, are dealt with by the introduction of the variable coefficient, $\phi(t)$, with which we are now familiar, where $1 \leq \phi(t) \leq \bar{\phi}$. Thus, we have: $c_s(x,t) = \phi(t)c_s(x)$. When $\phi(t) = \bar{\phi}$, we shall be at the point where $c_s(x) = c_p(x)$, the unit social cost of harvesting is equal to the unit private cost of harvesting. The social objective functional is given by:

$$PV = \int_0^\infty e^{-\delta t}(p - \phi(t)c_s(x))h(t)dt \tag{3.46}$$

and we have:

$$F'(x^*) - \frac{\phi(t)c_s'(x^*)F(x^*)}{p - \phi(t)c_s(x^*)} + \left[\frac{-\dot{\phi}(t)c_s(x^*)}{p - \phi(t)c_s(x^*)}\right] = \delta \tag{3.47}$$

where x^* denotes the socially optimal biomass level. Obviously, $x^* = x^*(t)$, i.e., the socially optimal biomass level will be a function of time.

What we are presented with is a 'myopic' investment decision rule. Only the immediate future has to be forecast.

Now let $x_{BE}(0)$ be the Bionomic Equilibrium at $t = 0$, as perceived by the fishing industry, i.e., prior to the introduction of the subsidy. Let $x^\dagger$ be the socially optimal biomass level when $c_s(x) = c_p(x)$. Recalling that the

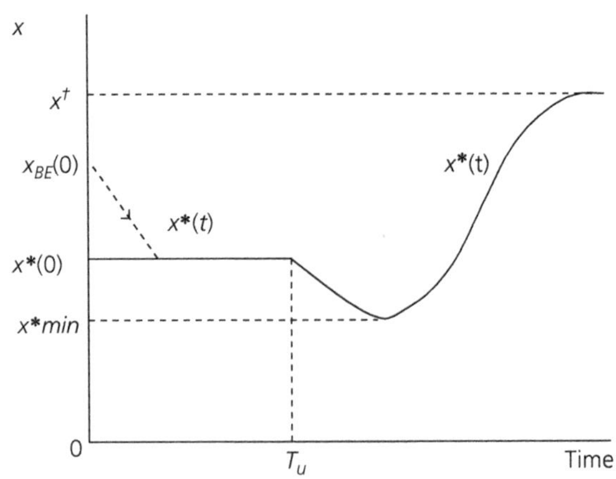

Figure 3.8. An unconstrained fishing subsidy programme

Source: Munro and Clark (1979, Figure A8)

We now consider case (i)—there are no constraints on fisheries subsidy policy; the authorities can increase or decrease the subsidies at will. It will be supposed that the authorities' efforts to dismantle the barriers to labour mobility commence at $t = T_u \geq 0$, and that these efforts continue until complete labour mobility is achieved.

The shifts in the social cost of harvesting, arising from the rising social costs of fishing labour, after $t = T_u$, are dealt with by the introduction of the variable coefficient, $\phi(t)$, with which we are now familiar, where $1 \leq \phi(t) \leq \bar{\phi}$. Thus, we have: $c_s(x, t) = \phi(t)c_s(x)$. When $\phi(t) = \bar{\phi}$, we shall be at the point where $c_s(x) = c_p(x)$, the unit social cost of harvesting is equal to the unit private cost of harvesting. The social objective functional is given by:

$$PV = \int_0^{\infty} e^{-\delta t}(p - \phi(t)c_s(x))h(t)dt \tag{3.46}$$

and we have:

$$F'(x^*) - \frac{\phi(t)c_s'(x^*)F(x^*)}{p - \phi(t)c_s(x^*)} + \left[\frac{-\dot{\phi}(t)c_s(x^*)}{p - \phi(t)c_s(x^*)}\right] = \delta \tag{3.47}$$

where x^* denotes the socially optimal biomass level. Obviously, $x^* = x^*(t)$, i.e., the socially optimal biomass level will be a function of time.

What we are presented with is a 'myopic' investment decision rule. Only the immediate future has to be forecast.

Now let $x_{BE}(0)$ be the Bionomic Equilibrium at $t = 0$, as perceived by the fishing industry, i.e., prior to the introduction of the subsidy. Let $x^\dagger$ be the socially optimal biomass level when $c_s(x) = c_p(x)$. Recalling that the

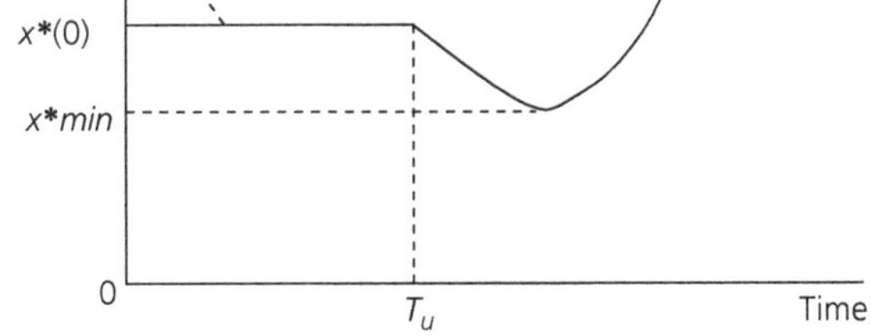

Figure 3.8. An unconstrained fishing subsidy programme

Source: Munro and Clark (1979, Figure A8)

economic model of the fishery to bear, in analysing this apparently straightforward policy issue.

The one difficulty with this straightforward optimal resource management policy is that it flies in the face of political reality. Throughout the world there are countless examples of 'temporary' subsidies (and taxes) that become permanent. In the case of subsidies, it commonly becomes politically very difficult to eliminate the subsidies once the economic justification for the subsidies has vanished. This caused the authors of the Malaysian study to raise a third question. What is the optimal resource management policy in the circumstances described thus far, if the authorities know full well that, upon introducing a fisheries subsidy, it will be very difficult to eliminate the subsidy in the future? The social benefits arising from the subsidy in the near to medium term could be followed by a permanent burden to society in the long term (Munro and Clark, 1979).

The analytical problem involves a solution that, as we shall see, involves myopic decision rules, and blocked intervals of permanent length.[18]

Munro and Clark (1979) proceed as follows. They address the problem by considering two polar extremes: (i) the case in which the authorities can effortlessly increase or decrease the fishing subsidy at will; (ii) the case in which, once a fishing subsidy is introduced, it can never be reduced.[19]

Munro and Clark (1979) consider both the case in which the economy outside of the fishing sector under consideration is enjoying full employment, and the case in which the outside economy is experiencing unemployment. We shall confine ourselves to the first case—full employment in the economy outside of the fishery.

Various simplifying assumptions are introduced. The underlying economic model of the fishery is the linear, Schaefer based model that we have now discussed extensively.

It is assumed further that produced capital is perfectly malleable with respect to the fisheries, while, initially, human capital is perfectly non-malleable—there are insurmountable barriers to labour mobility between the fishery and the rest of the economy. We commence with the fishery in Bionomic Equilibrium, as perceived by the fishing industry. The fishing industry will, it is assumed, always adjust to the point where the fishery is perceived by the industry to be in Bioeconomic Equilibrium, i.e., up to the point that the economic rent from the fishery, as seen by the private fishing industry, is equal to zero. The private industry's perception of the economic rent will obviously be affected by the subsidy, which can be negative as well as positive.[20]

It is assumed that unit private harvesting costs, $c_p(x)$, and the price of harvested fish, p, are independent of time. It is further assumed, if the barriers to labour mobility are fully dismantled—human capital is perfectly mobile—that $c_s(x) = c_p(x)$, where $c_s(x)$ denotes unit social harvesting costs. Finally, it is assumed that $\delta > 0$, where, as before, δ is the social rate of discount.

dismantling of the barriers to labour mobility commence at $t = T_u$, Munro and Clark show that the Singular Path can be seen as illustrated in Figure 3.8.

At $t = T_u$, there is actually a decline in $x^*(t)$ with $x^*(t)$ falling to $x^*(t) = x^*_{min}$. This is due to the impact of the Marginal Capital Gain Effect (*MCGE*), which we discussed in 3.7. In this particular model we have $MCGE \leq 0$.

We next consider the much more difficult, but also far more interesting, case (ii)—subsidies once introduced are irreversible. Thus the subsidy programme can only be used to decrease x. The programme can never be used to increase x.

The singular path as shown in Figure 3.8 can be thought of as showing the *optimum optimorum* through time. If, commencing at $t = 0$, the authorities set out straightaway to drive x to $x^*(t)$, the actual $x(t)$ would, in the more distant future, be far from the *optimum optimorum*. We would be faced with a blocked interval of immense proportions, with the blocked interval arising because of constraints on the subsidy rate.[21]

The problem then is to select what can be termed a constrained optimal biomass, which along with Munro and Clark (1979), we shall denote as x_z. The selection is to be made by balancing gains in the near future against losses in the more distant future. Once x_z is achieved, we are 'locked in'.

Return to Figure 3.8. It is possible that: $x_z \geq x^*(0)$, but it is also possible that: $x^*_{min} \leq x_z < x^*(0)$. If we have the second outcome, then a two-stage process may be involved (see Munro and Clark, 1979). We shall ignore this possibility, and suppose for the sake of simplicity, that we have: $x_z \geq x^*(0)$, which is shown in Figure 3.9.

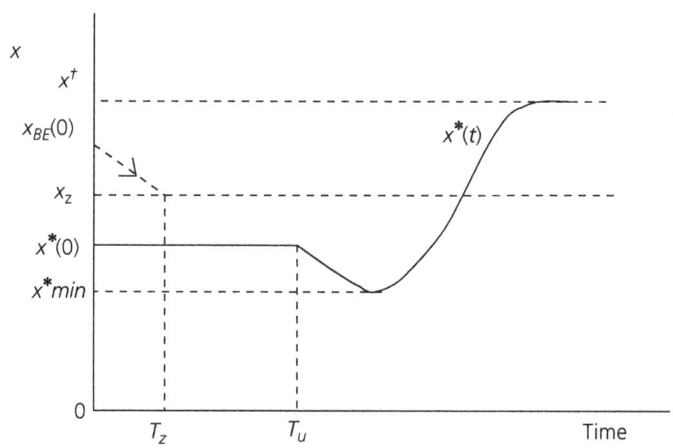

Figure 3.9. A constrained fishing subsidy programme

Source: Munro and Clark (1979, figure A8)

In passing, the selection of x_z is not restricted. It is possible that $x_z > x_{BE}(0)$. In this case, since we commence with subsidies equal to zero, and since subsidies cannot be reduced, the selected x_z would be infeasible. The least worst policy in this situation would be to introduce no subsidies and have the fishery remain at $x = x_{BE}(0)$.

Let the time at which x_z is achieved be denoted as $T_z \geq 0$. The objective functional can then be expressed as follows:

$$PV = \int_0^{T_z} e^{-\delta t}[p - \phi(t)c_s(x)]h(t)dt + \int_{T_z}^{\infty} e^{-\delta t}[p - \phi(t)c_s(x_z)]F(x_z)dt \tag{3.48}$$

We now differentiate PV with respect to x_z and set $dPV/dx_z = 0$. The resultant investment decision rule equation is, quite frankly, formidable. It can be shown that the formidable equation (which we shall not impose on the reader) can be simplified to the following elaborate version of the fundamental equation (see Munro and Clark, 1979 for details):

$$\omega\left\{F'(x_z) - \frac{\gamma_\delta(T_z)c_s'(x_z)Fx_z)}{p - \gamma_\delta(T_z)c_s(x_z)}\right\} = \delta \tag{3.49}$$

where

$$\gamma\delta(T_z) = \delta\int_0^{\infty} e^{-\delta t}\phi(t + T_z)dt \tag{3.50}$$

and

$$\omega = \frac{p - \gamma_\delta(T_z)c_z(x_z)}{p - \phi(T_z)c_s(x_z)} \tag{3.51}$$

Myopic decision rules must now be replaced by an 'omniscient' decision rule. It is necessary to project far, far out into the future. This 'omniscient' view of the world is captured in multiplier $\gamma_\delta(T_z)$, and the corrective ω. In the first instance, $\gamma_\delta(T_z)c_s(x_z)$ can be thought of as a weighted average of current and future harvesting costs. The larger are future harvesting costs, the sooner they are realized, and the less that future harvesting costs are discounted, the larger will be the Marginal Stock Effect, and the larger will be x_z. On the other hand, the lower is the current social cost of labour in the fishing industry, the lower will be x_z.

The $\gamma_\delta(T_z)$ multiplier actually does more than this. Given that: $x_z < x_{BE}(0)$, we have to consider the transition time from $x = x(0)$ to $x = x_z$. Recall that there is an upper bound on $h(t)$, $h(t) = h_{max}$. The economic benefits to be enjoyed from expanding the fishery are to be realized in the near future; the burden of expanding the fishery is to be felt in the more distant future. The longer the transition time from $x = x(0)$ to $x = x_z$, the smaller will be the size of the near future economic benefits arising from expanding the fishery.

Observe that the $\gamma_\delta(T_z)$ multiplier is an increasing function of T_z. The farther out in the future is T_z, the larger is $\gamma_\delta(T_z)$. The effect of using the $\gamma_\delta(T_z)$ multiplier is to translate the loss of near future economic benefits, arising from a transition phase being long rather than short, into future cost equivalents (Munro and Clark, 1979).

The corrective, ω, works in the opposite direction to $\gamma_\delta(T_z)$. If $\gamma_\delta(T_z) > 1$, then $\omega < 1$. The corrective, which is applied to the yield on the marginal investment in the resource, is accounting for the fact that future returns on the resource investment will be lower than current returns. The corrected marginal yield on the resource investment is analogous to the financial concept of 'yield to maturity',[22] although in this case, the maturity date is $t = \infty$.

Thus, the larger ω, the less will be the incentive to invest in the resource; the smaller will be x_z. Having said this, given the existence of a positive MSE, it can be shown that the impact of ω upon x_z can never exceed that of $\gamma_\delta(T_z)$ (see Munro and Clark, 1979, p. 24).

Not surprisingly, the 'omniscient' investment decision rule is very sensitive to the size of δ. Suppose, for example, that $\delta = 0$. It can be observed that (see Munro and Clark, 1979):

$$\lim_{\delta_s \to 0} \gamma_\delta(T_z) = \bar{\phi} \tag{3.52}$$

Thus, if $\delta = 0$, equation (3.49) reduces to:

$$F'(x_z) - \frac{\bar{\phi} c_s'(x_z) F(x_z)}{p - \bar{\phi} c_s(x_z)} = 0 \tag{3.53}$$

Future harvesting cost considerations swamp near future economic gains to be realized by expanding the fishery. Return to Figures 3.8 and 3.9. Equation (3.49) implies that $x_z = x^\dagger > x_{BE}(0)$. The solution is infeasible. The least worst policy for the authorities, the resource managers, is to set the fisheries subsidy rate equal to zero.

The case that we have discussed is specific to a particular developing fish state. Does it really have relevance to other parts of the world? The answer is most definitely. Copes (1972), writing about a fishing province in Canada subject to chronic unemployment, states that subsidy policies designed to expand fisheries beyond Bionomic Equilibrium are to be found in various parts of the world. The policies are typically justified by resource managers on the grounds of alleviating unemployment in fishing communities (Copes, 1972).

In any event, we can indeed conclude from this discussion that a case can be made for the use of subsidies, if the social cost of labour used in the fishery is far below the private cost of such labour. It must be recognized, however, that we live in a dynamic world, in which conditions are constantly changing. The resource managers, if they are to introduce subsidies must, therefore, be ruthless and uncompromising realists. Can the subsidies be reversed, they must ask, if conditions change? If the answer is no, then the authorities must be prepared to weigh carefully near term benefits against future losses.

The final comment to be made is on the use of dynamic economic models of the fishery. We have, in this chapter, attempted to emphasize the importance of dynamic, as opposed to static, models of the fishery. It can be stated, without fear of contradiction, that attempting to analyse the type of policy issue we have just explored, without the aid of dynamic models, is simply nonsensical.

Having now developed and analysed the basic dynamic economic model of the fishery, in the next chapter we turn our attention to applying dynamic bioeconomic models to real fisheries.

BIBLIOGRAPHY

Many of the books and articles referred to in the chapter 2 bibliography are of equal value to the examination of the basic dynamic economic model of the fishery, in particular: Munro and Scott (1985), Bjørndal and Munro (1998), Anderson and Seijo (2010), and Flaaten (2010). Of central importance are the books of Clark, *Bioeconomic Modeling and Fisheries Management* (1985) and *Mathematical Bioeconomics* (1990, 2010). In addition, the reader is directed to Conrad and Clark, *Natural Resource Economics* (1987).

NOTES

1. In Australia, where dynamic economic fishery models are being explicitly applied for policy purposes, one study carried out under the aegis of the Australian Government suggests that a δ in the order of 3 per cent (real) is in order (Kompas et al., 2009). In the United Kingdom, the UK Treasury uses a discount rate of 3.5 per cent for public investments, with the qualification that revenues and costs occurring after 30 years are to be discounted at a rate of 3 per cent.
2. Marine biologists do in fact use r as an indicator of whether the species is fast growing, slow growing, or in between (Clark et al., 2010a).
3. In chapter 7, we shall recount the story of the Alaskan pollock resource in the Bering Sea high seas enclave, known as the Doughnut Hole. From the early 1980s until the early 1990s, the Doughnut Hole Alaskan pollock fishery presented a

wonderful example of a Pure Open Access fishery. The exploitation of the Doughnut Hole Alaskan pollock resource during this period can best be described as a disorderly mining operation. In the words of the FAO, the resource was not just overexploited, it was plundered (FAO, 1994). The fishery resource has yet to recover.

4. In theory, this should guarantee that stock extinction will never occur. In reality, many stocks have been driven to near extinction even with harvesting costs depending on stock size. We shall return to this issue later.
5. In the empirical chapter to follow, chapter 4, a model of a real world example of such a fishery will be presented and discussed.
6. This is a necessary condition. Clark proves rigorously that the following slightly stronger condition is sufficient:

$$\delta > 2F'(0) = 2r(\text{Clark}, 1973).$$

7. The Clark and Munro (1978) model is linear, assuming, *inter alia*, that the demand for harvested fish is perfectly price elastic. If the demand price elasticity was finite, the decrease in the resource, and ultimate decline in harvests, could lead to a steady increase in the price of harvested fish, which would increase the threat of extinction.
8. An example is provided by a study of the Pacific halibut fishery undertaken by Crutchfield and Zellner (1962). The study, along with commentaries, was re-published in 2003 (Crutchfield and Zellner, 2003). The study is of particular interest because it contains the first fully developed dynamic economic model of the fishery. It discusses optimal management of the fishery in terms of a private 'sole owner'. Crutchfield and Zellner (2003, p. 19) state that 'it should be clear that the biological overfishing case, in which effort is pushed to the point where physical yields actually decline $[x < x_{MSY}]$, could not arise under private ownership of the resource'.
9. If it is possible that we shall have: $x^* < x_{MSY}$, how likely is such an occurrence? At the time of writing, there is an ongoing debate. Grafton et al. (2007) argue that such an occurrence is likely to be rare, even if the existence value of the resource is equal to zero. Not so, say Clark et al. (2010a, 2010b). The debate ranges over both theory and empirical models. Grafton et al. (2007) do, in their empirical models, assume constant price of harvested fish and constant unit fishing effort costs. They assume the standard harvest production function, $h = qE^{\alpha}x^{\beta}$. However, as well as relaxing the assumption that $\beta = 1$, they also relax the assumption that $\alpha = 1$. Life then becomes more complex. What Clark et al. then demonstrate, is that, given the assumptions employed by Grafton et al., if $\alpha + \beta > 1$ extinction will never be optimal, even if $r < \delta$. If, on the other hand, $\alpha + \beta \leq 1$, then there will exist some constellation of p, b, and q that will cause extinction to be optimal (Clark et al., 2010b). The relaxation of the assumption that $\alpha = 1$ greatly broadens the scope for extinction. If $\alpha = 1$, then obviously, we can only have $\alpha + \beta = 1$, if $\beta = 0$—a very demanding requirement.

In their response to Clark et al. (2010a), Grafton Hilborn (2010) focus on one of their empirical models. For that particular model it is to be found that $\alpha + \beta = 1$. Clark, Munro, and Sumaila confirm that, given the *p*, *b*, and *q* estimated by Grafton et al., the resource dealt with in that specific model is secure from extinction, as they claim, indeed even under Pure Open Access. No one can assume that the constellation of the relevant *p*, *b*, and *q* will be constant through time, however. Clark et al. demonstrate that a not unreasonable shift in the constellation would lead the Grafton et al. empirical model to predict that extinction of the resource in question would be optimal at a social rate of discount, δ, no greater than 3 per cent (Clark et al., 2010b). Ultimately what counts is empirical evidence from well specified bioeconomic models.

10. Implicitly it is here assumed that growth occurs prior to harvesting, Alternatively, one could specify growth as a function of escapement, i.e., $F(x_t\text{-}h_t)$.
11. Divide through both sides of equation (3.34) by $\alpha^t(p - c(x))$.
12. The AFZ is the Australian version of the EEZ.
13. The reversibility of investments, of lack thereof, is, of course, intimately linked to the concept of the malleability of capital discussed earlier.
14. The authors are grateful to Colin Clark for equation (3.45).
15. Having said this, if an environmental change is sudden and dramatic, then blocked intervals may be encountered. To a different issue, ideally, we should like to be able to investigate analytical models that are both non-autonomous and non-linear. This proves to be very difficult. One is generally forced to turn to numerical models (Clark and Munro, 1982).
16. Up to this point, it has been assumed that there is no divergence between the social and private cost of labour in the fisheries.
17. The Malaysian government programme of reducing barriers to labour mobility apparently had some success. About a decade and a half after the completion of the study, one of the authors (Munro) returned to Malaysia and had discussions with a senior fisheries official of the government of Malaysia. The official expressed some concern about the emerging 'shortage' of labour in the Malaysian inshore fisheries.
18. It took Clark and Munro about two years to arrive at the solution that we shall present. Following some considerable further delays, the solution was published by the Universiti Sains Malaysia in 1979 (Munro and Clark, 1979). A much shortened version of Munro and Clark (1979), also appears as a part of *Mathematical Bioeconomics*, first and second editions (see Clark, 1990, chapter 3, part 4). The part was prepared by Munro as a contribution to the book.

    It is worth noting that Clark and Munro (1982), to which we have given much attention, arose out of the preliminary work that the two authors had done in attempting to obtain a solution to the Malaysian policy problem.
19. It is assumed that the authorities, the resource managers, have only one fisheries management instrument available to them—subsidies/taxes. If other instruments were available, e.g., direct harvest controls, the case would become even more complex. Be that as it may, if subsidies are irreversible, they could, regardless of

In their response to Clark et al. (2010a), Grafton Hilborn (2010) focus on one of their empirical models. For that particular model it is to be found that $\alpha + \beta = 1$. Clark, Munro, and Sumaila confirm that, given the *p*, *b*, and *q* estimated by Grafton et al., the resource dealt with in that specific model is secure from extinction, as they claim, indeed even under Pure Open Access. No one can assume that the constellation of the relevant *p*, *b*, and *q* will be constant through time, however. Clark et al. demonstrate that a not unreasonable shift in the constellation would lead the Grafton et al. empirical model to predict that extinction of the resource in question would be optimal at a social rate of discount, δ, no greater than 3 per cent (Clark et al., 2010b). Ultimately what counts is empirical evidence from well specified bioeconomic models.

10. Implicitly it is here assumed that growth occurs prior to harvesting, Alternatively, one could specify growth as a function of escapement, i.e., $F(x_t\text{-}h_t)$.
11. Divide through both sides of equation (3.34) by $\alpha^t(p - c(x))$.
12. The AFZ is the Australian version of the EEZ.
13. The reversibility of investments, of lack thereof, is, of course, intimately linked to the concept of the malleability of capital discussed earlier.
14. The authors are grateful to Colin Clark for equation (3.45).
15. Having said this, if an environmental change is sudden and dramatic, then blocked intervals may be encountered. To a different issue, ideally, we should like to be able to investigate analytical models that are both non-autonomous and non-linear. This proves to be very difficult. One is generally forced to turn to numerical models (Clark and Munro, 1982).
16. Up to this point, it has been assumed that there is no divergence between the social and private cost of labour in the fisheries.
17. The Malaysian government programme of reducing barriers to labour mobility apparently had some success. About a decade and a half after the completion of the study, one of the authors (Munro) returned to Malaysia and had discussions with a senior fisheries official of the government of Malaysia. The official expressed some concern about the emerging 'shortage' of labour in the Malaysian inshore fisheries.
18. It took Clark and Munro about two years to arrive at the solution that we shall present. Following some considerable further delays, the solution was published by the Universiti Sains Malaysia in 1979 (Munro and Clark, 1979). A much shortened version of Munro and Clark (1979), also appears as a part of *Mathematical Bioeconomics*, first and second editions (see Clark, 1990, chapter 3, part 4). The part was prepared by Munro as a contribution to the book.

    It is worth noting that Clark and Munro (1982), to which we have given much attention, arose out of the preliminary work that the two authors had done in attempting to obtain a solution to the Malaysian policy problem.
19. It is assumed that the authorities, the resource managers, have only one fisheries management instrument available to them—subsidies/taxes. If other instruments were available, e.g., direct harvest controls, the case would become even more complex. Be that as it may, if subsidies are irreversible, they could, regardless of

wonderful example of a Pure Open Access fishery. The exploitation of the Doughnut Hole Alaskan pollock resource during this period can best be described as a disorderly mining operation. In the words of the FAO, the resource was not just overexploited, it was plundered (FAO, 1994). The fishery resource has yet to recover.

4. In theory, this should guarantee that stock extinction will never occur. In reality, many stocks have been driven to near extinction even with harvesting costs depending on stock size. We shall return to this issue later.
5. In the empirical chapter to follow, chapter 4, a model of a real world example of such a fishery will be presented and discussed.
6. This is a necessary condition. Clark proves rigorously that the following slightly stronger condition is sufficient:

$$\delta > 2F'(0) = 2r(\text{Clark}, 1973).$$

7. The Clark and Munro (1978) model is linear, assuming, *inter alia*, that the demand for harvested fish is perfectly price elastic. If the demand price elasticity was finite, the decrease in the resource, and ultimate decline in harvests, could lead to a steady increase in the price of harvested fish, which would increase the threat of extinction.
8. An example is provided by a study of the Pacific halibut fishery undertaken by Crutchfield and Zellner (1962). The study, along with commentaries, was re-published in 2003 (Crutchfield and Zellner, 2003). The study is of particular interest because it contains the first fully developed dynamic economic model of the fishery. It discusses optimal management of the fishery in terms of a private 'sole owner'. Crutchfield and Zellner (2003, p. 19) state that 'it should be clear that the biological overfishing case, in which effort is pushed to the point where physical yields actually decline $[x < x_{MSY}]$, could not arise under private ownership of the resource'.
9. If it is possible that we shall have: $x^* < x_{MSY}$, how likely is such an occurrence? At the time of writing, there is an ongoing debate. Grafton et al. (2007) argue that such an occurrence is likely to be rare, even if the existence value of the resource is equal to zero. Not so, say Clark et al. (2010a, 2010b). The debate ranges over both theory and empirical models. Grafton et al. (2007) do, in their empirical models, assume constant price of harvested fish and constant unit fishing effort costs. They assume the standard harvest production function, $h = qE^{\alpha}x^{\beta}$. However, as well as relaxing the assumption that $\beta = 1$, they also relax the assumption that $\alpha = 1$. Life then becomes more complex. What Clark et al. then demonstrate, is that, given the assumptions employed by Grafton et al., if $\alpha + \beta > 1$ extinction will never be optimal, even if $r < \delta$. If, on the other hand, $\alpha + \beta \leq 1$, then there will exist some constellation of p, b, and q that will cause extinction to be optimal (Clark et al., 2010b). The relaxation of the assumption that $\alpha = 1$ greatly broadens the scope for extinction. If $\alpha = 1$, then obviously, we can only have $\alpha + \beta = 1$, if $\beta = 0$—a very demanding requirement.

the resource management instruments available, still impose a burden upon society over the long run.

20. The optimal subsidy, s(t), on a per unit basis, can be expressed as follows: $[p - (c_p(x^*(t)) - s(t))] = 0$ (Munro and Clark, 1979).
21. Recall that blocked intervals arise when the constraints on the control variable become binding. In the case under discussion, the control variable, $h(t)$, is constrained by virtue of the fact that the subsidy rate can never be negative.
22. An example is a government or corporate interest bearing bond that is being traded in the secondary market. The buyer of the bond in the secondary market will commonly find that he/she is buying the bond at a price above or below the face value of the bond. If this is the case, then account must be taken of the fact that, if the bond is held to maturity, the holder will, as well as receiving interest over the life of the bond, experience either a capital loss or gain when the bond matures. This gives rise to the concept of 'yield to maturity'.

4 Empirical Applications of the Dynamic Economic Model

In this chapter, we will provide two empirical applications based on the dynamic economic model of the fishery introduced in chapter 3.

The two case studies are for the Western Channel sole fishery (section 4.1) and the North Sea herring fishery (section 4.2). Although there are many differences between the two fisheries, there will nevertheless be found a number of commonalities in the modelling.

For both fisheries, it will be assumed that they are managed by a resource manager whose objective is to maximize the present value of net economic benefits from the fishery. Although sole is harvested by several EU countries, the Total Allowable Catch (TAC) quota is determined at the EU level. North Sea herring is considered to be a shared stock, in that it is exploited by Norway and the European Union (EU). A TAC is jointly determined and is divided between the parties involved. Thus, the assumption of sole owner management can be justified by the way the TACs are set. In chapter 7, the implications of several countries participating in a fishery will be analysed.

As pointed out in chapter 3, a resource manager is faced with two main questions, namely: determining (i) the optimal stock level and (ii) the optimal rate of investment in the stock. The emphasis in the case studies will be on the first question, although the second will also be addressed.

We now employ discrete time dynamic models, where time is specified in years for the empirical applications (see chapter 3, section 3.4). This is natural, based on relevant characteristics of the fisheries in question. Moreover, only annual data are available.

A fish population can generally be divided into several sub-populations. As noted in chapter 1, in a bioeconomic, as opposed to a purely biological, model, the primary concern is to identify the harvestable population. Since for both Western Channel sole and North Sea herring this coincides with the spawning stock, the model will be formulated in terms of this variable.

4.1 The Western Channel sole fishery

The fishery analysed is sole (*Solea solea*) in International Council for the Exploration of the Sea (ICES) statistical area VIIe—the Western Channel.

Sole is a right-eyed flatfish that spawns in spring and early summer in shallow coastal water from May–June off the coast of southern England. Common sole become sexually mature at three–five years, when 25–35 cm long, the males being somewhat smaller than the females. Female sole grow larger than males, reaching up to 40 cm after five years and a maximum length of 60 cm. Males rarely grow longer than 40 cm (Marine Conservation Society, 2007).

Spawning stock biomass (SSB) and landings (harvest) per year for the period 1969–2006 are presented in Figure 4.1. SSB is seen to have increased from 2,800 tonnes in 1969 to a peak in excess of 5,400 tonnes in 1979. Subsequently, the trend has been negative, with 2007 stock level at 2,270 tonnes, the lowest ever recorded.

Landings peaked at 1,500 tonnes in 1982–3, and have in recent years averaged 1,000 tonnes. As the stock is in decline, the fishery does not appear to be sustainable. The TAC is shared among a number of EU countries. The UK has the largest share, with 59 per cent of the TAC.

Beam trawlers represent the most important fleet segment in the fishery, although sole is harvested also by other vessel groups. We focus on beam trawlers. The fishing vessels harvest other species, for example monkfish and megrim, as well as sole. As a consequence, we treat the vessel capital employed in the sole fishery as being highly malleable, in the sense that the vessels can readily shift from sole to other species. This greatly simplifies the

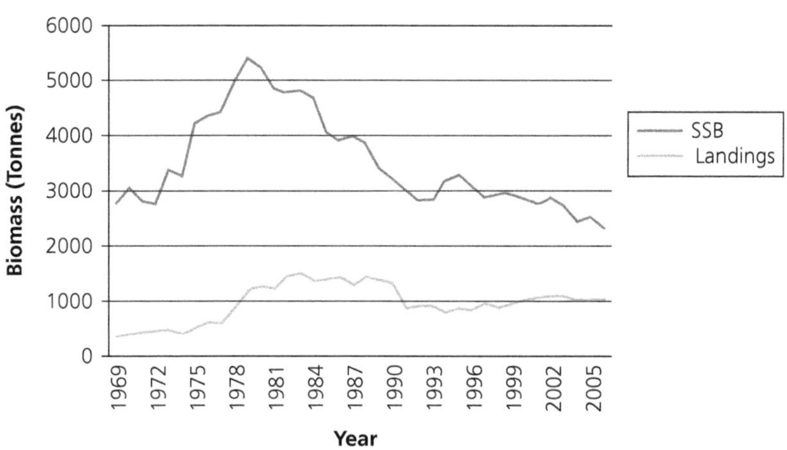

Figure 4.1. Spawning biomass and landings for Western Channel sole (1969–2006; tonnes)

analysis. In chapter 5, we shall discuss in detail the complexities that arise when the relevant vessel capital is other than highly malleable.

The EU established a multi-annual management plan for sole in the Western Channel in 2007, with the aim of rebuilding the stock.

As noted, the spawning stock biomass is considered the *fishable* biomass. This is because fishing and management measures in the main are directed at this stock.

Bjørndal and Bezabih (2008) analyse costs of harvesting sole based on Seafish reports for 2001, 2004, and 2005. Due to data limitations, it is impossible to relate unit harvesting costs to stock size. As a simplifying assumption, they assume that the cost of harvesting is independent of stock size, i.e., the Marginal Stock Effect is assumed to be equal to zero. It is also assumed that the unit cost of harvest is constant and invariant over time. The unit cost of harvest is estimated at £3 per kg, a cost which reflects both variable and fixed costs, including depreciation and a normal return on capital.[1]

The price of sole has shown a strong increase in recent years, with the price increasing from about £7.00/kg in 2005 to about £8.00/kg in 2006.

4.1.1 THE DYNAMIC BIOECONOMIC MODEL

With the simplifications discussed above, profits (π_t) in year t are given by:

$$\Pi_t = (p - b)H_t \tag{4.1}$$

Here, p is price per unit harvest and b the fixed unit cost of harvest. H_t is harvest in period t. Stock dynamics are given by

$$S_{t+1} = S_t + rS_t[1 - (S_t/G)] - H_t \tag{4.2}$$

Equation (4.2) presents the discrete time version of the Schaefer growth function, i.e., the model of population dynamics where S_t denotes the resource stock in time period t, and H_t denotes harvest at time t (see chapter 3, section 3.3). The term $rS_t[1-(S_t/G)]$ is the natural growth function, where r is the intrinsic growth rate of fish and G is the carrying capacity of the environment.

The present value (PV) of net revenues (rent) from the fishery, summed over time from $t = 0$ to $t = T$, is given by:

$$PV = \sum_{t=0}^{T} \left(\frac{1}{1+i}\right)^t (p - b)H_t \tag{4.3}$$

where $\alpha = 1/1+i$ is the discount factor and i the periodic (annual) discount rate.

The objective of a sole owner of the resource is seen to be that of maximizing the present value of (4.3) subject to the stock constraint given by equation (4.2). This optimization problem, within the context of discrete time models, was addressed in chapter 3, section 3.4

Given the assumption that harvesting costs are independent of stock size, the Marginal Stock Effect vanishes and we are left with the following solution for the optimal stock (cf. equation 3.36):

$$F'(S^*) = i \tag{4.4}$$

where S^* is the optimal stock level. By substituting the first derivative of the growth function for $F'(S^*)$ in equation (4.4), an explicit function for the optimal stock level can be derived:

$$S^* = (G/2)[1 - (i/r)] \tag{4.5}$$

According to equation (4.5), S^* is essentially determined by the relationship between the discount rate (i) and the intrinsic growth rate r.

The maximum sustainable yield (MSY) can be computed as follows:

$$MSY = H_{msy} = (rG/4) \tag{4.6}$$

The stock level corresponding to MSY is given by:

$$S_{msy} = (G/2) \tag{4.7}$$

The natural growth function has been estimated (Bjørndal and Bezabih, 2008), with results presented in Table 4.1. The intrinsic growth rate, r, is calculated to be 0.398 which indicates that when the spawning stock is at a very low level, it grows by almost 40 per cent yearly. This indicates high natural productivity. The carrying capacity, G, is estimated at 11,214 tonnes.

The natural growth function is illustrated in Figure 4.2. The stock level giving rise to MSY is 5,607 tonnes, with MSY equal to 1,116 tonnes. It can be noted that the growth curve is fairly 'flat' over a range of stock values around S_{msy}.

Table 4.1. Growth function characteristics for Western Channel sole

Parameter/Variable	Value
Intrinsic Growth Rate	0.398
Carrying Capacity (Tonnes)	11,214
Maximum Sustainable Yield (*MSY*) (tonnes)	1,116
Stock level giving MSY (S_{msy}) (tonnes)	5,607

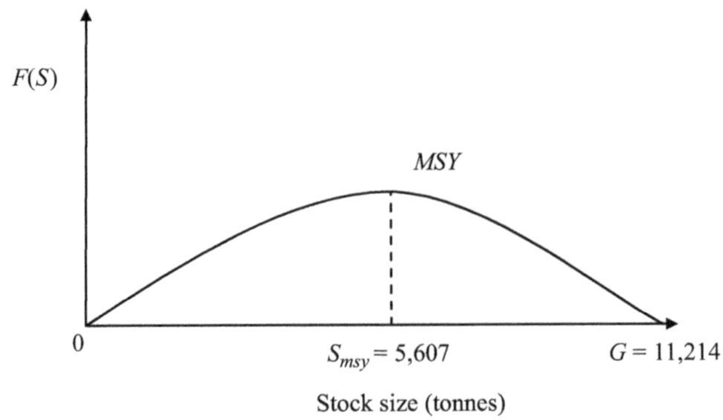

Figure 4.2. Natural growth $F(S)$ and spawning stock biomass for Western Channel sole (tonnes)

Table 4.2. Estimates of optimal SSB (S^*) and harvest (H^*) for Western Channel sole for different discount rates (tonnes)

Discount rate	S^*	H^*
0	5,607 (=S_{msy})	1,116 (=*MSY*)
0.035	5,114	1,107
0.05	4,903	1,098
0.10	4,198	1,045
0.20	2,790	834

Equation (4.5) can be solved for optimal stock level for different values of the discount rate. Corresponding harvests can be found from equation (4.2). Table 4.2 presents estimates of S^* and corresponding H^* for discount rates between 0 and 20 per cent. Optimal SSB varies between 5,607 tonnes ($i = 0$) and 2,790 tonnes ($i = 0.20$), with the corresponding annual harvest varying between 1,116 and 834 tonnes. It is noted that optimal harvest does not change much for discount rates between 0 and 10 per cent. This is due to the curvature of the estimated growth function, as pointed out above.

Above, it was noted that in recent decades there has been a very substantial *disinvestment* in the spawning stock biomass, which has been reduced from 5,421 tonnes in 1979 to 2,270 tonnes in 2007 (Figure 4.1). The results here call for a very substantial *investment* in the stock. The exact magnitude of the investment depends on the discount rate. For a discount rate of 3.5 per cent, as recommended by the UK Treasury for public investments, the optimal stock level is 5,114 tonnes. Essentially this calls for the stock to be rebuilt to its level prior to the start of the depletion process in the early 1980s.

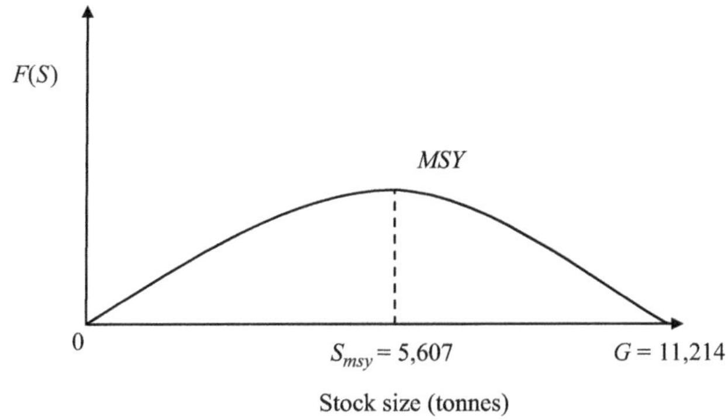

Figure 4.2. Natural growth $F(S)$ and spawning stock biomass for Western Channel sole (tonnes)

Table 4.2. Estimates of optimal SSB (S^*) and harvest (H^*) for Western Channel sole for different discount rates (tonnes)

Discount rate	S^*	H^*
0	5,607 (=S_{msy})	1,116 (=MSY)
0.035	5,114	1,107
0.05	4,903	1,098
0.10	4,198	1,045
0.20	2,790	834

Equation (4.5) can be solved for optimal stock level for different values of the discount rate. Corresponding harvests can be found from equation (4.2). Table 4.2 presents estimates of S^* and corresponding H^* for discount rates between 0 and 20 per cent. Optimal SSB varies between 5,607 tonnes ($i = 0$) and 2,790 tonnes ($i = 0.20$), with the corresponding annual harvest varying between 1,116 and 834 tonnes. It is noted that optimal harvest does not change much for discount rates between 0 and 10 per cent. This is due to the curvature of the estimated growth function, as pointed out above.

Above, it was noted that in recent decades there has been a very substantial *disinvestment* in the spawning stock biomass, which has been reduced from 5,421 tonnes in 1979 to 2,270 tonnes in 2007 (Figure 4.1). The results here call for a very substantial *investment* in the stock. The exact magnitude of the investment depends on the discount rate. For a discount rate of 3.5 per cent, as recommended by the UK Treasury for public investments, the optimal stock level is 5,114 tonnes. Essentially this calls for the stock to be rebuilt to its level prior to the start of the depletion process in the early 1980s.

The objective of a sole owner of the resource is seen to be that of maximizing the present value of (4.3) subject to the stock constraint given by equation (4.2). This optimization problem, within the context of discrete time models, was addressed in chapter 3, section 3.4

Given the assumption that harvesting costs are independent of stock size, the Marginal Stock Effect vanishes and we are left with the following solution for the optimal stock (cf. equation 3.36):

$$F'(S^*) = i \tag{4.4}$$

where S^* is the optimal stock level. By substituting the first derivative of the growth function for $F'(S^*)$ in equation (4.4), an explicit function for the optimal stock level can be derived:

$$S^* = (G/2)[1 - (i/r)] \tag{4.5}$$

According to equation (4.5), S^* is essentially determined by the relationship between the discount rate (i) and the intrinsic growth rate r.

The maximum sustainable yield (MSY) can be computed as follows:

$$MSY = H_{msy} = (rG/4) \tag{4.6}$$

The stock level corresponding to MSY is given by:

$$S_{msy} = (G/2) \tag{4.7}$$

The natural growth function has been estimated (Bjørndal and Bezabih, 2008), with results presented in Table 4.1. The intrinsic growth rate, r, is calculated to be 0.398 which indicates that when the spawning stock is at a very low level, it grows by almost 40 per cent yearly. This indicates high natural productivity. The carrying capacity, G, is estimated at 11,214 tonnes.

The natural growth function is illustrated in Figure 4.2. The stock level giving rise to MSY is 5,607 tonnes, with MSY equal to 1,116 tonnes. It can be noted that the growth curve is fairly 'flat' over a range of stock values around S_{msy}.

Table 4.1. Growth function characteristics for Western Channel sole

Parameter/Variable	Value
Intrinsic Growth Rate	0.398
Carrying Capacity (Tonnes)	11,214
Maximum Sustainable Yield (*MSY*) (tonnes)	1,116
Stock level giving MSY (S_{msy}) (tonnes)	5,607

4.1.2 INVESTING IN THE STOCK

Investing in the stock means that harvests must be reduced below natural growth for a period of time to permit the stock to grow. Let us proceed for the case of a 3.5 per cent discount rate, which implies $S^* = 5{,}114$ tonnes. Once the stock has increased to its optimal level of 5,114 tonnes, annual harvest can be increased to a sustainable level of 1,107 tonnes. We will now investigate alternative investment paths.

When the bioeconomic analysis of Western Channel sole was undertaken, stock data were available up to 2007. Thus, in the analysis of approach paths, the stock size as of 2007 was given with approach paths implemented as of 2008.

First, however, let us consider what will happen if present-day regulations are not changed. We assume that in this case the stock is subjected to a constant yearly harvesting of 958 tonnes, which is the actual harvest recorded in year 2006. Consideration of the impact of this management commences at the end of the year 2007. The results show that the stock will be driven to extinction by the year 2011, i.e., after four years (Figure 4.3). This may be a bit of an exaggeration, as catch per unit effort may decline as stock size is reduced to low levels, making stock depletion less severe.[2] Nevertheless, this result clearly indicates that if nothing is done, in the worst instance the stock may be driven to virtual extinction.

As pointed out above, an optimal policy will involve an investment in the stock. Hence our analysis of the approach paths to the steady state would involve building up the 2007 SSB level of 2,270 tonnes to the optimal steady state stock level of 5,114 tonnes, which will imply more than a doubling of the stock.

How quickly the optimal stock level will be reached depends on how much is harvested in the rebuilding period and the productivity of the stock. Three alternative approach paths are considered (Figure 4.3). The first one is the

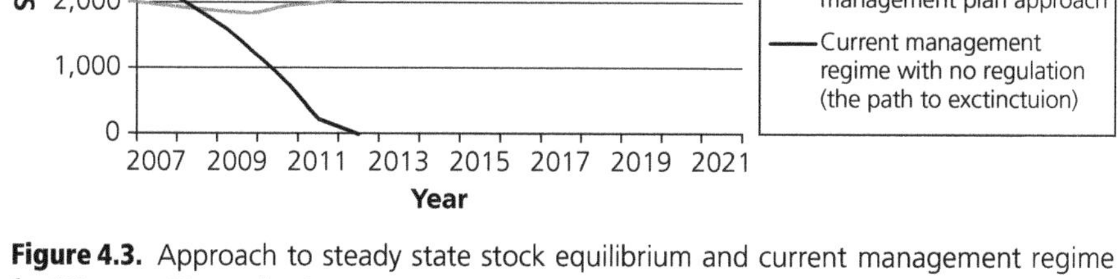

Figure 4.3. Approach to steady state stock equilibrium and current management regime for Western Channel sole

most rapid rate of investment in the stock, which involves a moratorium on the harvest until the optimal stock level is achieved. The second alternative involves an annual TAC of 500 tonnes until the optimal stock level is reached, while the third is based on the current management plan for sole.

In chapter 3, it was argued that, if our model is linear in the control variable, and if the produced (and human) capital in the fishery is perfectly malleable—the case we are now examining, the optimal policy involves maximum investment/disinvestment in the resource until the optimal stock level has been reached, i.e., a harvest moratorium, until the stock has recovered. This involves closing the fishery from 2008 until 2010. At the end of the year 2011, 853 tonnes will be harvested to get the stock to its steady state level of 5,114 tonnes. This level is reached in 2011, and a steady state TAC of 1,107 tonnes is introduced. Profit is estimated at £5 per kg; thus, in 2010 current profits represent £4.265 million, increasing to a steady state of £5.535 million in 2011.

This shows that, with a zero harvest, the stock will reach the optimal level in 2011. Subsequently, the steady state harvest will be 1,107 tonnes per year. This is a fairly short recovery period, a consequence of the good growth rate for sole.

Let us next test the claim that the most rapid approach path to the target stock is optimal, by taking a slower approach path. We do this by choosing an arbitrary yearly TAC value of 500 tonnes until the steady state is reached. A TAC figure of 500 tonnes is less than the TACs of recent years.

In this regime, the 500 tonne quota remains in effect for the period 2008–12, each year giving rise to net current profits of £2.5 million. In 2013 the quota is reduced to 407 tonnes so that the steady state can be achieved as from 2014. This shows that, compared with the most rapid approach, the steady state stock level is reached with a three-year delay.

As discussed above, a management plan for the Western Channel sole has been introduced. The third approach path is based on this management plans, and involves an even slower approach path. In the first two years of the recovery period, 2008 and 2009, the TAC is set at 765 tonnes. The TAC is then reduced to 600 tonnes per year until 2012. The TAC is further reduced to 550 tonnes in 2013 and maintained at this level until the steady state stock level has been achieved. In 2020, the last year before steady state is reached, the TAC is set to 868 tonnes.

Having considered three different approach paths, it is obvious that the size of the TAC in the rebuilding period has a great influence on how soon the steady state stock level is reached. With a most rapid approach, the fishery is reopened in 2010, and the steady state is reached in 2011. In the case of an approach path based on the current management plan, the steady state is reached in 2021, i.e., with a delay of ten years as illustrated in Figure 4.3.

4.1.3 ESTIMATION OF RETURNS FROM THE FISHERY

For the estimation of profits, it has been assumed that there is a net profit of £5 per kg. As both variable and fixed costs are included, this represents a return over and beyond a normal return on capital.

Consistent with our assessment of alternative paths to the steady states, the estimates of profits are based on the assumption that the year 2007 starts with an SSB of 2,270 tonnes, the target optimal stock level is 5,114 tonnes and a steady state annual harvest is 1,107 tonnes. Profits have been discounted to 2007[3] and, as before, the discount rate is 3.5 per cent.

In the case of a most rapid approach to the steady state there will be no harvests in the two-year period 2008–9. Thus, in these years there will be no revenue, and no profits will be generated. The situation changes in 2010, where 853 tonnes are harvested. With a profit of £5 per kg, the total profits will be £4,265 million. If the fishery is managed on a sustainable basis, as of 2011, the annual harvest level will be 1,107 tonnes and the associated profits of £5,535 million will be obtained on an annual basis ad infinitum. The present value of this stream of profits, measured as of 2007, is £140 million.

In the case with a 500 tonne TAC in the rebuilding period, profits are earned also in the rebuilding period. It does, however, take longer time to reach the steady state, which involves higher annual rents. In a sense, this represents a trade-off. The present value of rents for this alternative is £137 million, slightly less than for the most rapid approach.

The approach path based on the current management plan for sole gives a present value of £65 million, substantially lower than in the other two approach paths considered. The 'advantage' with this scenario is that it gives fairly high harvests and profits also while rebuilding the stock; however, this comes at the expense of delayed attainment of the steady state. The net result is, however, a much lower present value than for more rapid approach paths. Thus, given our assumptions, it is correct to state that the lower the rate of investment in the stock, the lower will be the net economic returns from the fishery, expressed in present value terms.

Continuation of the present policy gives rise to a present value of profits of £20 million. Fairly high profits are achieved in the first few years of the period, while the stock is being driven to extinction.

4.1.4 POLICY IMPLICATIONS

There are some important policy implications from the bioeconomic analysis that has been undertaken. First, there should be substantial investment in the stock to allow it to grow to an optimal level. Second, this can only be achieved by reducing harvest below natural growth for a period of years to allow the stock to recover. Third, the stock has the potential of yielding fairly substantial profits. With proper management and assuming a 3.5 per cent discount rate, after a rebuilding period the fishery could yield profits of more than £5 million per year and a present value of about £140 million.

To achieve the optimal level of profits, not only would the stock need to be allowed to grow to the optimal level, but the fleet would also need to be restructured in order to reduce excessive effort and harvesting costs. More detailed effort and cost data would be required to undertake a more thorough analysis of this issue.

An additional simplification has been to assume that the fishery is managed by a single owner. The implications of multi-country harvesting, for example on achieving an agreement for rebuilding the stock and quota compliance, are not considered part of this analysis. It should nevertheless be pointed out that improved management brings with it the promise of increased harvests, revenues, and rents for all countries participating in the fishery. These issues will be discussed further in chapters 6 and 7.

4.2 **The North Sea herring fishery**

We next consider North Sea autumn-spawning herring (*Clupea harengus L*) in ICES area IV and division VIId. Herring consists of three spawning stocks with spawning grounds east of Scotland, east of England, and in the English Channel. However, the three stocks mix on the feeding grounds and in the central and northern North Sea rendering it impossible to distinguish between catches from the three stocks. It is therefore customary to treat the three stocks as one.

Herring become sexually mature in age group two. Thus, the juvenile stock consists of age groups zero and one, while the spawning stock consists of age groups two and older. The herring fishery has a season running from May until September.

Spawning stock biomass (SSB) for the 60-year period 1947–2007 is illustrated in Figure 4.4. In 1947, right after the Second World War, the SSB estimated at 2.945 million tonnes is believed to have been close to its carrying capacity due to low fishing pressure during the war. What followed was a period with 'mining' of the stock, which was reduced to a level of 1.2 million tonnes in 1958. Subsequently, there have been substantial variations in stock size over time. A peak of 2.06 million tonnes was observed in 1959 which was followed by a dip, before the stock increased to almost 2.2 million tonnes in 1963.

The early 1960s saw the introduction of new technology, in particular the sonar and the power block,[4] which led to a technological revolution in the

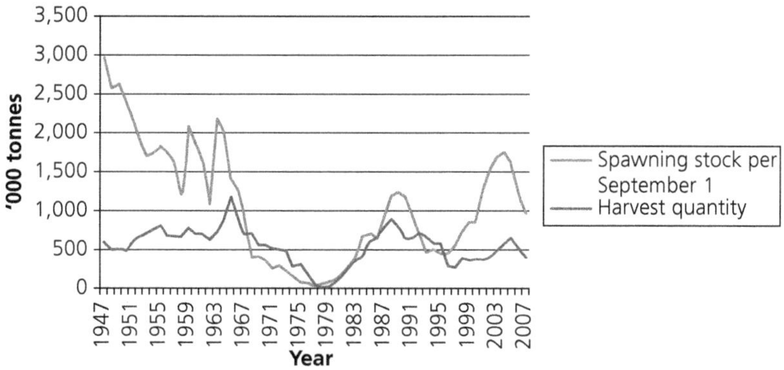

Figure 4.4. Spawning stock biomass and total landings for North Sea herring for 1947–2007 ('000 tonnes)

fishery with substantial increases in catches. This in turn caused rapid stock depletion, with the stock reaching a minimum of 47,000 tonnes in 1977, when a moratorium was introduced. Various regulations have been in effect ever since so as to allow for a sustainable fishery.

After the moratorium, the stock recovered reasonably quickly. For the years 1988–90, SSB averaged around 1.2 million tonnes. The stock was then reduced to a level around 0.5 million tonnes in the years 1993–7, later to increase again. In the period 2001–6, SSB varied in the range 1.3–1.8 million tonnes; the 2007 stock level was estimated at just below 1 million tonnes. The minimum safe SSB level, expressed as Blim,[5] is set at 800,000 tonnes, which implies that the current fishery is sustainable.

After the introduction of Extended Fisheries Jurisdiction (EFJ),[6] North Sea herring has been considered a shared stock, managed by Norway and the European Union (EU). A Total Allowable Catch quota is determined for the sustainable management of the stock and this TAC is divided between the parties involved, Norway receiving a 29 per cent share and the EU 71 per cent. The sharing is largely based on the zonal attachment of the stock to the Exclusive Economic Zones (EEZs) of Norway and the EU. Within the EU, the TAC is shared according to the principle of relative stability.[7]

Total landings of herring are also graphed in Figure 4.4. Landings increased from about 700,000 tonnes in 1960 to almost 1.2 million tonnes in 1965, and were maintained at a high level into the 1970s, despite a declining stock size. Note that in the years 1968–76, annual catches exceeded the size of the SSB.[8] Landings were reduced to 46,000 tonnes in 1977, when the moratorium was introduced, and stayed at a low level until the fishery was reopened in 1981. Catches then increased and reached almost 900,000 tonnes in 1988, after which they varied over time. Landings in 2006 were recorded at more than 500,000 tonnes, falling to 400,000 tonnes in 2007.

As for individual countries, Norway is seen to have the highest catch quantity most of the years, followed by Denmark. The Netherlands and the United Kingdom are also important harvesters of herring. The catch patterns for individual countries follow that of total catches.

The purse seine is the main technology used by several countries, including Norway and Denmark; however, in Norway trawlers and coastal vessels also participate in the fishery. In the UK, on the other hand, pelagic trawlers are most important.

Regulations of the fishery vary from country to county. In Norway, vessels are regulated with individual quotas that are not transferable. In Denmark, the purse seine fleet is managed by ITQs. Dutch vessels also have transferable quotas. In the United Kingdom, as a rule firms get their quota allocations via the producer organizations they belong to but one or two of the larger companies receive allocations directly from the government. The quotas are to a certain degree transferable.

The EU–Norway agreement on management of North Sea herring was updated in 2004. The revised rule specifies maximum fishing mortalities when the SSB is above 1.3 million tonnes, which is considered the B_{pa}.[9] In addition, there is now a rule specifying reduced fishing mortalities when the SSB is below 1.3 million tonnes.

4.2.1 UNITED KINGDOM FLEET AND VESSEL ECONOMICS

Although there have been substantial variations, overall the trend in prices is positive in recent years. For the years 1996–2000, price per tonne varied between £104 and £120 and this increased to between £317 and £289 for the years 2006–7. Herring is sold into an international market in competition with very close substitutes such as Norwegian spring spawning and Icelandic herring. It is the total supply of herring,[10] of which North Sea herring is a small part, that, in conjunction with demand, will determine price.

In the UK, herring is harvested by pelagic trawls for which we have available fleet and average vessel revenue and cost data.[11] The fleet harvests several species. In 2006, mackerel was by far the most important of these species, accounting for 67 per cent of the catch quantity. Herring accounted for 27 per cent of the catch quantity, with blue whiting accounting for the remaining 6 per cent.

Average value of the catch has increased considerably, from £306/tonne in 2004 to £402/tonne in 2006.[12] This is due to changes in the catch composition and/or increases in prices over time.

For the individual vessel, with an income of £4.14 million in 2006, profits are estimated at £566,900, or 13.7 per cent of income. Most of the profits come from mackerel.

4.2.2 THE BIOECONOMIC MODEL

Since for North Sea herring the harvestable population coincides with the spawning stock, the model will be formulated in terms of this variable.

In its most simple form, changes in the biomass of a fish stock over time will come from additions to the stock due to recruitment and natural growth, and deductions from the stock due to natural mortality and harvesting (chapter 1). The interactions between recruitment, natural growth, natural mortality, and harvesting have been fundamental in the development of the model of population dynamics, which has previously been used as part of bioeconomic analyses of North Sea herring (Bjørndal, 1988). This gives rise to a model that is infinitely more complex than the one considered hitherto.

The following delay-difference equation will be used to explain changes in the biomass over time:

$$S_{t+1} = (S_t - H_t)e^{G-M} + G(S_{t-\gamma}) \tag{4.9}$$

where S_{t+1} is spawning stock biomass in year $t+1$ and H_t is harvest in period t, both measured in tonnes. $G(S_{t-\gamma})$ is recruitment to the stock, taking place with delay of γ periods. G and M represent natural growth and mortality, respectively.

The first part of the right-hand side of equation (4.9) denotes stock changes due to natural growth, natural mortality, and harvesting. In the model, it is assumed that harvesting occurs in a short season at the beginning of the period. The escapement, $S_t - H_t$, is left to grow at the net instantaneous growth rate G–M. Therefore, in the absence of recruitment, changes in the spawning stock over time are given by

$$S_{t+1} = (S_t - H_t)e^{G-M}. \tag{4.10}$$

Alternatively, we could assume that the fishery took place at the end of the period, without affecting the qualitative nature of the model.

The second part of the right-hand side of equation (4.9) represents addition to the stock due to recruitment, which is assumed to occur at discrete time intervals. Moreover, recruits will normally join the parent population several years after spawning. We postulate that

$$R_{t+1} = g(S_t), \tag{4.11}$$

where R_{t+1} is the number of recruits to the juvenile population as a function of the previous period's spawning biomass. A certain fraction, λ, will survive the juvenile stage and join the spawning stock, so that

$$g(S_{t-\gamma})\lambda \tag{4.12}$$

is the number of recruits joining the spawning stock with a delay of γ periods. The delay occurs while the juveniles mature to spawning age. Letting w denote the weight of new recruits, we get

$$G(S_{t-\gamma}) = g(S_{t-\gamma})\lambda w, \tag{4.13}$$

where $G(S_{t-\gamma})$ denotes recruitment in weight to the spawning stock.

Herring spawn in September every year. The following year a number of recruits, called zero-group herring, join the juvenile population as indicated by equation (4.11). After another two years the survivors (equation 4.12) become sexually mature and join the spawning or adult population (equation 4.13). Thus for this species, $\gamma = 2$, and the delay between spawning and recruitment to the spawning stock is three years. The time periods are defined such that the beginning of one period coincides with the point of time when a year-class of new recruits enters the spawning biomass.

Next, the behaviour of the model under natural conditions will be considered. In the absence of fishing, equation (4.13) is reduced to

$$S_{t+1} = S_t e^{G-M} + G(S_{t-\gamma}) = S_t e^{\delta} + G\left(S_{t-\gamma}\right),$$

$$\delta \equiv G - M, \tag{4.13$'$}$$

where δ, in this case, is the mean net natural growth rate.[13] Under natural conditions, a fish population will grow towards its carrying capacity, which is the upper limit of the stock size as determined by environmental conditions. However, in equation (4.13′), it is assumed that G and M are constants. In reality, natural growth will be density-dependent because *ceteris paribus*, there will be relatively more food available to a small stock than to a large one. Natural mortality may also be density dependent, for example if the effectiveness of predation depends on stock size, or if cannibalism occurs or becomes more frequent at high stock densities. Hence one expects net annual growth to depend on stock density, with $\delta = \delta(S)$ and $\delta(\bar{S}) \leq 0$, where $\bar{S}$ is the carrying capacity of the stock; otherwise an upper limit to stock size would not exist under natural conditions. Equation (4.9) can now be restated as

$$S_{t+1} = (S_t - H_t)e^{\delta(St)} + G(S_{t-\gamma}). \tag{4.14}$$

For the purpose of this analysis, it is assumed that the resource in question is managed by a sole owner whose objective is to maximize the present value of net revenues from the fishery. The net revenue function is given by

$$\Pi(H_t, S_t). \tag{4.15}$$

We assume the manager of the fishery maximizes the present value of equation (4.15) subject to changes in the population level given by Equation (4.14). In addition, we have the feasibility constraint

$$0 \leq H_t \leq H_{max}. \tag{4.16}$$

This gives a discrete time dynamic bioeconomic model, with H_t and S_t as the control variable and the state variable respectively.

The method of Lagrange multipliers can be used to derive equilibrium conditions for an optimum (Clark, 1976, 1990), as we discussed in chapter 3, section 3.4. We define the Lagrangean expression:

$$L = \sum \left\{ \alpha^t \Pi(H_t, S_t) - q_t \left[S_{t+1} - (S_t - H_t)e^{\delta(S_t)} - G(S_{t-\gamma}) \right] \right\} \tag{4.17}$$

where $\alpha = 1/(1 + i)$ is the discount factor, i the periodic (annual) rate of discount and q_t is the discounted value of the shadow price.

First order necessary conditions for an optimum are

(i) $\partial L/\partial H_t = 0,\ t = 0, 1, 2,$
(ii) $\partial L/\partial S_t = 0,\ t = 1, 2, 3.$

Carrying out the derivations (Bjørndal, 1988), an implicit expression for the optimal spawning stock S^* is derived:

$$e^{\delta(S^*)}\{[\Pi_S + \Pi_H]/\Pi_H\} + \delta'(S^*)[S^* - G(S^*)] + \alpha^{\gamma} G'(S^*) - 1 = \mathrm{I} \tag{4.18}$$

As in the case of the linear dynamic Gordon-Schaefer model, the L.H.S. of equation (4.18) is the rate of return on marginal investment in the resource, or the 'own rate of interest' of the resource. It has two components. The first component is termed the Marginal Stock Effect (*MSE*), in this particular discrete time non-linear model given by the term $(\pi_s + \pi_H)/\pi_H$. It will be recalled that the MSE represents the impact of stock density on harvesting costs (see chapter 3, section 3.1). The second component represents that 'marginal product' of the natural component. In equilibrium this rate of return must be equal to the opportunity cost of capital as represented by the discount rate. The equation presented can be seen as a complex version of equation (3.35).[14]

4.2.3 DYNAMIC OPTIMIZATION

The model of population dynamics represents recruitment to the stock as well as net natural growth. In the Appendix, various functional forms for these functions are presented.[15] The functional structure of the model of

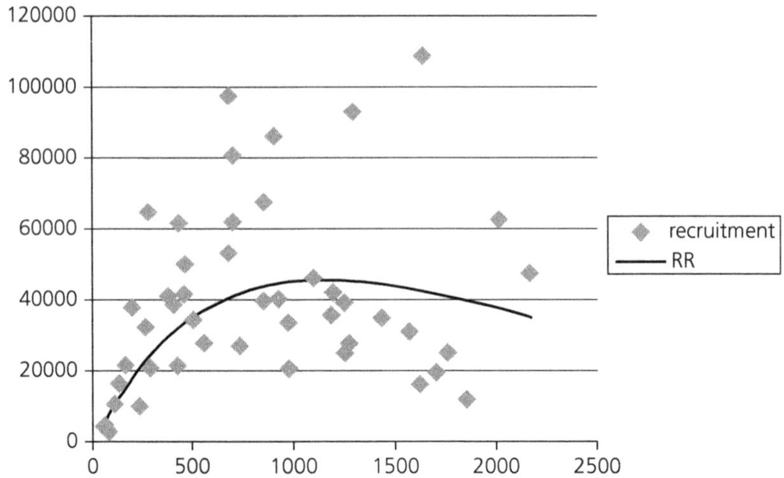

Figure 4.5. The Ricker stock recruitment function for North Sea herring
Recruitment in year (t+1) is measured in '000; spawning stock biomass is measured in '000 tonnes

population dynamics that will be used in the estimation of the optimal stock level is the combination of the Ricker stock-recruitment function and the linear net growth function.

In Figure 4.5 we have plotted recruitment in year t+1 against spawning stock biomass the previous year (year t). Visual inspection of the data points suggest recruitment is increasing in SSB up to a certain level but eventually levels off and declines. Moreover, it is noticeable that recruitment is particularly favourable in certain years. The Ricker stock-recruitment function, estimated on the basis of these data, is also illustrated.

The estimated model of population dynamics is thus:

$$S_{t+1} = (S_t - H_t)e^{(0.894 - 0.38.10^{-3}.S_t)} + 1.63.10^{-3}.S_{t-2}e^{(0.86.10^{-3}.S_{t-2})} \quad (4.19)$$

For this model

(i) $\bar{S}$ = 2,386,000 tonnes
(ii) S_{msy} = 1,284,000 tonnes with
(iii) MSY = 424,300 tonnes.

S_{msy} is the stock level corresponding to Maximum Sustainable Yield (MSY).

The model population dynamics is illustrated in Figure 4.6. To facilitate graphical representation of a function involving a time lag, a steady-state stock level is assumed. The harvest quantity is represented by the difference

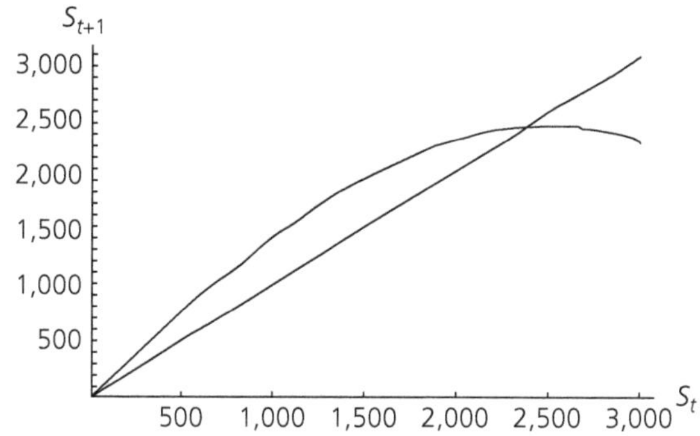

Figure 4.6. Stock dynamics (steady state stock levels) for North Sea herring ('000 tonnes)

between S_{t+1} and the 45° line. It is noteworthy that the steady-state harvest quantity is fairly constant over a wide range of stock values.

The harvest in period t will be given by the following Cobb-Douglas production function:

$$H_t = H(K_t, S_t) = AK_t{}^b S_t{}^g \tag{4.20}$$

where K_t is fishing effort in period t. Bjørndal and Conrad (1987a) found that, for North Sea herring, the number of participating vessels may be an appropriate measure of effort, an assumption that will be made in this study.

The standard Schaefer production function is a special case of equation (4.20), where $b = g = 1$. The schooling behaviour of the herring has permitted the development of very effective means of harvesting. With modern fish-finding equipment, harvesting can be viable even at very low stock levels. For this reason, we will expect $0 \leq g < 1$ for herring. The Cobb-Douglas production function will describe a 'pure' schooling fishery, where catch is independent of stock, when $g = 0$ (Bjørndal, 1988).

We assume the cost per unit of effort to be constant. Under this assumption, we can write the cost function as:[16]

$$C(S_t, H_t) = ck = c(H_t / AS_t{}^g)^{1/b} \tag{4.21}$$

where c is the cost per vessel per fishing season which includes a normal return on capital.

Industry profit is defined as:

$$\pi_t = ph(S_t, H_t) - ck_t = ph(S_t, H_t) - C(S_t, H_t), \tag{4.22}$$

where p is unit price of harvest. Estimated parameters for the Cobb-Douglas production function are taken from Bjørndal and Conrad (1987a),[17] where $a = 0.06157$, $b = 1.3556$, and $g = 0.5621$.

The parameter estimates show that the Schaefer production function is inappropriate for the North Sea herring fishery. The parameter g reveals, as expected, the output elasticity of stock size to be between zero and one. Thus, harvest will decrease with decreasing stock size, but relatively less than the change in stock size. The parameter b indicates an output elasticity of effort larger than one. This means that increased effort is met with increasing harvest. This may be the result of economies of scale in the search for schools of herring.

According to Anderson et al. (2008), annual operating and fixed costs for a UK pelagic trawler are £3,574,300. In 2006, herring represented 27 per cent of catches. We will assume that the costs in the herring fishery represent the same proportion out of total cost. Thus, the cost of operating one vessel in the herring fishery for one season is £965,000.

The analysis of potential net benefits should be based on the prices that will prevail in the future. We will set the price at £300/tonne in the analysis of optimal management and potential profits. This level is higher than prices observed in the period 1996–2007 except for 2006. We will, however, present a sensitivity analysis for changes in the herring price.

On this basis, we can establish the cost function as:

$$C(S_t, H_t) = Ck_t = c(H_t/AS_t^g) = 965{,}000\left(H_t/0.06157S_t^{0.5621}\right)^{1/1.3556} \quad (4.23)$$

Then, industry profit becomes:

$$\begin{aligned}\pi_t = ph(S_t, H_t) - cK_t &= 300{,}000H_t - 965{,}000(H_t/0.06157S_t^{0.5621})^{0.7377} \\ &= 300{,}000H_t - 7{,}544{,}139(H_t)^{0.7377}S_t^{-0.4147}\end{aligned} \quad (4.24)$$

It is important to note that an underlying assumption is that price and cost conditions for all countries participating in the North Sea herring fishery are similar to those in the United Kingdom. Herring is sold into an international market, so the assumption of a fixed European price is appropriate.

Table 4.3. Estimates of optimal stock level S^* and the corresponding harvest H^* for North Sea herring ('000 tonnes; price = £300/tonne; cost per vessel per season = £965,000)

Discount rate	S^*	H^*
0.0	1,375.3	421.6
0.035	1,340.0	423.3
0.05	1,325.0	423.8
0.10	1,275.8	424.3

The following alternative recruitment functions were estimated by Bjørndal (1988) and Bjørndal et al. (2011):

$$R_{t+1} = S_t e^{a(1-St/b)},$$

$$R_{t+1} = aS_t/(b+S_t),$$

$$R_{t+1} = aS_t{}^b,$$

$$R_{t+1} = aS_t - bS_t{}^2.$$

Here, R_{t+1} is the number of recruits, which is a function of stock size. The first three functional forms are known as the Ricker, Beverton-Holt, and Cushing stock-recruitment functions, respectively, while the fourth specification is a quadratic function. Under an appropriate transformation each function above can be estimated using least squares procedures (Bjørndal, 1988).

In the model of population dynamics, we will employ the Ricker function. First, the explanatory power of this function is good. Second, predictions of S_{max} and R_{max} appear to be very reasonable. Third, while recruitment is continuously increasing in stock size for the Cushing function, it reaches a maximum and subsequently declines for the Ricker function, which is in line with theoretical properties for the recruitment function.

The above regressions relate zero-group herring in year $t + 1$ to the previous year's spawning stock. As noted above, $\gamma = 2$, and not all recruits survive to age-group two. The survival rate of equation (4.12) is defined as $\lambda = e^{-(M_0 + M_1)}$, where M_0 and M_1 are the natural mortality rates of year-class zero and one, respectively. We make use of the following estimates: $M_0 = M_1 = 1.0$.[20] In addition, we add fishing mortality, using the following parameters: $F_0 = 0.046$, $F_1 = 0.076$.[21] The mean weight of age-group two herring (w) is estimated to be 125 g.[22] We then have estimates of all parameters of the stock-recruitment equation (4.13) or $G(S_{t-2}) = g(S_{t-2})\, e^{-(M_0 + M_1)} w = g(S_{t-2})\, e^{-2.122} 0.125$.

Two functional forms for the relationship between δ and S are specified:

$$\delta(S) = a + bS$$

$$\delta(S) = a + bS + cS^2$$

These functions have been estimated by Bjørndal (1988) and Bjørndal et al. (2011). In this chapter, the linear version is used.

NOTES

1. In chapter 2 a distinction was made between salvageable and non-salvageable vessel costs. All of the vessel 'fixed' costs in this fishery can be deemed to be salvageable by virtue of the high degree of vessel capital malleability.

A number of countries fish for herring in the North Sea and regulations and harvesting technologies vary from country to country. Although this is ignored in the current analysis, this is likely to have an impact on current rent and the ability to attain potential rent. It would be useful and interesting to extend the analysis to take account of different regulatory and technological features on rent generation for herring in the North Sea.

BIBLIOGRAPHY

Biological data, including harvest and stock data, for both Western Channel sole and North Sea herring are available from the International Council for the Exploration of the Sea (www.ices.dk).

To the best of our knowledge, the only bioeconomic analyses of Western Channel sole are provided by Bjørndal and Bezabih (2008, 2010).

The delay-difference discrete time bioeconomic model for North Sea herring was developed and estimated by Bjørndal (1988). Bjørndal (1987; reprinted in 2002) estimated a harvest supply function for herring and derived the underlying structure of the production function, representing an early application of duality theory to fisheries. Bjørndal and Conrad (1987a) analysed the open access dynamics of North Sea herring, while Bjørndal and Conrad (1987b) estimated capital dynamics in the fishery.

To the best of our knowledge, there are not many game theoretic studies pertaining to North Sea herring. One exception is provided by Bjørndal and Lindroos (2004) who analysed the sharing of the stock between Norway and the EU.

Nøstbakken and Bjørndal (2003) developed a discrete time bioeconomic model which was used to derive supply curves for the open-access and the optimally managed fisheries. This is one of the few examples in the literature of empirical supply functions for fisheries.

As noted above, to save the herring stock from total depletion under open access, the fishery was closed from 1977 until 1981. Nøstbakken (2008), in the context of a discrete-time stochastic bioeconomic model, analysed the North Sea herring fishery under alternative management regimes. It is shown how catches and harvest policies change with the price of herring. Feedback policies are found for the optimally managed fishery. The management of the North Sea herring, after the moratorium was lifted in 1981, is evaluated. The results indicate that the management has been sub-optimal. Under optimal management, the fishery should have stayed closed until 1983.

Bjørndal et al. (2010) estimated current and potential rent in the herring fishery.

APPENDIX: THE MODEL OF POPULATION DYNAMICS FOR NORTH SEA HERRING

In Figure 4.5, spawning stock biomass and recruitment the following year are graphed. Notice the big spikes in recruitment in 1961, 1986, and 2001.

2. In other words, our assumption that the Marginal Stock Effect is equal to zero might have to be modified.
3. Thus, profits in 2008 are discounted one year, rents in 2009 two years, etc.
4. A mechanical winch that was used to pull in the seines, allowing for much larger nets and eventually larger vessels.
5. See Horwood (1999), <http://www.cefas.co.uk/publications/miscellaneous-publications/the-precautionary-approach.aspx>.
6. That is, the implementation of the EEZ regime.
7. This principle says that the relative (percentage wise) shares of the different countries involved in the fishery remain constant over time.
8. Catches include juvenile herring that are part of the total biomass but not the SSB. As well, there is growth in the SSB during the year. As an example, in 1974 total landings were about 275,000 tonnes. Total biomass and SSB, at the beginning of the year, were 912, 000 and 162,000 tonnes, respectively. By the time the fishing season started, both total biomass and SSB would have grown. Thus, the harvest of 275,000 tonnes would be from a larger stock, and a substantial part of it would be immature herring.
9. B_{pa} is the minimum biomass level according to the precautionary approach.
10. By total supply of herring is meant world supply; however, most of this comes from the North Atlantic.
11. Anderson et al. (2008).
12. Average value per tonne is defined as total value divided by total landings. It is worth pointing out that, in 2006, average value of the landings of this fleet was £402/tonne, while the herring price was £317/tonne. The reason for this is that mackerel, which represents about two thirds of the catch, receives a higher price than herring, which in turn receives a higher price than blue whiting.
13. This is, of course, different from the continuous time discount rate.
14. The model that we are considering here is far more complex than the general model that we discussed in chapter 3, section 3.4.
15. Various functional forms both for stock recruitment and the net growth function have been estimated (see Appendix). Different alternatives may be combined in the model of population dynamics.
16. See chapter 2. If $b = g = 1$, this reduces to equation (2.17).
17. To the best of our knowledge, more recent estimates of the production function do not exist. Moreover, data do not exist that would enable us to estimate a production function for the UK pelagic fleet.
18. A somewhat larger discount rate is used in this example than when Western Channel sole was highlighted. This is to illustrate that the Marginal Stock Effect is important even at a reasonably high discount rate of 5 per cent.
19. These outcomes are fully consistent with having a significantly positive *MSE*, and a reasonable but modest discount rate, as shown in chapter 3.
20. Source: ACFM (2008).
21. These are average fishing mortalities for 2000–5.
22. This is an average for 2006 for area IVa(E).

5 Extensions to the Basic Dynamic Economic Model of the Fishery

In chapter 3, we discussed, in some detail, the linear and non-linear versions of the basic dynamic economic of the fishery. We then followed this discussion with empirical applications of the basic model in chapter 4. In this chapter, we shall explore some of the many extensions to the basic model. The extensions, to a large degree, arise from relaxing the highly restrictive assumptions that we employed in chapters 3 and 4.

The extensions to be considered in this chapter are:

(a) the incorporation of the processing sector into the model (5.1);
(b) the introduction of multi-species, as opposed to single species, fisheries (5.2);
(c) the introduction of non-malleable produced and human capital into the model (5.3); and
(d) the introduction of uncertainty (5.4).

It will be found that extensions (a) and (b) can be addressed with relative ease. Extension (c), on the other hand, is much more challenging. The extension has major policy implications, particularly now that attention is being directed more and more towards the rebuilding of hitherto overexploited capture fishery resources, and the threats posed to such rebuilding programmes by the emergence of fishing overcapacity.[1]

The final extension, (d), the introduction of uncertainty, is, of course, important, by virtue of the fact that uncertainty is a fundamental fact of life in fisheries management. That said, introducing uncertainty does not have dramatic effects upon the analytical economic model of the fishery. Uncertainty, we shall argue, has to be addressed under the heading of policy issues, and is thus a major topic for a later chapter.

5.1 The processing sector included

All of the economic models of the fishery considered up to this point, static and dynamic, ignore the processing sector. It is as if all fish harvested is sold

5 Extensions to the Basic Dynamic Economic Model of the Fishery

In chapter 3, we discussed, in some detail, the linear and non-linear versions of the basic dynamic economic of the fishery. We then followed this discussion with empirical applications of the basic model in chapter 4. In this chapter, we shall explore some of the many extensions to the basic model. The extensions, to a large degree, arise from relaxing the highly restrictive assumptions that we employed in chapters 3 and 4.

The extensions to be considered in this chapter are:

(a) the incorporation of the processing sector into the model (5.1);
(b) the introduction of multi-species, as opposed to single species, fisheries (5.2);
(c) the introduction of non-malleable produced and human capital into the model (5.3); and
(d) the introduction of uncertainty (5.4).

It will be found that extensions (a) and (b) can be addressed with relative ease. Extension (c), on the other hand, is much more challenging. The extension has major policy implications, particularly now that attention is being directed more and more towards the rebuilding of hitherto overexploited capture fishery resources, and the threats posed to such rebuilding programmes by the emergence of fishing overcapacity.[1]

The final extension, (d), the introduction of uncertainty, is, of course, important, by virtue of the fact that uncertainty is a fundamental fact of life in fisheries management. That said, introducing uncertainty does not have dramatic effects upon the analytical economic model of the fishery. Uncertainty, we shall argue, has to be addressed under the heading of policy issues, and is thus a major topic for a later chapter.

5.1 The processing sector included

All of the economic models of the fishery considered up to this point, static and dynamic, ignore the processing sector. It is as if all fish harvested is sold

2. In other words, our assumption that the Marginal Stock Effect is equal to zero might have to be modified.
3. Thus, profits in 2008 are discounted one year, rents in 2009 two years, etc.
4. A mechanical winch that was used to pull in the seines, allowing for much larger nets and eventually larger vessels.
5. See Horwood (1999), <http://www.cefas.co.uk/publications/miscellaneous-publications/the-precautionary-approach.aspx>.
6. That is, the implementation of the EEZ regime.
7. This principle says that the relative (percentage wise) shares of the different countries involved in the fishery remain constant over time.
8. Catches include juvenile herring that are part of the total biomass but not the SSB. As well, there is growth in the SSB during the year. As an example, in 1974 total landings were about 275,000 tonnes. Total biomass and SSB, at the beginning of the year, were 912, 000 and 162,000 tonnes, respectively. By the time the fishing season started, both total biomass and SSB would have grown. Thus, the harvest of 275,000 tonnes would be from a larger stock, and a substantial part of it would be immature herring.
9. B_{pa} is the minimum biomass level according to the precautionary approach.
10. By total supply of herring is meant world supply; however, most of this comes from the North Atlantic.
11. Anderson et al. (2008).
12. Average value per tonne is defined as total value divided by total landings. It is worth pointing out that, in 2006, average value of the landings of this fleet was £402/tonne, while the herring price was £317/tonne. The reason for this is that mackerel, which represents about two thirds of the catch, receives a higher price than herring, which in turn receives a higher price than blue whiting.
13. This is, of course, different from the continuous time discount rate.
14. The model that we are considering here is far more complex than the general model that we discussed in chapter 3, section 3.4.
15. Various functional forms both for stock recruitment and the net growth function have been estimated (see Appendix). Different alternatives may be combined in the model of population dynamics.
16. See chapter 2. If $b = g = 1$, this reduces to equation (2.17).
17. To the best of our knowledge, more recent estimates of the production function do not exist. Moreover, data do not exist that would enable us to estimate a production function for the UK pelagic fleet.
18. A somewhat larger discount rate is used in this example than when Western Channel sole was highlighted. This is to illustrate that the Marginal Stock Effect is important even at a reasonably high discount rate of 5 per cent.
19. These outcomes are fully consistent with having a significantly positive *MSE*, and a reasonable but modest discount rate, as shown in chapter 3.
20. Source: ACFM (2008).
21. These are average fishing mortalities for 2000–5.
22. This is an average for 2006 for area IVa(E).

into the fresh fish market. Since in many, if not most, fisheries the harvested fish are processed to some degree, it behoves us to ask what difference, if any, it will make to include the processing sector in our model. This question was examined by Clark and Munro (1980). It is their model that we shall consider.[2]

The general conclusions of Clark and Munro are that, if both the harvesting and processing sectors are characterized by atomistic competition and both are free of non-malleable produced and human capital, the processing sector can be safely ignored when studying the consequences of Pure Open Access. The processing sector cannot be ignored, however, when analysing the socially optimal management of the fishery. In addition, since processing sectors are commonly characterized by oligopoly/oligopsony, rather than atomistic competition, an interesting policy issue is seen to arise (Clark and Munro, 1980).

Clark and Munro commence by observing that in many fisheries, particularly groundfish fisheries, the unit cost of processing and the price of finished product are functions of the average size of fish. The larger the fish, the lower is the unit cost of processing raw fish, and the higher is the unit value of the finished product.[3] Clark and Munro incorporate the possible effects of size of fish on unit harvesting cost and price of finished product into the basic dynamic economic model of the fishery by making the ad hoc assumption that the larger the biomass, the larger will be the average size of harvested fish (Clark and Munro, 1980).

The resultant Clark and Munro model is as follows. To begin, the authors assume that the processing sector faces a perfectly elastic demand for finished product and assume further that the processing firms are subject to constant returns to scale. The harvesting sector is modelled exactly as in chapter 3. It is assumed initially that the unit cost of fishing effort is a constant, and it is assumed further that harvest is given by:

$$h(t) = qE(t)x(t).$$

In the processing sector, we first take note of the recovery factor, which we shall here denote as α, expressed as the fraction of a unit of raw fish that can be transformed into finished product. We have: $0 \leq \alpha \leq 1$ and further:

$$\frac{d\alpha}{dx} \geq 0 \tag{5.1}$$

Next let $c_2(x)$ denote the unit cost of processing raw fish. We have:

$$c_2'(x) \leq 0 \tag{5.2}$$

Finally, we have the price of finished product, p_m, assumed to be constant.[4]

The net revenue received by the processing sector, per unit of raw fish processed and sold (excluding the cost of acquiring the raw fish), is:

$$p_2(x) = \alpha(x)p_m - c_2(x) \tag{5.3}$$

(Clark and Munro, 1980).

From the point of view of the social manager, the fishery must be viewed on an integrated basis. Hence, to the social manager the net economic return from the fishery at any given point in time is given by:

$$\pi_s = [p_2(x) - c_1(x)]h \tag{5.4}$$

where $c_1(x)$ denotes unit harvesting costs, and where by assumption: $c_1(x) = b/qx$. As before, b, a constant, denotes unit fishing effort costs and q, a constant, is the catchability coefficient.

At a later point we are going to consider the possibility of the management of the fishery being under the control of a processing sector monopsony. Thus, we are careful to denote net economic return from the fishery as the net economic return to the social manager as π_s.

In any event, the resource manger's objective functional can be expressed as follows:

$$\begin{aligned} max\ PV &= \int_0^\infty e^{-\delta_s t}\pi_s(h(t), x(t))dt \\ &= \int_0^\infty e^{-\delta_s t}[p_2(x) - c_1(x)]h(t)dt \end{aligned} \tag{5.5}$$

subject to the usual constraints (see equation 3.7), and where δ_s denotes the social rate of discount.

By a routine application of the Maximum Principle, we are led to the following version of the Fundamental Equation for the social manager:

$$F'(x_s^*) + \frac{[p'_2(x_s^*) - c'_1(x_s^*)]F(x_s^*)}{p_2(x_s^*) - c_1(x_s^*)} = \delta_s \tag{5.6}$$

Thus, if $p_2(x)$ is, in fact, completely independent of the size of x, $p_2(x) = p_2$, then explicitly incorporating the processing sector adds nothing to our model. If, in fact, $p'_2(x) > 0$, then the effect of explicitly recognizing the processing sector will cause the social manager to be more conservationist than would otherwise be the case.[5] Be that as it may, we are wholly unjustified in assuming that we shall always have: $p'_2(x) = 0$ (Clark and Munro, 1980).

Clark and Munro, drawing upon the fact that the typical processing sector is an oligopsony, investigate a claim made at an earlier date that, if the firms of an oligopsonistic fish processing sector were not merely allowed, but actively encouraged, to merge and form a monopsony, a socially optimal outcome would be forthcoming. This would be true, even if the harvesting sector is highly competitive (Crutchfield and Pontecorvo, 1969). As is so often the case, it all depends.

Clark and Munro investigate a case in which monopsony will not result in a socially optimal outcome. This is a case in which harvesting costs are non-linear in $h(t)$ (resulting from the total cost of fishing effort, $C(E)$, being such that $C'(E) > 0$, while assuming that $C''(E) = 0$).

They assume to begin with that $\delta_2 = \delta_s$, where δ_2 denotes the monopsonistic processor's rate of discount.

The Fundamental Equations of the social manager and the monopsonist can be expressed as follows:

$$F'(x_s^*) + \frac{\partial \pi_s / \partial x_s^*}{\partial \pi_s / \partial h}_{|h = F(x_s^*)} = \delta_s \tag{5.7a}$$

$$F'(x_2^*) + \frac{\partial \pi_2 / \partial x_2^*}{\partial \pi_2 / \partial h}_{|h = F(x_2^*)} = \delta_2 \tag{5.7b}$$

It is found that for any given level of x, we shall have:

$$\frac{\partial \pi_2 / \partial x}{\partial \pi_2 / \partial h} > \frac{\partial \pi_s / \partial x}{\partial \pi_s / \partial h} \tag{5.8}$$

Why? Without going into the full details of the model, the reason is that the monopsonist controls harvesting through its control of the ex-vessel price of fish, which we can denote as p_1. The total cost of harvested fish to the monopsonist is thus equal to: $p_1 h$. The harvesting sector will be in equilibrium when p_1, which is the marginal revenue from harvested fish perceived by individual harvesters, is equal to marginal harvesting costs. In the circumstances described, marginal harvesting costs exceed average harvesting costs.[6] As a result, the total cost of harvested fish, as perceived by the monopsonist, incorporates producer's surplus of the infra-marginal harvesters. Note that the price, p_1, applies to all fishermen, high cost and low cost alike. Hence, the monopsonist *overestimates* the true cost of harvested fish. On the other hand, the cost of harvested fish, as perceived by the social manager, is equal to the true cost.[7] Consequently, the monopsonist is more sensitive than is the social manager to the harvesting and processing cost reducing benefits of having a larger, rather than smaller, fish stock.

Given that $\delta_2 = \delta_s$, we shall thus have:

$$x_2^* > x_s^* \tag{5.9}$$

The monopsonist will, from a social standpoint, be excessively conservationist! If it were the case that $\delta_2 > \delta_s$, then we might get $x_2^* = x_s^*$, or something close to it, because of offsetting distortions (Clark and Munro, 1980).[8]

Clark and Munro discuss how the monopsonist might be guided towards the social optimum through the use of taxes and subsidies, a discussion that we shall not investigate. One glaring problem with the Clark and Munro (1980) approach to socially optimal fisheries management via a processor monopsony is a political one. The bulk of the net economic returns from the fishery would flow into the hands of the processing monopsony. In their article, Clark and Munro acknowledge that the resource manager would have to find a means of capturing some of the resource rent, without upsetting the resource manager's attempt to guide the monopsonist along the optimal path (Clark and Munro, 1980).[9]

Clark and Munro also acknowledge that it strains one's credulity, in the situation described, to believe that the fishermen would continue to act competitively, thereby allowing themselves to be 'exploited' by the monopsonist. In a follow-up paper, Munro (1982) assumes that the fishermen react and form a fully functioning cooperative. The monopsony situation is thus transformed into a bilateral monopoly.

Munro analyses the bilateral monopoly within the framework of Nash's model of a two-player cooperative game. The discussion of game theory and fisheries will be put off to a later stage. Be it sufficient to say that, if the monopsonist and fishermen cooperate, and employ the same discount rate, the outcome is straightforward, and is very similar to a standard wage bargaining model. The two 'players' should be expected to cooperate to maximize the present value of the net economic returns from the fishery, and then bargain over the division of net economic returns through time (Munro, 1982).

Complications can, of course, arise. Suppose that the monopsonist's and the fishermen cooperative's discount rates differ, and suppose, further, that both differ from the social rate of discount. The analysis can become daunting.

Prior to leaving the topic of the processing sector, and the processing sector as a monopsony, we cite one example. The example comes from the developing world, and involves a developing coastal state, where something close to the monopsony model was applied, and was applied with apparent success. The coastal state is Mauritania, in Northwest Africa. The monopsonist, in this case, was a government monopsonist, so that there were no concerns about divergences between private and social discount rates, or about a private entity absorbing most of the resource rent.

The introduction by Mauritania of a 200-mile EEZ had important economic consequences for the country. The EEZ encompassed potentially valuable fishery resources. There was no dispute over the need for the public sector to capture most of the resources rent.

The overwhelming bulk of the fish landed in Mauritania is exported. The government established the Mauritanian Fish Marketing Authority designed to give the country some degree of market power in exporting the fish. The government body was also a monopsonist, however, since all fish landed in Mauritania had to be sold to the Authority, which did, by the way, do a modest amount of processing. There are few landing points, so that enforcement was reasonably easy (Cunningham et al., 1995).

Cunningham et al. (1995) concede that the state monopsony system did not work perfectly. The authors argue persuasively, however, that significant resource rent was forthcoming from the fisheries, and that the system worked far better than any of the alternatives (Cunningham et al., 1995).

Finally, one might ask the following. What is the real difference between the Mauritanian government monopsony scheme and management of the fishery through the use of harvest taxes? The answer is: what indeed! We shall refer to the Mauritania case again, when we come to discuss the use of taxes in fisheries management.

5.2 Multispecies fisheries: an introduction to ecosystem resource management

Every economic model of the fishery that we have examined thus far has been a single species one. In the real world multispecies fisheries are commonplace. Vessels in a fishery may be harvesting several different species at the same time. Relevant species may interact with one another, by competing for food sources, or because there are predator–prey relationships between, or among, them.

The attempt to address the multispecies question can be seen as our first attempt at examining ecosystem-based approaches to fisheries management. A simple dictionary definition of an ecosystem is: a biological community of interacting organisms and their physical environment (Oxford Dictionaries Online, 2011). The American National Research Council (1999) notes that any attempt to manage an entire marine ecosystem would be an impossible undertaking. A reasonable first approach to an ecosystem-based approach to fisheries management, the Council argues, is to recognize explicitly interaction between and among species, and their interaction with the marine environment (National Research Council 1999, p. 15).[10]

In any event, multispecies fisheries should cause us no conceptual difficulties. Rather than thinking in terms of managing a single natural resource asset, we now think in terms of managing a portfolio of such assets. What matters is not the economic return on a single fishery resource asset, but rather the economic return on the portfolio as a whole. This means that, *inter alia,* we must be concerned with portfolio balance—the mix of fishery resource assets.

We illustrate by taking the two-species predator–prey model of Clark (1990, chapter 10), e.g., capelin and cod. Let x denote the biomass of the prey species and y denote the biomass of the predator species. It is assumed that the two species are independent of all other species, and that the species are exploited independently.

It is then assumed that the fisheries can be modelled as follows:

$$\begin{aligned} F(x,y) &= rx\left[1-\frac{x}{G}\right]-\alpha xy \\ M(x,y) &= sy\left[1-\frac{y}{N}\right]+\beta xy \end{aligned} \tag{5.10}$$

where r, s, G, N, α, and β are constants. The species interaction is captured by αxy and βxy, where α and β, in this context, are to be seen as 'interaction coefficients'.

We have further:

$$\begin{aligned} dx/dt &= F(x,y)-h_1(t) \\ dy/dt &= M(x,y)-h_2(t) \end{aligned} \tag{5.11}$$

where $h_1(t)$ and $h_2(t)$ are the harvest rates of the prey and predator species respectively. It is assumed in addition that the demands for the harvested fish of the two species are independent and that they are both perfectly elastic. It is also assumed that unit costs of harvesting the two species are decreasing functions of the respective biomasses, but are independent of the rates of harvest (Clark, 1990). The joint rent, or net economic return, function can be expressed as follows:

$$\pi_J(t)=[p_1-c_1(x)]h_1(t)+[p_2-c_2(y)]h_2(t) \tag{5.12}$$

where subscripts 1 and 2 denote x and y fisheries respectively.

There is now a set of interdependent Fundamental Equations, which can be expressed as (Clark, 1990, p. 325):

$$F_x+\frac{(\partial\pi_J/\partial h_2)M_x+\partial\pi_J/\partial x}{\partial\pi_J/\partial h_1}\Bigg|_{\substack{h_1=F(x,y)\\ h_2=M(x,y)}}=\delta \tag{5.13}$$

$$M_y + \frac{(\partial\pi_J/\partial h_1)F_y + \partial\pi_J/\partial y}{\partial\pi_J/\partial h_2}\Bigg|_{\substack{h_1 = F(x,y)\\ h_2 = M(x,y)}} = \delta$$

where F_x, F_y, M_x, and M_y denote partial derivatives, and where δ denotes the social rate of discount.[11]

Observe first that optimality calls for both the own rate of interest of x and that of y to be equal to δ, which means that they are equal to one another. This satisfies a fundamental condition for optimal portfolio balance, namely that assets in the portfolio of a common risk class must be found to have equal yields.

Next observe the interactive terms, $(\partial\pi_J/\partial h_2)M_x$ and $(\partial\pi_J/\partial h_1)F_y$. The expression $(\partial\pi_J/\partial h_2)M_x$ can be interpreted as the impact of an incremental change in x upon the flow of net economic returns from the y fishery, while the expression $(\partial\pi_J/\partial h_1)F_y$ can be interpreted in a similar fashion. Since x is the prey species and y is the predator species, we have $M_x > 0$; $F_y < 0$.

The optimal portfolio balance will, of course, depend, *inter alia*, on the net unit values of h_1 and h_2. If, for example, the prey species is high valued and the predator species low valued—e.g., Pacific salmon vs. dogfish (a.k.a. lemon shark), it might be optimal to drive the predator species below the Bionomic Equilibrium level. This can be illustrated as follows.

Rewrite the second of the two Fundamental Equations (5.13) as

$$\frac{1}{\delta}\left[\left(\frac{\partial\pi_J}{\partial h_2}\right)M_y + \left(\frac{\partial\pi_J}{\partial h_1}\right)F_y + \frac{\partial\pi_J}{\partial y}\right] = \frac{\partial\pi_J}{\partial h_2} \tag{5.14}$$

One interpretation of the L.H.S. of (5.14) is that it represents the marginal 'user cost' of the resource,y, which can be defined as the present value of the stream of net economic returns from the resource that would be lost, if there was an incremental reduction in the resource.

Now consider the consequences of the resource being driven to Bionomic Equilibrium,$y = y_{BE}$. At Bionomic Equilibrium we have: $\partial\pi_J/\partial h_2 = 0$, which means that the L.H.S. of (5.14) reduces to $1/\delta\left[(\partial\pi_j/\partial h_1)F y_{BE} + \partial\pi_J/\partial y_{BE}\right]$. Given that $F y_{BE} < 0$, it is quite possible that at $y = y_{BE}$, we would have the following inequality:

$$\frac{1}{\delta}\left[\left(\frac{\partial\pi_J}{\partial h_1}\right)F y_{BE} + \frac{\partial\pi_J}{\partial y_{BE}}\right] < 0 \tag{5.15}$$

The implication is that at $y = y_{BE}$, the marginal 'user cost' of y is negative, which in turn implies that it would be optimal to reduce$y$ below y_{BE}. Fishermen would not harvest y at a loss, so that they would have to be offered a bounty to harvest$y$. The y resource would acquire the status of 'pest'.

An example is provided by the Barents Sea, the fisheries of which have been studied extensively on a multispecies basis (Flaaten, 1988, 1991; Eide and Flaaten, 1998). Flaaten (1988) examines a three species model consisting of cod, capelin, and sea mammals. The three species are in fact an aggregation of: (a) demersal species, (b) pelagic species, and (c) seals and whales. The demersal species predate on the pelagic species, while the sea mammals predate on both demersal and pelagic species. The harvests of both cod and capelin have significant economic value, while the harvests of sea mammals do not.

The optimal mix of cod and capelin in the resource asset portfolio will, not surprisingly, depend upon the relative net economic returns from the harvests of the two species. Unpalatable to many is the conclusion that, if this multispecies fishery were managed optimally, the sea mammal biomass would be driven down below the level that would occur under Pure Open Access (Flaaten, 1988, 1991).[12]

While neither Eide nor Flaaten mention the fact, the study provides an example of the law of unintended consequences. There are in many countries as well as the EU severe restrictions on imports of whale and seal products, with the objective being to protect seals and whales. This may have benefits under Pure Open Access fisheries. Where sea mammal fisheries, and other interacting fisheries, are managed, however, the low prices for sea mammal products, arising from the ban, could result in the sea mammals being placed in the 'pest' category for management purposes.

A special version of the multispecies problem—which strictly speaking is not multispecies at all—involves sub-populations of the same species, meta-populations, that are spread out spatially, and in which there is dispersal, or diffusion, among the sub-populations. A particularly clear-cut case is that of a sink–source relationship, in which fish flow, unidirectionally, from one sub-population—the source—to the other sub-population—the sink. In any event, the interaction, in this case, arises from dispersion among the sub-populations (Sanchirico and Wilen, 1999, 2005).

The unidirectional sink–source case has been recently modelled by Clark (2010). It is to his model that we now turn. In the Clark model, the source sub-population will grow to an upper limit, or saturation level, G. Once G is achieved, any further 'surplus' production by the source sub-population will be instantaneously transferred to the sink subpopulation. If the source sub-population is below G, transfers to the sink sub-population will be zero. The sink sub-population is wholly dependent upon the source. If transfers from the source are zero, the sink sub-population will ultimately vanish.

Denote the source sub-population as x_1, and the sink sub-population as x_2. Let r denote the intrinsic growth rate of x_1 and let m denote the rate of natural mortality of x_2. We then have the following, which looks very different indeed from the Schaefer model to which we have become so accustomed:

$$\frac{dx_1}{dt} = F(x_1) = \begin{cases} rx_1 \text{ if } x_1 < G \\ 0 \text{ if } x_1 = G \end{cases} \tag{5.16}$$

$$\frac{dx_2}{dt} = \begin{cases} -mx_2 \text{ if } x_1 < G \\ rG - mx_2 \text{ if } x_1 = G \end{cases} \tag{5.17}$$

With regards to natural capital management, the key resource is obviously x_1. If our resource management is such that the optimal level, x_1^*, is less than G, then x_2 becomes irrelevant. If it is optimal to let x_1 grow to G, then the question becomes how to exploit rG.[13] One will harvest at the source, or at the sink, but not both (Clark, 2010).

Clark allows for density dependent harvesting costs in both sink and source fisheries, and, of course, must allow for the fact that fishing effort costs will be different between the two fisheries. So we have the following unit harvesting costs: $c_1(x_1)$ and $c_2(x_2)$ respectively. It is assumed that the price of harvested fish, p, is identical for the two fisheries. Where to harvest will depend, *inter alia*, upon relative costs in the sink and source fisheries, but will also depend upon the social rate of discount. Look at the source fishery, where capital theoretic decisions have to be made. Clark shows that the following version of the Fundamental Equation is relevant for the source fishery, given that it is feasible economically to exploit that fishery (i.e., $p > c_1(G)$):

$$r - \frac{c'(x_1^*)rx_1^*}{p - c(x_1^*)} = \delta^{14} \tag{5.18}$$

If $r > \delta$, then equation (5.18) has no solution, implying that it is optimal to stabilize x_1 at G. Then the question becomes how to harvest rG, period after period.[15] If, on the other hand, $r < \delta$, then it is quite possible that we will have: $x_1^* < G$. Then the sink would be irrelevant.

At this point, we should look ahead to coming discussions of uncertainty. The preceding discussion of sink–source cases is entirely in terms of deterministic models, i.e., models in which uncertainty is ignored. Allowing for uncertainty can have a drastic impact upon our conclusions. Suppose, in an uncertain world, the social manager estimates that the *expected* economic returns from the fishery would be maximized by fishing at the source alone. The rational social manager could still opt to concentrate harvesting in the sink alone. In light of uncertain data and uncertain modelling of the resource, the manager has the assurance that by concentrating harvesting on the sink alone the future of the resource will not be endangered, in spite of management mistakes that may be made. The social manager has no such assurance, if the harvesting is concentrated on the source alone. We shall return to this point in a later chapter.

5.3 Non-malleability of produced and human capital in the fishery

The term 'malleability' of produced and human capital in a fishery, which we encountered briefly in chapter 2 in our discussion of Regulated Open Access, refers to the ease with which such capital can be shifted in and out of the fishery. If the capital can be shifted easily and costlessly in and out of the fishery, it is deemed to be highly 'malleable'. The malleability of produced/human capital is analogous to the concept of liquidity in the world of finance.[16]

The seemingly esoteric concept of malleable, as opposed to non-malleable, produced and human capital does in fact have important policy implications. This is particularly the case when we come to address the issue of rebuilding hitherto overexploited fish stocks.

The produced and human capital, with which we are concerned, consists of fleet capital—vessels and gear—processing capital, and finally the human capital embodied in fishermen and those employed in the processing sector. In the analysis to follow, however, we shall focus exclusively on fleet capital. We would, nonetheless, argue that the analysis is applicable, *mutatis mutandis*, to both processing and human capital.

In the literature, there are three articles that address the issue of non-malleable fleet capital head on. These are the article by Clark et al. (1979) and the two articles by McKelvey (1985, 1987). The models in all three articles are single species in nature. We commence with Clark et al. (1979)—CCM from hereon in.

To begin, the CCM model looks very similar to the linear autonomous dynamic economic model of the fishery discussed in chapter 3, but with two differences—one major; the other minor. The major difference is that the chapter 3 model implicitly assumes that all produced and human capital is perfectly malleable. This assumption is now relaxed. The minor difference is that it is found more convenient to let $E(t)$ be a control variable, rather than $h(t)$.

The CCM model, resting upon the Schaefer biological model, is as follows:

$$\frac{dx}{dt} = F(x) - h(t) \tag{5.19}$$

where by assumption: $F(x) = rx[1 - x/G]$. Moreover,

$$h(t) = qE^{\alpha}(t)x^{\beta}(t) \tag{5.20}$$

where by assumption $\alpha = \beta = 1$. So far, so good.

Now denote the stock of fleet capital by $K(t)$, where $K(t)$ can be thought of in terms of the number of 'standardized' fishing vessels. We then have:

$$0 \leq E(t) \leq E_{max} = K(t) \tag{5.21}$$

which asserts that the maximum fishing effort capacity is determined by the number of vessels in the fleet, and that actual fishing effort cannot exceed E_{max}. Effort capacity may, or may not, be fully utilized.

Net investment in fleet capital is given by:

$$\frac{dK}{dt} = I(t) - \gamma K, K(0) = K^0 \tag{5.22}$$

where $I(t)$ denotes the rate of gross investment in K at time t, and where γ, a constant, is the rate of depreciation, $\gamma \geq 0$.

It is assumed that the price of harvested fish, p, is a constant, as is b_{oper}, operating costs per unit of fishing effort. The replacement unit cost (purchase price) of fleet capital is denoted by λ, while the 'scrap' value, or resale price, of fleet capital is denoted by λ_s. Obviously λ_s can reflect the returns to be enjoyed by employing capital in other fisheries. Both λ and λ_s are deemed to be constant.

It is assumed that 'pulse' investment in fleet capital is possible, meaning that the upper bound to $I(t)$ is: $I(t) = +\infty$ (CCM, 1979).

The resource is assumed to be under the control of an all-powerful resource manager whose objective functional can be expressed as:

$$PV = \int_0^{\infty} e^{-\delta t}(ph(t) - b_{oper}E(t) - \phi(I(t))dt \tag{5.23}$$

where, as before, δ is the social rate of discount, and where ϕI is:

$$\phi I = \begin{cases} \lambda I \text{ if } I > 0 \\ \lambda_s I \text{ if } I < 0 \end{cases} \tag{5.24}$$

CCM then define various degrees of 'malleability' of fleet capital as follows:

(i) $\lambda_s = \lambda$
(ii) $\lambda_s = 0; \gamma > 0$
(iii) $0 < \lambda_s < \lambda; \gamma \geq 0$
(iv) $\lambda_s = 0; \gamma = 0$

Consider first case (i). This is the case of perfect 'malleability' of capital. Any unused, unwanted, fleet capital can be promptly disposed of at its purchase price. There is no threat of a capital loss. No sane resource manager would allow any unused capacity to exist.[17]

CCM demonstrate that, in this instance, the manager's objective functional would reduce to:

$$PV = \int^{\infty} e^{-\delta t}\{ph(t) - [b_{oper}+(\delta+\gamma)\lambda]E(t)\}dt \qquad (5.25)$$

where $(\delta+\gamma)\lambda$ can be viewed as the per unit 'rental cost' of vessel capital (CCM, 1979). $K(t)$ effectively disappears from the model. We thus have a one state variable, $x(t)$ and one control variable, $E(t)$, optimal control problem. Indeed, what we have is simply the linear, autonomous dynamic economic of the fishery described in chapter 3.

This leads to an important conclusion. The H. Scott Gordon model of the fishery, and its dynamic versions, which we have discussed up to this point, implicitly assume that *all* produced and human capital used in the fishery is perfectly malleable. This in turn means, as we shall come to see, that these models are quite inadequate for the purpose of analysing certain important policy issues.

Now consider case (iv), the polar opposite to case (i). Once the fleet capital is acquired, it is there for ever. It cannot be sold off at a positive price, and it never depreciates. $K(t)$ most certainly does not disappear from the model. On the contrary, we are now confronted with a two state variable, $x(t)$ and $K(t)$, and two control variable, $E(t)$ and $I(t)$, problem.

Cases (ii) and (iii) are intermediate cases, which we can refer to as involving quasi-malleable fleet capital. In case (ii) the fleet capital, once acquired, has no resale value. It can, however, be disposed of over time, through depreciation. In case (iii) fleet capital, once acquired, can be sold off, but only at a capital loss, $\lambda_s < \lambda$. Once again, we are faced with a two state variable; two control variable problem.

In cases (ii), (iii), and (iv), the mathematics becomes daunting. The standard optimal control theory, underlying the dynamic economic models of the fishery discussed thus far, is not applicable. The Maximum Principle does not apply when the permissible value of a control variable depends upon current value of a state variable, which is the situation that now confronts us, i.e., $E(t) \leq K(t)$ (CCM, 1979). We shall discuss the analytical results forthcoming from the CCM model, but we shall not examine the underlying mathematical proof.[18]

Let us now explore the consequences of non-malleability of fleet capital for optimal management, by considering the more realistic case of quasi-malleable fleet capital. We shall, for the sake of simplicity, focus on case (ii), $\lambda_s = 0$; $\gamma > 0$. Let us assume that we commence with a virgin fish stock that is commercially exploitable. Consider now Figure 5.1, which can be described as a feedback control, state space diagram, and which is adapted from CCM (1979).

Consider first x^* and $\tilde{x} < x^*$, both of which are to be explained momentarily. Associated with x^* and $\tilde{x}$ are K^* and $\tilde{K} > K^*$ respectively. These represent the minimum levels of K required to harvest x^* and $\tilde{x}$ on a sustained yield basis. Given the underlying Schaefer model, the lower is x, the larger will be the K required to harvest the resource on a sustained yield basis.

2008). We reiterate: in the circumstances described, the policy of investing in the resources at the most rapid rate would not only be bad policy on social grounds, it would as well be bad policy on economic grounds.[23]

5.4 Uncertainty introduced

We have to this point employed 'deterministic' economic models of the fishery, in which we have assumed the future to be known with absolute certainty. These models are useful as a first approximation, but it is now universally recognized that fisheries management has to take place under great uncertainty. The uncertainty has been given formal recognition through the Precautionary Approach to fisheries management, a concept which is now embedded in international treaty law (UN, 1995, Article 6).

Writing over 30 years ago, Walters and Hilborn (1978) listed three broad classes of uncertainty in fisheries management. While their classification is concerned largely with environmental uncertainty, it applies with equal force to economic uncertainty. The Walters–Hilborn classification, which has lost none of its relevance over time, is as follows:

(a) random effects whose future frequency of occurrence can be determined from past experience;
(b) parameter uncertainty that can be reduced by research and acquisition of information through future experience; and
(c) ignorance about the appropriate variables to consider and the appropriate form of the model.

Almost all attempts to incorporate uncertainty into analytic economic models of the fishery focus on (a), the least demanding form of uncertainty. What is normally done is to produce a stochastic version of the standard dynamic economic model of the fishery. We take an example from Clark (2010).

The Clark stochastic model is based upon a discrete time, as opposed to continuous time, model, which is distinct from the very general discrete time model discussed in chapter 3. Clark's model is a stock recruitment model, in which we focus on recruitment to the fishery, harvesting during the fishing season, and escapement of fish at the end of the season. We simplify by assuming a single lifetime reproductive episode.

In the model's deterministic formulation, we have the following. Recruitment to the fishery in period t is denoted as R_t, the harvest in period t is denoted as H_t, and escapement at the end of the period as S_t. We thus have:

$$S_t = R_t - H_t \tag{5.33}$$

and

$$R_{t+1} = F(S_t) \tag{5.34}$$

It is assumed that net natural growth during the fishing season is zero, so that any net natural growth, which might occur, does so off season.

Clark then shows that, with density-dependent harvesting costs, and with our usual assumptions of perfectly elastic demand for harvested fish and perfectly elastic supply of fishing effort, the optimal escapement, S^*, is given by:[24]

$$F'(S^*)\cdot\frac{p - c(F(S^*))}{p - c(S^*)} = \frac{1}{\alpha} \tag{5.35}$$

where α is the per season discount factor, $\alpha = 1/1+i$, where i is the per period social discount rate (Clark, 2010, chapter 2).

We next turn to Clark's stochastic model, which is based in turn on the model of Reed (1979). Environmental uncertainty is added to the foregoing deterministic discrete time model, by re-expressing equation (5.34) as follows:

$$R_{K+1} = Z_K F(R_K - H_K) \tag{5.36}$$

where K is a specific period, and where: $0 \leq H_K \leq R_K$, and where the Z_K are i.i.d. random variables with a common density $\phi(z)$ and mean $\bar{z} = 1$.

The objective is that of maximizing the *expected* present value:

$$EPV = E\left\{\sum_{K=0}^{T} \alpha^K \pi(R_K, S_K)\right\} \tag{5.37}$$

where T is the terminal period, where α is the discount factor, and where:

$$\pi(R_K, S_K) = \int_{S_K}^{R_K} [p - c(x)]dx \tag{5.38}$$

It is assumed that there is perfect foresight with regards to p and fishing effort costs (Clark, 2010, chapter 7).

Through the application of stochastic dynamic programming, an equation for the optimal S, S^*, emerges, which can be expressed as follows:[25]

$$F'(S^*)\cdot\frac{p - E_z\{Zc(ZF(S^*))\}}{p - c(S^*)} = \frac{1}{\alpha} \quad (5.39)$$

(Clark, 2010, p. 277). The L.H.S. of equation (5.39) is to be seen as the *expected* own rate of interest of the resource. Thus, the equation can be interpreted as stating that one should invest/disinvest in the resource up to the point that the *expected* yield, or rate of return, on the marginal resource investment, is equal to the appropriate rate of interest.

Note the following: if we had $Z \equiv 1$, i.e. complete certainty, then equation (5.39) would reduce to equation (5.35). Hence, equation (5.39) can be seen as a generalization of the deterministic case (Clark, 2010, p. 277).

The question now arises as to whether the optimal escapement level in the stochastic model is greater than, less than, or equal to the optimal escapement level in a deterministic world. There is no definite answer. Reed (1979), and also Lewis (1981), imply that, under reasonable cost and revenue assumptions, the stochastic S^* will be larger than the deterministic S^*.

The argument is that under uncertainty there are asymmetric upside and downside risks. The upside risk is that the actual escapement may prove to be larger than anticipated; the downside risk that actual escapement may prove to be smaller than anticipated. Dealing with the downside risk consequences can be expected to be much more difficult and painful than dealing with the consequences of the upside risk. Even if the resource manager is risk neutral, prudence requires that the escapement target should be somewhat larger than it would be under deterministic conditions (Reed, 1979).

That said, this really does not get us very far. As Clark points out, the difference between the stochastic and the deterministic S^* may be minor (Clark, 2010, p. 278).

What is needed is protection against true irreducible uncertainty (Walters and Hilborn category c), which could imperil the resource. This really should be discussed under the heading of policy. The question is: which policies will provide the fishery with resilience, the ability to withstand and react quickly to shocks arising from, say the environment, or from management errors arising, in turn, from the irreducible uncertainty? What we know, from experience in other fields, is that one common way of mitigating risk is through portfolio diversification. This will lead us into a discussion of the use of marine protected areas (MPAs) and fishing opportunity portfolio diversification at the fisherman level, for example by ensuring that fishermen can fish for many different species.

We can, at this point, make one comment arising from our discussion of non-malleable capital. It should be obvious that the greater is the degree of non-malleability of produced and human capital in the fishery, the less flexible, the less resilient will the fishery be. Any policy that enhances the malleability of produced/human capital in the fishery will strengthen the fishery in the face of irreducible uncertainty.

BIBLIOGRAPHY

Along with the many articles and books cited in this chapter, the interested reader is encouraged to turn to Andersen and Sutinen (1984), Charles (1983, 1992, 2007), Charles and Munro (1985), Sumaila (1995), Lane and Stephenson (1998), Doyle et al. (2007), and Flaaten (2010). The articles by Charles and Doyle et al. are of particular interest in that they combine both investment in non-malleable fleet capital and uncertainty. Nøstbakken (2006) develops a bioeconomic model of a fishery that encompasses both stock uncertainty and price uncertainty. Nøstbakken and Conrad (2007) review the body of literature dealing with uncertainty in bioeconomic modelling of fisheries.

NOTES

1. Recall our references in chapter 1 to the World Bank/FAO report, *The Sunken Billions* (World Bank, 2009). The report argues that, if world capture fishery resources are to yield their maximum contribution to the world economy, they will need to be doubled in size—implying a massive resource investment programme.
2. The motivation for the Clark and Munro (1980) article came from research on the fishing industry of Newfoundland and Labrador that Munro had undertaken for the then Economic Council of Canada. Newfoundland fisheries at the time were dominated by groundfish. It became obvious to Munro that to ignore the processing sector in the economic analysis of the fisheries was nonsensical (Munro, 1980).
3. See Munro (1980).
4. There is no reason, as we have just argued, why p_m should not be a function of x as well.
5. Suppose that $c_1'(x_s) = 0$. Even in the absence of an existence value, the Marginal Stock Effect could be positive by virtue of the fact that $p_2'(x_s) > 0$.
6. If we had $C'(E) = 0$, then marginal and average harvesting costs would, of course, be equal.
7. See Clark and Munro (1980, part 4) for a detailed discussion of this point.
8. Suppose that the monposonist were to integrate backwards. This would be socially desirable, if $\delta_2 = \delta_s$, but would be undesirable, if $\delta_2 > \delta_s$ (Clark and Munro, 1980).
9. This would essentially be in the nature of a principal-agent problem.
10. The ecosystem approach to resource management does carry with it the flavour of the economist's concept of general equilibrium—everything depends upon everything else. We shall have much more to say about ecosystem approaches to fisheries management in chapter 6.
11. Since, in contrast to the discussion on the processing sector, we do not have to make a distinction between a social and private resource manager, we can return to denoting the social rate of discount simply as δ.
12. Importantly, the studies ignore existence values, which, if included, might bring forth more 'palatable' results.

13. This raises one question. Suppose that harvesting is zero. x_1 would be stabilized at $x_1 = G$. What about x_2? Look at equation (5.17), x_2 would be stabilized at: $x_2 = rG/m$.
14. Note that $F'(x_1) = r$ (see equation (5.16)).
15. Relative fishing effort costs are obviously of key importance, but that is not all. If exploitation is done at the source, then all of rG is taken, period after period. If harvesting is done at the sink, then in the Clark model, some part of rG is lost through natural mortality (see equation (5.16)). Moreover, since harvesting costs in the sink fishery are density dependent, by assumption, this will limit the extent to which it is optimal to harvest $[rG - mx_2]$ in the sink (Clark, 2010, p. 307).
16. For a financial asset to be deemed to be highly liquid two conditions must be met. First, it must be possible to convert the asset cash quickly. Secondly, there must be no risk of a capital loss when converting the asset into cash. A United States federal government three-month treasury bill is highly liquid. Common shares of a private company, traded actively on a recognized stock exchange, meet the first condition, but not the second.
17. One can see the link to the concept of liquidity, in that the asset—fleet capital—can be disposed of quickly, and without risk of a capital loss.
18. The reader interested in the mathematical proof is directed to CCM (1979, part 7).
19. This is by assumption. As will be pointed out, the parameters could be such that the optimal investment in K at $t = 0$ will give rise to a fleet that is not sufficiently large over time to reduce x to $\tilde{x}$.
20. The economic intuition behind this optimal policy will be discussed at a later point.
21. The biological growth functions in our models have been characterized by pure compensation, which implies that there is not a positive minimum viable population level. If the biological growth function is characterized by critical depensation, then a positive minimum viable population level will exist (see Clark, 1990, chapter 1 for further details). McKelvey (1985) points out that, if a positive minimum viable population level exists, the existence, in turn, of highly non-malleable fleet capital in a Pure Open Access fishery could put the survival of the resource at serious risk.
22. The temporarily 'free' fleet capital can be seen as yielding a Marshallian quasi-rent (Marshall, 1927; Munro, 2010).
23. One should also refer back to the American 1996 Sustainable Fisheries Act and the fact that the act imposes a maximum stock rebuilding period of 10 years (unless biological considerations dictate otherwise). This has proven to be very controversial. It is argued that the 10-year maximum is unrealistic, and threatens excessive consequences for fishing communities (Khwaja and Cox, 2010, p. 62).
24. This is similar to the continuous time and discrete time versions of the Fundamental Equation discussed in chapter 3. Clark shows that equation (5.35) can be re-written as: $F'(S^*) - c(F(S^*)) - c(S^*)/p - c(S^*).F'(S^*) = 1/\alpha$. The link is now much more transparent (Clark, 2010).
25. Assuming that recruitment is always greater than escapement (Clark, 2010, p. 278).

6 The Economic Management of Capture Fisheries at the National/Regional Level

In the preceding chapters, we have discussed and analysed the economic consequences of marine capture fisheries being 'common pool' in nature. We are reminded of *The Sunken Billions* report's estimate that, due to such consequences, the world economy is losing in the order of US$50 billion per year (World Bank, 2009). We now have to examine what can be done to eliminate, or at the very least to mitigate, the 'common pool' nature of such fisheries through effective economic management. In so doing, we make a distinction between management at the national, or regional, level and management at the international level. The management at the national/regional level implies intra-EEZ management of the fisheries.[1]

While a distinction is made between national and international fisheries management, let it be immediately acknowledged that management at the national level and the management at the international level do not stand in isolation from one another. There is, of course, an interaction between the two, upon which we shall comment in a later chapter.

We shall argue in this chapter that the management of capture fisheries at the national level has gone through a process of evolution, which is continuing at the present time. The root cause of the management problem is the perverse (from society's point of view) incentives facing fishermen in 'common pool' fisheries. It will be seen that resource managers initially relied upon what the FAO calls incentive blocking measures, in which fishermen are hopefully blocked from responding to the perverse incentives. It will then be seen that disappointment with these measures has led to greater emphasis being placed on what the FAO terms incentive adjusting measures, in which the management measures are designed to adjust the incentives facing fishermen, so that these incentives are more compatible with the goals of society.

In this chapter, we shall cover the following:

(i) a discussion of the goals of fisheries management, ecosystem considerations, the evolution of fisheries management, and a Principal-Agent framework for analysing fisheries management (6.1–6.4);

(ii) incentive blocking approaches to fisheries management at the national level (6.5);
(iii) incentive adjusting approaches to fisheries management at the national level (6.6);
(iv) the role of marine protected areas (6.7);
(v) the impact of subsidies upon fisheries management (6.8); and
(vi) uncertainty considerations (6.9).

6.1 The goals of fishery resource management at the national (regional) level

With regards to the basic goal of fisheries management at the national (or regional) level, we return to chapter 1. In following the lead of the World Bank (2005), we argued that the capture fishery resources of a nation are to be seen as a subset of that nation's portfolio of real capital assets. The basic goal of management, then, is to manage this subset of real capital assets in such a manner that they make the maximum contribution to the nation's economy through *time.*[2] In the first instance, we measure this contribution in terms of resource rent, broadly defined. The definition explicitly includes non-market, as well as market, components.

Sustainability of the fishery resources is often put forward as a key goal of resource management. While accepting the importance of sustainability, we would argue that the goal of sustainability is subsumed under our aforementioned goal. The goal, let it be re-emphasized, is to maximize resource rent *through time.* The goal cannot be realized if the fishery resources are exploited on an unsustainable basis. No nation can hope to prosper through time if it allows its capital assets, of whatever form, to be steadily depleted over time.

While we adopt a broad definition of resource rent, it does seem faintly crass to focus so heavily upon resource rent. It is, therefore, worthwhile being reminded of the importance of resource rent to the world community in general and to developing nations in particular. Recall that the World Bank (2005) publication, *Where Is the Wealth of Nations?*, maintains that natural resources are of key importance to the real capital portfolios of developing states.

Thus, if resource rent is dissipated or squandered, on a worldwide basis, it is the developing states that will suffer the most.

What about equity? There is no question that, within a nation, equitable distribution is an important issue. Indeed, we shall argue that the viability of some of the fisheries management schemes to be described will be undermined if the distribution of the forthcoming resource rent is perceived to be grossly inequitable. Having said this, first things first. There is no sense in talking about equitable distribution of resource rent until we have in place a resource management regime that ensures the ongoing flow of positive resource rent. We must first be certain that we have something worthwhile to distribute.

Resource management regimes are not costless. Hence, we require some concept of 'efficient' vs. 'inefficient' resource management schemes. A simple working definition of an efficient fisheries management scheme is one in which the resource rent (gross) flowing from the fishery through time exceeds the cost to society of administering the scheme. A change in the management regime that leads to an increase in net resource rent (net of management costs) is to be seen as efficiency improving.

6.2 **Ecosystem considerations**

The FAO reports that, since the early 1990s, all international agreements pertaining to fisheries stress the need for adopting ecosystem approaches to fisheries management (FAO, 2011). Examples are given by the FAO Code of Conduct for Responsible Fisheries (FAO, 1995) and the 2001 Reykjavik Declaration on Responsible Fisheries in the Marine Ecosystem, signed by 57 states (FAO, 2001, appendix I). We commented briefly on ecosystem management in chapter 5. Recall the standard definition of an 'ecosystem': 'a biological community of interacting organisms and their physical environment' (Oxford Dictionaries Online, 2011). The definition is so broad as to be of very limited practical value for management purposes.[3]

We also mentioned in passing the fact that the question of an ecosystem based approach to fisheries management was addressed in detail in the late 1990s by the National Research Council of the United States, through its Committee on Ecosystem Management for Sustainable Fisheries (National Research Council, 1999). In its report, the Committee, at the outset, makes two fundamental points. First, given the state of scientific knowledge, managing a large marine ecosystem comprehensively and reliably is quite simply infeasible. Secondly, humans (e.g., fishermen) are not to be regarded as just users of the ecosystem. They are integral components *of* the ecosystem (National Research Council, 1999, p. 15).

To the Committee, it will be recalled, an ecosystem based approach to fisheries management means, at a minimum, abandoning strictly single species approach to fisheries management, and recognizing explicitly species interaction, and the interactions of the species with the shifting aquatic environment (National Research Council, 1999). As noted, this causes us no conceptual difficulties, and is not contrary to our stated goal of economic management of fisheries resources.

First, with regards to species interaction, we have already seen in the previous chapter that our dynamic economic model of the fishery can readily accommodate such interactions. As stated earlier, we then think, not in terms of managing a single 'natural' capital asset, but rather in terms of managing a portfolio of interacting 'natural' capital assets.

If species interactions and the interactions of these species with the surrounding aquatic environment are ignored, we shall not succeed in achieving the goal of maximizing resource rent (broadly defined) through time. Consider the following simple example. Suppose that a particular species, y, predates upon a species x. Suppose further that y is a high valued species, while species x has negligible value in the market place. One cannot sensibly engage in the economic management of species y, without reference to the seemingly valueless species x. While species x has no direct value in the market, it has indirect economic value by virtue of the fact that it supports y.[4]

The interactions of the fish species and the surrounding aquatic environment, and the severe difficulties in measuring these interactions, let alone predicting changes in these interactions, have profound economic implications. They relate directly to the irreducible uncertainty encompassing the management of the fishery resources that must be confronted in the economic management of fishery resources.

6.3 The evolution of capture fisheries management: an overview

Capture fisheries management did not suddenly emerge. It has evolved gradually and is continuing to evolve at this time. At the beginning of the evolution, the most efficient management of capture fisheries was seen to be no management whatsoever.

Return again to chapter 1. In the discussion of the evolution of the legal framework surrounding world capture fisheries, it was pointed out that well into the twentieth century there was a belief that the great ocean fishery resources were free from the threat of serious overexploitation. While there were a few attempts at significant resource conservation in the early twentieth century, such as the North Pacific fur seal fishery (1911) and the Pacific halibut fishery (1923), the management of ocean fisheries remained minimal until the close of the Second World War. This, we argued, could be ascribed to economics. Given the state of fisheries technology, it was in the past too costly to engage in massive exploitation of capture fishery resources in the high seas.

In chapter 1, we also made reference to a speech given by the famous nineteenth-century British biologist, Thomas Huxley, in 1883, at which time Huxley held the post of Inspector of Fisheries for the United Kingdom. Let us now quote more fully from Huxley's 1883 speech:

> 'The cod fishery, the herring fishery, the pilchard fishery, the mackerel fishery, and probably all of the great sea fisheries, are inexhaustible: that is to say that nothing we do seriously affects the number of fish. And any attempt to regulate these fisheries seems consequently... to be useless (cited in Gordon, 1954, p. 126).

The implications arising from Huxley's speech are clear enough. In economics parlance, the fisheries natural capital is to be seen as 'free' capital. Thus, the most 'efficient' form of fisheries management is no management at all.

As discussed in chapter 1, the economic protection given to ocean capture fishery resources was beginning to fray, even as Huxley spoke. By the end of the Second World War, it was clear to all but a few that much of this economic protection had been stripped away by ongoing advances in fisheries technology, which had led to steadily falling harvest costs. With this stripping away, the perverse fishermen incentives[5] brought forth by the 'common pool' nature of the fishery resources began to manifest themselves. We analysed in detail, in chapters 2 and 3, the consequences of these perverse incentives.[6]

What then is to be done? A conceivable option is to attempt to turn the clock back to a world in which it is too costly to exploit heavily the great ocean fishery resources. One gets a sense of this from some marine biologists who rail against 'industrialized' fisheries, and wish for a world of small vessels, with primitive equipment. As commonly observed, however, technological advances cannot be uninvented. Also, from an economic standpoint, there is something utterly perverse in turning to high cost methods of production. It is almost as if we were to argue that agriculture ought properly to revert to the era of oxen and wooden ploughs.

6.4 A Principal-Agent framework

If we are not to return to the past, where the 'common pool' nature of the resources did not seem to matter all that much, then the problems presented by the socially perverse fishermen incentives must be addressed. A useful framework, within which to consider this problem, is provided by Principal-Agent analysis. While the rigorous application of Principal-Agent analysis to fisheries economics is in its very early stages (Vestergaard, 2010), the analysis is nonetheless useful as a broad framework within which to organize our thinking about the problem at hand.

One can think of the coastal state, as represented by the government of the state, constituting the legal owner of the intra-EEZ fishery resources (McRae and Munro, 1989b), as the Principal. The fishermen exploiting the intra-EEZ fishery resources constitute the Agents.

The basic nature of Principal-Agent (PA) analysis can be thought of in the following manner. The Principal, be it a person, a firm, or a state, wishes to see undertaken certain tasks that it is unable or unwilling to do itself, and so acquires the services of one, or more, Agents to undertake these tasks. Classic examples are an owner of a firm hiring a manager, and a landlord leasing farm land to a tenant farmer. The PA analysis has application far beyond these simple examples, for example industry regulators and the firms being regulated (Sappington, 1991). More formally, the PA paradigm can be formally described as follows (see Clarke and Munro, 1987, pp. 83–6).

A strict hierarchical relationship exists in which the Principal chooses an incentive scheme (e.g., set of regulations) to be applied to the Agents. The Principal's incentive scheme, along with the actions taken by the Agents, determines both the returns to the Agents and to the Principal. As seen from the perspective of the Principal, a first-best situation exists when the Principal can, at minimal cost, contractually and enforceably, specify the actions of the Agents. Wishes, urges, and desires of the Agents, contrary to the best interests of the Principal, are entirely suppressed. The Agents are essentially robots.

In the normal second-best situation, the Principal lacks the power, or more to the point finds it too costly, to force a set of actions upon the Agents. The Agents thus have some freedom of choice. The Principal can hope to influence the Agents' choices only indirectly through the incentive scheme. This gives rise to the concept of an *agency cost* (Clarke and Darrough, 1983), which is the difference between the actual return to the Principal, and what it would receive under a first-best situation. It reflects the insufficiency of the Principal's incentive scheme in compensating for its inability to monitor perfectly the Agents' actions and enforce its will upon the Agents. At the heart of the PA problem is monitoring and enforcement imperfection (Clarke and Munro, 1987).

For our purposes we can think of the coastal state government—resource managers—as the Principal attempting to maximize the resource rent (broadly defined) from the fisheries through time. The Principal's incentive scheme has to be such as to counter effectively the perverse fishermen incentives.[7]

The FAO (1998), in its discussion of fisheries management, makes a distinction between:

I. Incentive blocking approaches.
II. Incentive adjusting approaches.

Incentive blocking approaches to fisheries management involve the coastal state, as Principal, doing the obvious in dealing with perverse fishermen (Agent) incentives. Regulations, and accompanying threats of punishment for regulation violations, are imposed with the hope of blocking fishermen from responding to the undesirable incentives.

It will be seen that these approaches have led in many instances to the emergence of massive *agency costs*. This experience has, in its turn, led to incentive adjusting approaches, which involve implementing management measures that will hopefully alter fishermen (Agent) incentives in such a manner that they are made more consistent with society's goals.

6.5 Incentive blocking approaches to fisheries management

The term incentive blocking approach to resource management implies, as we have indicated, that the resource manager deals with the perverse incentives of the fishermen by blocking the fishermen as they attempt to respond to these incentives. Thus, if fishermen have the incentive to overexploit fishery resources, block the fishermen from responding to these incentives by imposing strictly enforced season by season global harvest quotas, for example TACs. In a given year, once the global harvest quota has been taken, the fishing season is brought to an end, and the fishery is closed until the following year. Any fisherman found to be fishing after the season closure is to be punished.

Fishermen incentive blocking approaches were adopted in the early days of extensive fisheries management, because, as we have said, they constituted the obvious response. Libecap (2007) reports that incentive blocking approaches were also the first to be used in other 'common pool' industries. They are, as Libecap points out, less complex than the incentive adjusting approaches to be discussed. Furthermore, the implementation of such approaches makes it apparent that public resources (such as fishery resources) remain firmly with the public domain (Libecap, 2007).

Historically, the first major thrusts towards ocean capture fisheries management after the Second World War took place through international organizations, because the coastal state jurisdiction over capture fishery resources was so limited. The International Commission on Northwest Atlantic Fisheries (ICNAF), 1949–77, and the North East Atlantic Fisheries Commission (NEAFC) provide examples. With the coming of the EEZ regime much of the resource management passed to individual coastal states.

The development of incentive blocking approaches can be seen as moving in two stages:

1. Global harvest restrictions on the fishery, possibly accompanied by gear restrictions, for example mesh size. No restrictions are placed on fleet size. We will refer to this as the TAC (or equivalent) alone approach.
2. Limited entry, plus Olympics style TACs.

In stage 2, the TACs (or equivalent) are supplemented with a programme of restricting the number of fishing vessels allowed into the fishery. The vessels granted access to the fishery are then to compete for shares of the TAC, hence the term Olympics style TAC.[8]

There also exists what we might term a third stage involving marine protected areas (MPAs), areas in which harvesting is forbidden outright, or is severely restricted. We will put off discussing MPAs until after the discussion of incentive adjusting approaches to fisheries management. No one seriously advocates relying upon MPAs alone. It will be argued, however, that MPAs can serve as a useful ancillary management instrument.

Stage 1—TACs alone—we now know from chapter 2 as class II open access or regulated open access. The focus is on the resource alone, with the TAC (or equivalent) being a measure to block the fishermen from responding to the incentive to overexploit the resource. The drawback to this resource management approach, learned painfully by a generation of resource managers, lies first in the underlying economics of the situation. The restricted season by season harvest becomes the 'common pool'. Competition for shares of the valuable restricted harvest leads almost invariably to excess fishing capacity, which leads, in turn, to the dissipation of resource rent.

We explored the consequences of regulated open access in chapter 2, with the aid of a simple static model. Now that we have become familiar with dynamic economic models of the fishery and have a greater understanding of the concept of non-malleable fleet capital, we can re-visit the topic and hopefully provide a more detailed and sharper explanation. The static model in chapter 2 hinted at non-malleable vessel capital, but did not discuss the concept in any detail.

In the following model, based upon Clark (2010) and Munro (2010b), it is assumed, as we did in chapter 2, that the resource manager exercises full control over the season by season TAC. It is assumed further that the vessels and crews are identical, that technology is frozen, that there is no 'crowding', and that all harvested fish is sold into the fresh market, with the industry facing a perfectly elastic demand for that fish. It was seen in chapter 2 that, given these assumptions, Regulated Open Access presents us with no economic problem, if the vessel capital is perfectly malleable, i.e., the unit re-sale value of vessels is equal to their purchase price($\lambda_s = \lambda$).[9] If the vessel capital is not perfectly malleable, then economic waste can arise through the emergence of unequivocal vessel capital redundancy.

We now attempt to drive these points home by turning to two cases with admittedly extreme assumptions. In the first case, we assume that the rate of depreciation of the vessel capital, γ, is equal to zero, but that the vessel capital is perfectly malleable, with respect to the fishery under consideration, by virtue of the fact that $\lambda_s = \lambda$. In the second case, we assume that the vessel capital is perfectly non-malleable, which, as we have seen in an earlier chapter implies that: $\gamma = \lambda_s = 0$.

In the model to follow, we suppose that we have:

K	–	the number of 'standardized' vessels in the fishery
q	–	catchability coefficient expressed as tonnes of harvested fish/day/vessel
Q	–	the fixed Total Allowable Catch (TAC) per season, expressed in tonnes
D	–	the length of season in days, with D_{max} denoting maximum season length
b_{oper}	–	daily operating costs per vessel, a constant
p	–	price of harvested fish, a constant
λ	–	purchase price of a vessel, a constant
λ_s	–	per unit resale value of a vessel, a constant

During the fishing season the existing fleet is fully utilized, so that we always have $E = K$. It is assumed that the constraint in terms of the TAC is binding. The seasonal harvest is thus: $qKD = Q$. Fleet operating profits over the season D are given by: $(p - b_{oper}/q)\cdot Q$.[10]

In our first case, let us recall, we assume that vessel capital is perfectly malleable: $\lambda_s = \lambda$. Now let ρ denote the annual rate of interest (Clark, 2010), and assume no compounding of interest within the year. Since we assume that the rate of depreciation is zero, the rental cost of the fleet capital on a per annum basis can be expressed as $K\lambda\rho$, and the rental cost on daily basis by $(K\lambda\rho)/365$. The fleet rental cost on a seasonal basis can be expressed as$((K\lambda\rho)/365)$, or $K\phi D$, where $\phi = (\lambda\rho)/365$. Denote seasonal fleet profits as ω. We thus have:

$$\omega = (p - b_{oper}/q)\cdot Q - K\phi D \tag{6.1}$$

Given our assumptions, fleet annual profits are independent of the fleet size, essentially because the fleet rental cost is really an additional form of operating costs. To see why, let us first recall that we have $qKD = Q$, where q and Q are, by assumption constants. Suppose, for the sake of argument, that initially $K = 100$ and $D = 200$. Then suppose that the fleet is doubled, with $K = 200$. It follows that the season length must be reduced by one half to $D = 100$. Return to equation (6.1). The first term on the R.H.S. is obviously independent of K. With respect to the second term any increase (decrease) in K will be fully offset by a reduction (increase) in D.

Next turn to our second case, in which we assume that vessel capital is perfectly non-malleable: $\gamma = \lambda_s = 0$. The situation is now changed dramatically. The cost of vessel capital is a once and for all cost, not an operating cost. As a consequence, we must now look at the cost of the fleet capital in relation to the stream of operating profits from the fishery over time and think in terms of the Net Present Value of the fishery (NPV).

We can express the PV of operating profits from the fishery, from season one to infinity, as:

$$PV_{oper} = \frac{[p - b_{oper}/q] \cdot Q}{\rho} \tag{6.2}$$

(see Clark, 2010). The cost of the fleet capital, given our assumptions, is simply λK. Thus the aforementioned NPV is

$$NPV = PV_{oper} - \lambda K \tag{6.3}$$

Obviously, if we are to maximize NPV, we must ensure that K is the minimum size. Recall that we have: $qKD = Q$, and thus $K = Q/qD$, which makes it obvious, in turn, that minimizing K implies maximizing D, $D = D_{max}$.[11] Denote the minimum K as K_{opt}.

Now let us suppose that we commence with $K = K_{opt}$, and that $\lambda K_{opt} < PV_{oper}$. It is assumed that the TAC is fully enforced, but it is also assumed that there is no control over fleet size. We are now confronted with a disequilibrium situation, even though any additional vessel entering the fishery will be strictly redundant. The reason is straightforward, and is as follows.

Since, by assumption, the vessels are identical, an owner of a single vessel can be expected to enjoy an average share of PV_{oper}. Thus, the owner of a prospective new addition to the fishing fleet will have an incentive to join the fleet, if:

$$\lambda < PV_{oper}/K \tag{6.4}$$

It can thus be argued that investment in fleet capacity will continue up to the point that:

$$\lambda K_{ROA} = PV_{oper} \tag{6.5}$$

where K_{ROA} denotes regulated open access equilibrium fleet size. It can be said further that $K_{ROA} - K_{opt}$ represent unequivocal fleet overcapacity and that the economic loss—a redundancy deadweight loss—is equal to: $\lambda(K_{ROA} - K_{opt})$. Let it also be noted that, given the assumption of perfect non-malleability of fleet capital, the deadweight loss, once incurred, can never be recovered (Munro and Clark, 2003).[12]

When the fishery is in regulated open access equilibrium, the season length will obviously be below the maximum. Indeed, it is quite possible that the actual season length, D, will be: $D \ll D_{max}$. Does this matter? In the context of the model just presented, the answer is no, given the assumptions of the absence of 'crowding' externalities, the sale of all harvested fish into the fresh market, the perfect elasticity of the demand for harvested fish and iron maintenance of the TAC by the resource manager, and assuming that the maintenance of the TAC is at little or no cost. The entire economic burden of overcapacity is captured in: $\lambda(K_{ROA} - K_{opt})$ (Clark, 2010; Munro, 2010b).

Once the restrictive assumptions of the model are relaxed, the shortening of the season does matter very much in economic terms. If 'crowding', congestion, does exist—vessels impeding one another, inadvertently destroying one another's gear—then it takes little imagination to see that the problem will intensify the shorter the season becomes (see Brown, 1974). If the fish is not sold in the fresh market, but is rather sold to processors with non-malleable processing capital, then economic inefficiencies can arise from the glut/famine cycles being faced by the processors. Next, if the demand facing the industry has finite price elasticity, then the consequences of the short season are obvious.[13]

Finally, the control that the resource managers have over the global season by season harvest is often less than iron like. With less than iron control, the larger the fishing capacity, the greater is the resource manager's surveillance and enforcement problem. Thus, the build-up of capacity can become a threat to the stability of the resource.[14]

The threat can arise even if the vessel capital is perfectly malleable. An example is provided by the herring fishery of British Columbia. The fleet capital that was employed in the fishery was malleable with respect to the fishery. Nonetheless, the intensity of the fishing was so great in the short openings allowed by the resource managers, that the resource managers lost control of the fishery and seriously thought of closing the fishery. Ultimately a solution was found by turning to an incentive adjusting approach (Munro and Clark, 2003).

Let us now review the situation. What has been seen up to this point is that implementing harvest controls may be all that is required. The resource rent may be atomized—spread over a large number of vessels—but it is not lost, as is implied by the static Gordon model. In terms of our Principal-Agent analysis, we could say that the Principal (the coastal state) has achieved a first-best outcome—the *agency cost* is reduced to zero.

See, however, what conditions must be satisfied for the first-best outcome to be achieved. The demand for harvested fish products must be perfectly elastic. There must be no 'crowding'. There must be perfect malleability of produced and human capital involved in both fishing and processing activities. The resource manager must, at little or no cost, be able to enforce with rigour the controls over the season by season harvests, regardless of how short the season, and regardless of how large the fleet may be. So demanding is the sum total of the conditions to be met that we can regard the case in which harvest controls *alone* will lead to a first-best outcome as being of theoretical interest only.

In the real world, second-best outcomes, with large *agency costs*, were and are the norm for coastal states, adopting the stage 1 incentive blocking approach to fisheries management.

In the evolution of capture fisheries management, the unsatisfactory outcomes of the stage 1 approach led to stage 2—limited entry with Olympics style TAC. If under the stage 1 approach, there is a build-up of excess fishing capacity, then the obvious response of the resource manager is to block the fishermen's incentives leading to the emergence of excess capacity by restricting entry into the fishery. The objective would be to ensure that the restrictions are such that $K = K_{opt}$.

This obvious response gave rise to what are commonly referred to as limited entry, or licence limitation, schemes. The latter term arises from the fact that limited entry is achieved first by the resource manager decreeing that only vessels carrying a specific licence will be granted entry to the fishery. The resource manager then proceeds to place a strict limit on the number of the specific licences that are issued.

The root of the regulated open access problem, as we have seen, lies in the fact that vessel capital is other than perfectly malleable. The one immediate problem with limited entry schemes is that they are normally introduced after there has already been a build-up of excess fleet capacity. Thus, at the commencement of the scheme at $t = 0$, we have $K(0) >> K_{opt}$. Normally, it is politically impossible simply to order vessels to leave the fishery.[15]

The problem of initial excess capacity is usually addressed through government financed decommissioning, 'buyback', schemes. Milazzo (1998), in a detailed World Bank study on subsidies in world capture fisheries, succinctly states the case for government subsidized decommissioning schemes as follows: '... many commentators have noted how difficult it is to induce the exit of capital from fishing because these assets ... have little other uses ... disinvestment in fisheries has to be promoted with ... subsidies' (Milazzo, 1998, p. 65). While Milazzo does not use the term non-malleable vessel capital, it is obvious that the vessel capital to which he is referring is decidedly non-malleable.

In any event, the rationale for the combined limited entry plus decommissioning/buyback schemes is straightforward. Remove excess fleet capacity by bribing vessel owners to leave the fishery, and then prevent the re-emergence of excess fleet capacity by limiting entry to the fleet.

The pioneering set of limited entry/buyback schemes occurred in the British Columbia Pacific salmon fishery in the late 1960s. At the time, Pacific salmon was the most important fishery resource for the fishing industries of British Columbia, the American Pacific Northwest, and Alaska.

We noted earlier in the chapter that serious interest in managing the world's ocean fishery resources did not emerge until the end of the Second

World War. Pacific salmon was one of the exceptions. Pacific salmon is an anadromous resource that is produced in fresh water, moves out into the ocean, and then after a cycle of two to four years, returns to its origin of birth to spawn and die. The fish are normally caught as they approach rivers on their return journey. Examples of Pacific salmon rivers in North America are the Columbia, marking the boundary between the American states of Oregon and Washington, and the Fraser River in southern British Columbia.

Since the fish are normally caught as they approach their rivers of origin, they are very vulnerable to overexploitation. In British Columbia, regulation of the salmon fisheries can be traced back to the 1880s, when the population of the province was very modest indeed (Fraser, 1979). The initial regulations took the stage 1 form just described.

Due to the nature of the resource, TACs are not employed. Rather the resource managers implement an escapement policy—the number of salmon to escape harvesting to go up river to spawn. Fish in excess of the escapement targets are open to exploitation, administered through declared 'openings'.

The resource managers relied upon the escapement policy combined with weak controls over capacity through gear restrictions. The regulated open access syndrome played itself out. While the conservation of the Pacific salmon stocks was reasonably successful, the growth of fleet capacity in British Columbia seemed unstoppable and reached crisis proportions by the late 1950s. The resource managers became worried about the stability of the resource. Given the magnitude of the fleet capacity, even minor errors in management could have had disastrous consequences (Fraser, 1979).

In response to the crisis, the Canadian federal government commissioned an economist (Sol Sinclair) to study the problem and then to come up with recommendations for solving it. Drawing upon the then recent economic theory of H. Scott Gordon (1954), Scott (1955), and others, Sinclair (1960) identified the problem as one arising from perverse economic incentives confronting fishermen. In his report, Sinclair recommended a limited entry scheme, a recommendation that was endorsed by the industry and the major fishermen's union (Fraser, 1979). For political reasons, the Sinclair recommendations were not implemented until the late 1960s (Fraser, 1979).

The limited entry scheme, pioneered in the British Columbia salmon fishery, was to be copied many, many times. Moreover, at the outset, the scheme was greeted with enthusiasm by a considerable number of economists (see, for example, Crutchfield, 1979). It was only over time that disillusionment began to set in.

One can now say, after the fact, that, if the vessels had been uniform, if there had been fixed factors of production (inputs) proportions in the vessels (i.e., non-substitutability of inputs), if technology had been frozen, and if the accompanying buyback scheme had taken the industry by surprise, the combined schemes might well have been successful. This can be seen by returning to our model, in which it is assumed that vessels and crews are uniform and that technology is frozen.

Suppose for the moment that the buyback scheme is unanticipated, is successful and that K is reduced toK_{opt}. This is the bare minimum size fleet required to take the TAC. With the vessels uniform, and technology frozen, vessels cannot effectively out-compete one another. The outcome should, at the very least, be satisfactory.

The problem in retrospect is that, in many fisheries with stage 2—limited entry schemes—none of these conditions were, or are, met. In many, input substitution is feasible, technology is anything but frozen over time, and decommissioning/buyback schemes are not unanticipated by the industry.

We begin by first turning to the issue of decommissioning/buyback schemes. These have proven to be highly controversial (see, for example, Curtis and Squires, 2007). The criticisms of such schemes fall into two parts. The first criticism is that they have proven to be ineffective. Once fleet capacity has been reduced in a fishery, the capacity has a tendency to seep back into the fishery over time. We shall refer to this as the 'seepage' problem. It will be seen that the 'seepage' problem is a part of the general problem of controlling capacity in a fishery with limited entry with Olympics style TAC. The second criticism is that, if the 'buyback' scheme is anticipated by the industry, the scheme will end up by making a bad situation worse. We shall refer to this as the time consistency problem.

We put the 'seepage' problem to one side for a moment and turn to the time consistency problem. The term 'time consistency', now widely used (particularly by macro-economists), can be traced back to a famous article by Kydland and Prescott (1977). The term has caused some confusion, because in the way first employed by the two authors, time consistency is not a 'good thing'.

In any event, the concept is easily explained with reference to the authors' introductory two period model, a model that will prove to be very useful to our case. A policy, in each time period, is deemed by the authors to be consistent, in an optimal control context, if it maximizes the social objective function, taking as given the decisions made by the economic agents in previous time periods, and given that future policy decisions are arrived at in the same manner (Kydland and Prescott, 1977, p. 475).

Thus, in their two period model (t = 1; 2), the consistent policy in $t = 2$ ignores the impact of that policy upon economic decisions made by economic agents in $t = 1$. If economic agents in 1 are rational and anticipate the policies introduced in $t = 2$, the chances are excellent that the time consistent policy rule will be sub-optimal. Writing years later, Prescott (2004) stated that 'time consistent policy is not [the] best'. The essence of the time consistency problem is caught in the following quote: 'there is no mechanism to induce future policy makers to take into consideration the effect of their policy, via the expectations mechanism, upon current decision of [economic] agents' (Kydland and Prescott, 1977, p. 481).

Return to our model and continue to assume that the vessel capital is perfectly non-malleable, that the vessels are identical, that there are no 'crowding' externalities, that all fish is sold into the fresh market, and that the demand for harvested fish is perfectly elastic. We assume that we commence at $t = 0$, with $x(0) = x_{opt}$ and $K(0) = K_{opt}$. We assume further that $\lambda < PV_{oper}/K_{opt}$. The resource manager exercises complete control over x, but at $t = 0$ exercise no effective control over K.

With no initial restrictions on fleet size, there is, for reasons that we have discussed, a build-up of K. We end up with $K_{ROA} > K_{opt}$. Now let it be supposed that the *ROA* equilibrium is achieved at $t = 10$. We further suppose that, with the *ROA* equilibrium being achieved at $t = 10$, the resource managers see the apparent error of their ways, and introduce a 'sudden death' decommissioning/buyback scheme to reduce K from K_{ROA} to K_{opt}. We assume that they are fully effective in blocking seepage of capacity back into the fishery from thereon in. If the vessel owners' expectations at $t = 0$ are myopic, with respect to the resource manager's future policies, so that they do not anticipate the 'sudden death' buyback at $t = 10$, it can be shown that the decommissioning/buyback scheme will do no harm, although given our assumptions, the scheme will really do no good either[16] (Munro and Clark, 2003; Munro, 2010b).

We now change one of our key assumptions and suppose that at $t = 0$, the vessel owners' expectations with respect to the resource managers' future policies are rational, rather than myopic. We shall take rational expectations to mean perfect foresight, an admittedly extreme assumption upon which we shall comment later.[17]

Under a normal decommissioning/buyback scheme, the vessel owners cannot be forced out of the fishery. At $t = 10$, when the scheme is implemented, all vessels remaining in the fishery will expect to enjoy an equal share of the PV_{oper} of the fishery discounted back to $t = 10$. No vessel owner will allow him/herself to be bought out at a price less than the equal share of PV_{oper} to be enjoyed by each remaining vessel. We shall suppose (unrealistically perhaps) that the decommissioning/buyback price is equal to the aforementioned equal share of the PV_{oper}. Denote the buyback price as λ', i.e., $\lambda' = PV_{oper}/K_{opt}$.

Thus, in effect, every vessel owner at $t = 10$ will receive λ', either as a cash payment or as a share of PV_{oper}. Let us now denote the regulated open access fleet at $t = 0$, assuming that the vessel owners' expectations are rational, as K'_{ROA}. It can be shown that K'_{ROA} will be given by the following equation:

$$\lambda K'_{ROA} = \sum_{i=0}^{10} \frac{(p - b_{oper}/q)Q}{(1+\rho)^i} + \frac{\lambda' K'_{ROA}}{(1+\rho)^{10}} \tag{6.6}$$

or

$$\left(\lambda - \frac{\lambda'}{(1+\rho)^{10}}\right) K'_{ROA} = \sum_{i=o}^{10} \frac{(p - b_{oper}/q)Q}{(1+\rho)^i} \tag{6.6a}$$

(Munro and Clark, 2003; Munro, 2010b).

The R.H.S. of equation (6.6) consists of the present value of fleet operating profits from $t = 0$ to $t = 10$, plus λ' times the fleet size discounted back to $t = 0$. Recall that at $t = 10$, each vessel owner will receive λ', either as a cash payment or as a share of PV_{oper} discounted back to $t = 10$. Equation (6.6) states that investment in fleet capacity will expand up to the point that the cost of investment in fleet capacity at $t = 0$ is equal to the PV of both anticipated future operating profits from the fishery and 'buyback' subsides.

When we examine equation (6.6a), the problem becomes immediately apparent. Few dispute the fact that subsidies for the acquisition of new vessels can undermine effective fisheries management. Decommissioning/buyback subsidies that are anticipated by the industry act like vessel acquisition subsidies granted today.

This is not idle academic speculation. An empirical study on EU fisheries and fisheries policy by Jörgensen and Jensen (1999) concludes that decommissioning/buyback subsidies in such fisheries do in fact stimulate investment in new fishing vessels. Furthermore, the authors add that would-be investors in new vessels find their bankers to be much more accommodating than they would have been had such subsidies not existed.

Of course the assumption of perfect foresight on the part of the vessel owners is an extreme one. The point remains, nonetheless. If vessel owners anticipate the possibility of future decommissioning/buyback schemes, this will influence their vessel investment decisions, as Jörgensen and Jensen (1999) clearly indicate. Furthermore, if the perfect foresight assumption is extreme, an even more extreme assumption is that the vessel owners' expectations pertaining to resource managers' policies remain persistently myopic (Clark, Munro, and Sumaila, 2005).

We turn now to the issue of 'seepage'. This is really about the problem of dealing with the fishermen's incentives to expand fishing capacity under limited entry with Olympics style TAC programmes. In other words, if we commence with a limited entry fishery in which the capacity happens to be appropriate at the beginning of the programme, we will have exactly the same problem that will be confronted, if capacity is reduced through a 'buyback' scheme.

The basic problem is that the fishermen's incentives to compete for shares of the limited harvest remain. As we noted at the earlier point, this might not matter a great deal, if there were fixed factor proportions and if technology was frozen. In many (if not most) fisheries, neither of these conditions are met. Fishing capacity is best seen as a bundle of inputs. Resource managers find it difficult, if not impossible, to regulate all inputs. Rational fishermen will, in the first instance, attempt to substitute uncontrolled for controlled inputs.

In the early 1980s, Peter Pearse was appointed as a one-person Royal Commission to examine Canada's fisheries policy off the country's Pacific coast. Pearse stated the nature of the problem very clearly in his final report:

> This approach [limited entry and Olympics style TAC] has a fundamental weakness: when one or more inputs in the fishing process are restricted, the capacity of the fleet can continue to expand by adding other, unrestricted inputs. As a result, this technique has consistently failed to achieve the desired results.... The basic problem is that the design and structure of a fishing unit is flexible, and restrictions on one or two dimensions cannot, in the long run, prevent increased investment in other dimensions. While, theoretically, restrictions could be placed on all dimensions of fishing effort simultaneously, such restrictions would have to be so numerous and diverse (covering vessel size, power, crew, time spent fishing, gear for finding, catching and holding fish, and so on) that they would be virtually impossible to administer and enforce. (Commission on Pacific Fisheries Policy, 1982, p. 83)

There are two aspects of the problem that Pearse does not stress. First, it is not just a matter of substituting existing uncontrolled for controlled inputs. The competition among the fishermen will act as a spur to technological developments, with the common result that a shrinking number of vessels in the fleet will continue to retain the same level of fishing capacity.

Secondly, it has come to be recognized that the competition among fishermen is more intense than heretofore realized, because of strategic interaction among the fishermen. Up to this point, we have assumed, by and large, that the fishing industry is perfectly competitive. Strategic interaction among firms in perfect competition is not an issue.

By definition, perfect competition does not exist in limited entry fisheries. Strategic interaction among the limited number of fishermen can be expected to emerge. We are thus presented with a competitive fishermen game. Even if each and every fisherman recognizes that competition among the fishermen will dissipate economic returns from the fishery, each fisherman has no choice but to compete aggressively. Not to compete aggressively is to look forward to losing one's share of the TAC—or equivalent. What we are presented with is a classic Prisoner's Dilemma game.

In the late 1990s, a detailed study was undertaken on 'buyback' programmes in the United States, Canada, Australia, Japan, Taiwan, Norway, and the European Union (Holland, Gudmundsson, and Gates, 1999). The overall conclusion of the study is that decommissioning/buyback schemes have very limited effect. The authors of the study remark that, in order for such schemes to have a lasting effect, the schemes must be accompanied by the constraining of '...increases in effective effort due to various forms of input stuffing by existing fishermen...constraining the growth of effective effort requires very strict conditions that hardly are the case in any real world situation' (Holland et al., 1999, p. 68).

A striking example is provided by the pioneer in combined limited entry and decommissioning buyback schemes, the British Columbia Pacific salmon fishery. The first decommissioning/buyback scheme in the fishery was introduced in 1970. Over the ensuing decades, there have been four more buyback schemes in the fishery (Grafton and Nelson, 2007).

At first glance, the combined buyback schemes seemed to have had considerable success. By the first decade of the new millennium, the number of vessels in the British Columbia salmon fleet had been reduced to less than one third of the number of vessels at the time of the first buyback in 1970 (Fraser, 1979, p. 761; Grafton and Nelson, 2007, p. 197). Thus, there had been a dramatic decline in fleet capacity over the 30+ year period, or so it appeared.

Appearances, as we keep being reminded, can be deceptive. Fishing capacity, as we have argued, can be thought of as a bundle of inputs. Consequently, simply counting the number of vessels in the fleet does in fact give a wholly inadequate gauge of fleet capacity. A 'workable' definition of actual fleet capacity used by the FAO is as follows. Given a fishery resource of a certain size and age structure, the actual capacity of the fleet is measured by the total harvest that the fleet could take during a specified period of time, if fully utilized. Denote the actual fleet capacity by Y_c.

A measure of fleet overcapacity can be obtained by comparing Y_c with the 'target' fishing capacity, which we shall denote as Y_T. Y_T represents the harvest of a fully utilized fleet during a specified period of time, where the harvest is equal to that desired by the resource managers. From this, we can derive what can be termed the Excess Capacity Coefficient, θ, which is given simply by: $\theta = (Y_c - Y_T)/Y_T$ (FAO, 1998; Munro, 2010b).[18]

Grafton and Nelson (2007) in their study of the British Columbia Pacific salmon fishery maintain that the dramatic decline in fleet size between 1970 and the early 2000s was not accompanied by commensurate decline in fleet capacity. Indeed, they express serious doubt that the British Columbia Pacific salmon fishery excess capacity coefficient in the early 2000s differed significantly from the excess capacity coefficient for 1970[19] (Grafton and Nelson, 2007, p. 195).

The case of the British Columbia fishery points to a link between the 'seepage' and time consistency problems. At an earlier point, we stated that decommissioning/buyback scheme could be beneficial, or at least do no harm, if the industry is completely taken by surprise. We should now qualify the statement to say so long as the industry is taken by surprise and 'seepage' can be effectively suppressed. If 'seepage' is not suppressed, then a new 'buyback' programme will be required. The fishing industry is not going to go on being taken by surprise.

The British Columbia Pacific salmon fishing industry was probably taken by surprise in 1970, because the decommissioning/buyback scheme had never been tried before. It is difficult to believe that any rational British Columbia Pacific salmon fisherman would now ignore the possibility of future 'buyback' schemes.

Given 'seepage,' once the expectation of future buybacks becomes embedded in the industry, the 'seepage' will be intensified. The reason is straightforward. The perceived cost of input/capital 'stuffing' will be reduced by the anticipated 'buybacks'. The resource managers can then easily become enmeshed in a decommissioning/buyback treadmill, so evident in the case of the British Columbia Pacific salmon fishery.

In any event, there is nothing unique about the experience of the British Columbia Pacific salmon fishery. In Principal-Agent parlance, the stage 2 incentive blocking approach to fisheries management has led to large, and indeed in many cases unsustainable, *agency costs.* The disappointment with incentive blocking approaches to fisheries management in general has led to greater and greater attention being given to Incentive Adjusting approaches to fisheries management.

6.6 Incentive adjusting approaches to fisheries management—taxes

The basic idea behind the incentive adjusting approach to fisheries management is that, if it is not possible to prevent fishermen from responding to socially perverse incentives, then design management schemes that will adjust and modify fishermen's incentives appropriately, or so the argument goes. The economist's classic prescription for correcting perverse incentives of any form, dating back to the work of the British economist Pigou (1912), is to use taxes, positive or negative (i.e., subsidies). The call for carbon taxes to address the problem of global warming is one current example. It is to taxes, then, that we first turn.[20]

We shall first look at the use of taxes within the context of the static Gordon-Schaefer model. This will provide us with a broad, and uncomplicated, view of the impact of taxes. We shall then attempt to be somewhat more precise by examining the impact of taxes within the context of the dynamic economic model of the fishery.

Suppose now that all of the assumptions underlying the Gordon-Schaefer model, as analysed in chapter 2, hold. We suppose further that x_{MEY} (as defined by Gordon) is indeed the optimal biomass, i.e., we have: $x^* = x_{MEY}$.

Consider now Figure 6.1, which is based upon figure 2.4a. As explained in chapter 2, total sustainable revenue from the fishery (TR_{s1}) is simply: $pF(x)$, where p, a parameter, is the price of harvested fish. Now introduce a per unit harvest tax, τ. TR_{s2}, then, is the sustainable revenue from the fishery, as perceived by the fishing industry, i.e., $[(p-\tau)F(x)]$.

From chapter 2, we have: $TC_{min} = C(F(x), x)$. Now select a tax τ, such that the following hold true:

$$(p-\tau)F(x_{MEY}) - C(F(x_{MEY}), x_{MEY}) = 0 \tag{6.7}$$

Thus, with the fishing industry unconstrained, Bionomic Equilibrium will be achieved at the optimal biomass level: $x = x_{MEY}$. Bionomic Equilibrium is benign![21]

This happy outcome has occurred because the 'common pool' has been eliminated. The state, as owner of the resource, is exercising its full property rights by absorbing the entire resource rent. Re-write equation (6.7) as:

$$pF(x_{MEY}) - C(F(x_{MEY}), x_{MEY}) = \tau F(x_{MEY}) \tag{6.7a}$$

The L.H.S. of equation (6.7a) is the true sustainable resource rent as seen from society's point of view. The R.H.S. of the equation is the sustainable harvesting tax revenue, equal to the true sustainable resource rent.

There is no need to regulate the fishing industry. If x should be driven down below $x = x_{MEY}$, the fishing industry will experience negative economic profits and contract. The contraction will continue until x is restored to x_{MEY}. Gordon did, in fact, in an earlier article (Gordon, 1953) state that one possible

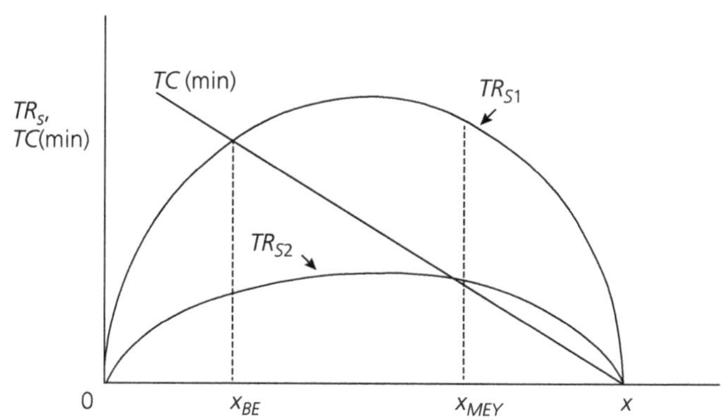

Figure 6.1. A tax on harvests
TR_{S1} = Total Sustainable Revenue without the Tax on Harvests
TR_{S2} = Total Sustainable Revenue Net of the Tax on Harvests

solution to the 'common pool' fishery problem was to have the fishery resource come under public ownership, and then have the public owner tax away the resource rent.

Determining the perfect τ would, of course, be very difficult. The resource manager would have to have perfect knowledge about the resource and about the fishing firms' cost functions. Furthermore, it would be necessary to allow for the fact that in the real world conditions are constantly changing, which would mean that the tax would have to be continually adjusted. This has led to a considerable debate in the literature, with some authors going so far as to argue that the information demands that would be imposed upon the resource manager are so immense as to make the task impossible (e.g., Arnason, 1990). On the other hand, Weitzman (2002) argues that under the inescapable environmental uncertainty facing resource managers, taxes will perform better than will harvesting rights programmes, to be discussed in the following section.

One can also see the possible political difficulties that could be encountered if the tax scheme is implemented after the fishery is in pure open access Bionomic Equilibrium. The imposition of the tax will push the industry from a position of zero true economic profits to one of negative economic profits.

In the Gordon-Schaefer model, this is not a serious problem because of the implicit assumption of perfect malleability of produced and human capital in the fishery. If the assumption does not hold, then the imposition of the tax would result in economic hardship for the industry, and would invite intense political opposition.

Now consider the use of taxes as a resource management instrument, within the framework of the dynamic economic model of the fishery. In so doing, the implementation of the tax scheme is viewed in strict Pigouvian terms—the internalizing of an externality.

As mentioned in an earlier chapter, we can, in the context of pure open access, think of a fishing firm that depletes the resource, by however small an amount, as imposing an externality upon all of the other fishing firms. The impact of the resource depletion upon the firm in question is so small that it is reasonable to suppose that the firm will ignore it.

We shall examine the argument in the context of the basic linear, autonomous dynamic model of the fishery. From chapter 3, we have the following equation for the Hamiltonian (equation (3.8)), which we re-state as:

$$H = e^{-\delta t}\left\{ [p - c(x)]h(t) + \psi(t)F(x) - h(t) \right\} \tag{6.8}$$

In chapter 3, it was stated that the current adjoint, or co-state, variable, $\psi(t)$, can be seen as the shadow price of the marginal addition to the stock of 'natural' capital in the form of the fishery resource.

It will be recalled from chapter 3 that the Fundamental Equation can be expressed as follows (see: equation (3.14)):

$$p - c(x^*) = \left[\frac{d}{dx^*}\left\{(p - c(x^*))F(x^*)\right\}\right] \cdot \frac{1}{\delta} \tag{6.9}$$

Clark and Munro (1982) demonstrate that, along the singular path, $\psi(t)$ is equal to the R.H.S. of equation (6.9).

The R.H.S. of equation (6.9), as well as being seen as the shadow price of the marginal addition to the resource stock, can also be seen as the 'marginal user cost' of the resource, a concept that owes its origins to John Maynard Keynes (Keynes, 1936, p. 135; Scott, 1955; Clark and Munro, 1982). As such, it represents the cost of an incremental depletion of the resource, in terms of the lost present value of future resource rents.[22]

Re-express equation (6.9) as:

$$p = c(x^*) + \left[\frac{d}{dx^*}\left\{(p - c(x^*))F(x^*)\right\}\right] \cdot \frac{1}{\delta} \tag{6.9a}$$

Seen in this fashion, optimality is realized by satisfying a type of marginal revenue-marginal cost rule, in the sense that we shall have achieved our resource goal, x^*, when the marginal revenue from harvesting is equal to the sum of the marginal harvesting cost and the marginal user cost of the resource.

We have seen, in the context of the linear, autonomous dynamic economic model of the fishery, that pure open access can be equated to setting $\delta = \infty$. If $\delta = \infty$, then the perceived marginal user cost goes to zero, with the implication that, under pure open access, the rational fisherman sets the marginal user cost equal to zero—hence the resource externality.

The optimal tax policy is to be seen as one that will internalize the externality (à la Pigou). Set a unit harvest tax equal to the marginal user cost evaluated at $x(t) = x^*$ (Brown, 1974; Clark and Munro, 1982, pp. 41–2; Arnason, 1990). If we denote the optimal unit harvest tax as τ^*, then, with pure open access, Bionomic Equilibrium will be achieved at: $x(x) = x^*$, i.e.,

$$(p - \tau^*) - c(x^*) = 0 \tag{6.10}$$

which we can re-express as:

$$\tau^* = p - c(x^*) \tag{6.10a}$$

The resource rent will not be dissipated at Bionomic Equilibrium. Rather it will be appropriated by the resource managers via the tax.

Let us observe that, in theory, exactly the same result could be achieved by imposing a tax on fishing effort. In so doing, we could be said to be ensuring that the costs faced by the fishermen reflect the true social costs of exploiting the resource. Let us continue to denote unit fishing effort cost as b, and suppose, as usual, that we have: $c(x^*) = b/qx^*$. Now let us observe that we can re-express equation (6.10) as:

$$p - [c(x^*) + \tau^*] = 0 \tag{6.11}$$

or as:

$$p - \left[\frac{b + \tau^* qx^*}{qx^*}\right] = 0 \tag{6.11a}$$

Hence, $\tau^* qx^*$ can be seen as the optimal per unit effort tax, which we can denote simply as ϕ^*. We thus end up with:

$$p - \left[\frac{b + \phi^*}{qx^*}\right] = 0 \tag{6.11b}$$

and

$$\phi^* = (pqx^* - b) \tag{6.11c}$$

There is a major problem with the harvest tax scheme, as we have presented it. As pointed out by Clark and Munro (1980), the tax scheme, as presented, depends critically upon harvesting costs being sensitive to the size of the biomass—i.e., the Marginal Stock Effect must be positive. This can be seen most clearly if we revert to the case of the harvest tax. Return to equation (6.10a).

Suppose that $x(t) > x^*$. We would then have $\tau^* < p - c(x)$. The fishing industry would enjoy true economic profits, the industry would expand, the resource would decline, and as the resource declined unit harvesting costs would gradually rise. Equilibrium would be achieved at: $x(t) = x^*$. If $x(t) < x^*$, the reverse would occur. The fishing industry would contract, the resource would increase, unit harvesting costs would gradually fall, and equilibrium would be achieved at $x(t) = x^*$.

Now suppose that unit costs are wholly insensitive to the stock size. The automatic corrective mechanism that we have described becomes non-existent. The tax scheme would essentially be infeasible.[23] Given the increasing evidence that many fisheries may exhibit weak Marginal Stock Effects (Clark et al., 2010a), the drawback to the tax scheme is indeed a major one.[24]

Implementing the perfect tax scheme may well be an impossible undertaking. While conceding this, we must not allow the best to become the enemy of the good. A tax scheme, imperfect though it may be, could well offer benefits greater than alternative fisheries management schemes.[25]

Examples in which taxes are used extensively in fisheries management are hard to come by. There is, however, one case that has been reported upon in detail. This is Mauritania in Northwest Africa. We made a passing reference to this case in the last chapter under the heading of processor monopsony, and pointed out that the case could also be viewed as one of resource management by taxation.

In any event, to repeat, the coastal state borders the Atlantic, and has rich fishery resources lying offshore. These resources are of prime economic importance to the country with much of the country inland consisting of desert. The major species of commercial importance consist of cephalopods (octopus, in particular), along with some demersals. These major species are harvested by national fleets, with some using chartered foreign vessels. The overwhelming bulk of the harvest is exported, largely to European and Japanese markets (Cunningham et al., 1995; Cunningham et al., 2005).

In 1984, the Mauritanian government established the Société Mauritanienne Commercialisation de Poissons (SMCP). All vessels harvesting cephalopods and demersals were required to sell their harvests to the SMCP. The SMCP was established ostensibly for the purpose of enhancing the country's bargaining power in the export markets. The SCMP was, in fact, much more than this. It was a government monopsony. According to Cunningham et al. (1995), the monopsony was feasible because the vessels had but one port at which to land their catches.

While the monitoring of harvests was undoubtedly less than perfect, the SMCP was able to use its monopsony position to extract a significant amount of resource rent through the use of a set of taxes, the most important being an export tax. In addition, since the SMPC was a monpsony, it was able to set the ex-vessel price for fish, which enabled it to enjoy a profit over and above the explicit taxes. The profits could be seen as a type of implicit tax.

The taxes enabled the government to extract a significant amount of resource rent from the fisheries. The taxes did more than that, however. They provided an instrument for fisheries management, by guiding fishing effort in the direction that it deemed to be appropriate (Cunningham et al., 1995).

Cunningham et al. conclude by stating that:

> in the fisheries economic literature it is often recognised that taxation provides an ideal method of correcting externalities, but one that is too difficult to implement in practice. This [study] has shown how a very successful [tax] system can be developed. (Cunningham et al., 1995, p. 10)[26]

6.7 Incentive adjusting approaches to fisheries management: harvesting rights programmes

If taxes, for reasons right or wrong, are not to be used as a means of adjusting fishermen incentives, then some other approach must be found. The most common 'other' approach consists of harvesting rights, popularly referred to as 'catch shares.' Under the heading of harvesting rights/'catch' shares, we include:

(i) individual harvest quotas (IQs)
(ii) community based fisheries management programmes, also referred to as territorial use rights fisheries—'TURFs'
(iii) fishermen cooperatives
(iv) sectoral TAC programmes.

Far from being mutually exclusive, one can find blends of the four. We shall, moreover, argue that there is a convergence among the four.

6.7.1 INDIVIDUAL HARVEST QUOTAS

The original form of limited entry programmes, in which the limited number of vessels granted entry to the fishery are invited to compete for shares of the TAC (or equivalent), can, as we have now seen, readily lead to an uncontrollable expansion of actual fishing capacity. The expansion of capacity, in turn, leads to a dissipation of resource rent and can readily undermine the resource managers' ability to conserve the resource.

A response to this problem emerged in the 1970s, soon after the introduction of the original limited entry programmes. In 1973, American economist Francis Christy (Christy, 1973), to be followed by the Canadian economists Maloney and Pearse (Maloney and Pearse, 1979; see also Commission on Pacific Fisheries Policy, 1982), argued that the competition among the vessel owners in a limited entry could be eliminated, or at least sharply reduced, by 'stinting', by giving individual vessel owners quotas to harvest specified shares of the TAC (or equivalent). To quote from Pearse's report on Canada's Pacific fisheries policies:

> The outstanding advantage of this approach is that it eliminates the basic cause of overcapacity in the fishing industry by removing the incentives of individual fishermen to protect and increase their share of the catch...(Commission on Pacific Fisheries Policy, 1982, p. 84)

Pearse points out that there is nothing particularly novel about this approach, and that, within his home country, Canada, this type of 'stinting' approach has been used to regulate other publicly owned renewable natural resources, such as timber, water, grazing rights, etc. (Commission on Pacific Fisheries Policy, 1982, p. 84).[27]

Christy, and Maloney and Pearse, differed in some details, but were in general agreement. Christy said nothing about the transferability of such individual quotas; Maloney and Pearse stressed the importance of such transferability. Christy (1973) recommended that the individual quotas be expressed as percentages of the TAC; Maloney and Pearse (1979) urged that they be expressed as fixed quantities. All three economists agreed that the individual harvest quota (IQ) scheme would not work unless enforcement is effective. No one of the three claimed that IQs are a universal panacea.[28]

The IQ schemes, when first introduced, were not greeted with universal enthusiasm by economists. Some fishery economists argued that IQ schemes were too complex to be feasible, that they would lead to equity problems, or that they were inherently unnecessary, or all of the above (e.g., Copes, 1986; McConnell and Norton, 1978; Crutchfield, 1979). In any event, IQ schemes were only of minor interest prior to the advent of the EEZ regime, given that the management control of coastal states over fishery resources was extremely limited.

All of this changed with the coming of the 200-mile EEZ regime and the massive transfer of fishery resources from the high seas to coastal state management and control. IQ schemes then ceased to be little more than a theoretical curiosity and achieved practical significance. It is worth noting that Maloney and Pearse (1979) published their paper two years after several coastal states, including Canada, had unilaterally introduced their EEZ regimes, prior to the conclusion of the Third UN Conference on the Law of the Sea.[29]

While IQ schemes now became of practical significance, the implementation of IQ schemes in the late 1970s and early 1980s proceeded at a slow pace. There remained ongoing scepticism, and ongoing objections—many of which continue to the present day—that IQs mean the privatization of fishery resources, the 'giving away' of public resources to private interests (e.g., Copes, 1986; Bromley, 2009).

In more recent years, however, IQ schemes have begun to spread rapidly. At the beginning of the 1980s, only two coastal states had introduced IQs, Iceland and the Netherlands.[30] At the end of the decade, the number of adopting states had increased, but only to five. By mid-2010, the number of states adopting increased to 22, accounting for almost 25 per cent of global capture fishery harvests, as measured by the FAO (Arnason, 2010). There is no evidence that the spread of IQ schemes is abating.[31]

IQs, which as we have seen, were originally designed to curb the 'race for the fish', are now seen as doing much more. If properly designed, and administered, it is argued, they will serve to give the fishermen an incentive to invest in the resource, and provide a means of mitigating fishermen risk.

Before proceeding further, there is one point that we must keep uppermost in our minds. The limited entry with Olympics style TACs was described as a competitive game leading to a Prisoner's Dilemma type of outcome. The question that has to be asked is, whether under a given IQ scheme, the strategic interaction among fishermen disappears. If it does not, then it can be argued that the IQ scheme will be fully effective, if and only if, it succeeds in transforming the fishermen competitive game into a cooperative one.[32] This proves to be true for the other rights based resource management schemes as well.

The IQ scheme facilitates a fishermen cooperative game by providing a means of sharing the TAC among the fishermen. The limited entry with Olympics style TAC scheme almost compels the fishermen to compete. Having said this, the IQ TAC sharing mechanism is not enough— in, of, and by itself—to ensure a cooperative fishermen game that is stable through time.[33]

With this in mind, let us consider the following aspects of IQs:

(a) IQs as property rights to the fishery resource.
(b) IQs, short term or long term in nature.
(c) IQs expressed in fixed quantities over time, or as percentages of the TAC.
(d) IQs, non-transferable or transferable—individual transferable quotas, ITQs.
(e) IQs and the implications for equity.
(f) IQs and multiple species fisheries.
(g) IQs as cooperative games, and problems of surveillance and enforcement.

(a) IQs as resource property rights

This is a question that has caused a great deal of controversy, and charges of IQ schemes leading to public resource 'give aways' to undeserving private interests. The position taken by these authors is that IQs can be seen as short term, or long term, claims to shares of the TAC, and as such they are to the IQ holders clearly assets. What they are not are property rights to the fishery resource itself. Within the EEZs, these rights are held by the state. Gray (1991) maintains that the essence of property rights lies in the ability to control access to what one deems to be property The control over access to the intra-EEZ fishery resources lies ultimately with the state.

The two fishing states that most aggressively promoted IQs, in the early days of these schemes, Iceland and New Zealand, both make it explicit in their legislation that IQs/ITQs do not constitute property rights to the fishery resources.[34] The Icelandic Fishery Management Act of 1988, and all subsequent revisions to the act, for example, states that 'the fish stocks around Iceland are the property of the Icelandic people' (cited in Matthiasson and Agnarsson, 2010, p. 303).

Aspects (b), (c), (d), and (e) of IQs are interdependent. Keeping this in mind, we turn to aspect (b): IQs, short term, or long term.

(b) IQs, short term or long term

As well as mitigating, if not eliminating, the 'race for the fish', one hope for IQs is that they will give fishermen incentive to invest in the resource, or at least to cease regarding the resource as a non-renewable resource to be 'mined' with all possible speed. Obviously, if the IQs are truly short term in nature, the holder, if rational, will discount future economic returns from the fishery heavily, thereby undermining the fishermen's incentive to invest in the resource. This becomes a serious issue, when a fishery resource is to be rebuilt. The rebuilding of a fishery resource, as we have stated over and over, is a programme of positive investment in natural capital. Every positive investment entails a current cost undertaken in the hope of a future payoff. If the fishing industry is called upon to bear part, or all, of the investment cost, through reduced current harvests, and is given no assurance that it will enjoy some part of the future payoff, the industry, if rational, will resist the investment programme.

In the United States, as we have already indicated, such fish stock rebuilding is mandated by federal law. Hanna (2010) provides contrasting examples of two such fish stock rebuilding programmes, Gulf of Maine cod stocks and Bering Sea snow crab. In the Gulf of Maine case, the fishermen were given no effective assurance that they would enjoy a part of the future payoff; in the Bering Sea snow crab case, the fishermen were. As Hanna (2010, p. 158) concludes, in the Gulf of Maine case, stakeholder resistance to the fish stock rebuilding programme was maximized; in the Bering Sea case stakeholder resistance was minimized.

(c) IQs expressed in fixed quantities over time, or as percentages of the TAC

The question of IQs being, over the long term, expressed as fixed quantities, or as percentages of the TAC, pertains to the sharing of the risk between fishermen and resource managers. If the IQs are expressed as fixed quantities, the risk is borne by the resource managers. If the TAC has to be reduced, IQs have to be withdrawn, or at least reduced. The fishermen will demand compensation.

The experience of New Zealand provides a good example of the consequences of fixed quantity IQs. With the introduction of the New Zealand Quota Management System (QMS) in the mid-1980s, the IQs, ITQ in nature, were issued in perpetuity and were expressed in terms of fixed quantities. If

TACs had to be reduced, the government had to intervene in the quota market to buy back quota.

Matters were brought to a head by the quota for deep water species—orange roughy in particular—that the New Zealand government issued and sold to the fishing industry[35] in the 1980s. It was discovered that a major reduction of the orange roughy TAC was necessary, which in turn meant that the government would have to buy back quota. The government faced the prospects of spending hundreds of million of NZ dollars in so doing. In the 1990s, the government changed the relevant legislation so that the quotas were thenceforth to be expressed as percentages of the TAC (Sharp and Batstone, 2007; Connor and Shallard, 2010). There was, needless to say, a strong reaction from the fishing industry.[36]

It is reasonable to say that, after the New Zealand experience, the policy for other than very short-term IQs is to have them expressed as percentages of the TAC, i.e., to have the industry bear the risk. In terms of creating the desired fishing industry incentives, this policy has much to recommend it. With the risk falling on the industry, the industry has a powerful incentive to invest in the resource and to foster good resource management. If the resource management is flawed, and the resource is subject to undue depletion, the fishermen will suffer the economic consequences.

(d) IQs, non-transferable or transferable—individual transferable quotas, ITQs

With regards to the issue of transferability of quota, i.e., allowing quota to be leased and/or sold, the traditional economist's argument for making the quota transferable—ITQs—is one of efficiency. Vessels and vessel owners are not homogeneous. Transferability of quota will enable the shift of quota from inefficient to efficient fishing 'firms', although fears of undue concentration in the industry may limit the extent of the transferability. The argument is certainly valid, but there are two additional arguments to be made on behalf of transferability of quota. The first pertains to the incentive to invest in the resource.

If IQs are ITQs, then quota markets can and do emerge. Suppose now that the ITQs are long term, in fact if not in law, and that the ITQs are expressed in terms of percentages of the TACs. The ITQs then become claims to shares of the economic returns from the fishery now and in the future, whatever those economic returns might be.[37]

If participants in the quota market are rational, the quota prices should reflect the present value of the expected economic returns from the fishery. If the resource management is flawed and causes market participants to believe that TACs will have to be reduced in the future, these beliefs should be promptly reflected in today's quota prices.

Suppose a quota holder expects to exit from the fishery, long before the expected reductions in the TAC occurred. If the quota market participants are rational, this will do him/her no good. The consequences will have a negative impact on quota prices today.[38]

The second additional argument for transferable IQs pertains to the question of multiple species fisheries. We examine that argument at a later point. We turn first to the difficult issue of equity.

(e) IQs and the implications for equity

IQs, and in particular ITQs, have been roundly criticized on equity grounds. Let it be conceded from the start that the equity issue is far from settled.

The equity based criticisms of ITQs can be seen as having at least two strands. There is first the argument that the economic returns from the fishery with an ITQ scheme, based upon a public resource, come into the hands of the 'lucky few'. The immediate response to this argument is that in the evolution of fisheries management, the 'lucky few' came into being before the advent of ITQs. This, after all, is what limited entry was, and is, all about.

The 'lucky few', under the limited entry with Olympics style TACs, were not all that 'lucky', due to the dissipation of resource rent. If the ITQ schemes lead to the generation of positive resource rent from the fishery, this is an important step forward. There is not much point in worrying about equitable distribution of net economic returns from the fishery, if those returns are non-existent.

Having said all of this, if we accept the proposition that the fishery resources constitute state property then, of course, the argument can be made that some of the resource rent forthcoming from the fisheries under ITQ schemes should accrue to the state, through auctioning off quota,[39] by imposing royalties, or some other means. This then raises the obvious question of the extent to which the state can extract resource rent from the fisheries, without undermining the positive fishermen incentives the ITQ scheme is designed to foster. The question, at the time of writing, remains unanswered and is the focus of intense debate (see, for example, Libecap, 2007, 2010).

What we can say is the following. Go back to the argument that intra-EEZ fisheries management is best considered within the framework of Principal-Agent analysis. The question, or problem, that we have posed is a classic Principal-Agent one (Sappington, 1991). The debate continues. Much additional research remains to be undertaken.

The second strand of the argument runs as follows. If IQs are transferable, fortunate quota holders may effectively withdraw from the fishery and lease out their quotas as the rentiers that they most assuredly are—'armchair fishermen'. Worse yet, the quota may be sold to non-fishermen. What is not known at this point, with any accuracy, is the extent of this 'rentier class', or the consequences. If it should be demonstrated that a serious problem exists, then the problem would have to be corrected by adjusting the regulations surrounding ITQs. The existence of the problem—in, of, and by itself—provides no justification for abandoning ITQs.

(f) IQs and multiple species fisheries

In the not so distant past, many pro-ITQ economists held the view that, while ITQ schemes could be expected to work well in single species fisheries, they could not be expected to do so in multiple species fisheries. The complexities would be overwhelming. This view of the ITQ world was set forth forcefully by Squires et al. (1998). If this view is correct, then the usefulness of ITQ schemes is limited, perhaps severely limited, in spite of the increase in their adoption. Within finfish fisheries, at least, there are few genuine single species fisheries.[40]

This pessimistic view of ITQs has been proven, by experience, to be wrong. Two of the authors of the aforementioned 1998 article, Grafton and Turris, explicitly recanted and came to argue that, if anything, ITQs come into their own in multiple species fisheries (Grafton et al., 2007). The motivation for the recantation was the experience with the groundfish trawl fishery off British Columbia, Canada.[41]

The fishery is an extremely complex one, with a large number of species, and over 60 different stocks. It was, and is, a multiple species fishery par excellence. Upon Canada's introduction of the EEZ regime in the late 1970s, the British Columbia groundfish trawl fishery was subject to a limited entry programme with Olympics style TACs. The usual economic pattern developed, combined with an inability to implement effectively the TACs in several of the key sub-fisheries. The *agency cost* ultimately became so massive that the resource managers—the Department of Fisheries and Oceans—felt compelled to shut the fishery down in the autumn of 1995—a historical first for the Department (Munro et al., 2009).

In 1997, an ITQ scheme was established for the fishery. Under the scheme, 55 stock TACs were established. The Department proceeded to allocate quotas to all 55 stocks to all vessels licensed to operate in the fishery, and then hoped for the best (Munro et al., 2009). The transferability of quota was to be essential, if the scheme was to have any chance of success.

The complexities of the fishery did not prove to be a daunting obstacle. A sophisticated quota market quickly emerged. The licence holders, working through the market, solved the allocation problem.

The licence holders became, in effect, quota portfolio managers. Many of the vessels specialized to some degree by species and/or area, and/or gear. Such specialization required the licence holders to rebalance their quota portfolios. The required portfolio balancing was, and is, effected through the quota market. As of 2009, up to 3,000 ITQ transfers occurred annually in the fishery. The Department of Fisheries and Oceans has to approve each transfer, but this can be done speedily through electronic means (Munro et al., 2009).

Along with the groundfish trawl fishery, there are other groundfish fisheries off British Columbia, the most important ones being a Pacific halibut fishery and a sablefish fishery, both of which operate under ITQ schemes. The fisheries are linked through bycatches, in that vessels operating in one fishery can readily have bycatches of fish from the other two fisheries.

Prior to 2006, the three groundfish fisheries were segmented legally. The consequence was that a groundfish trawl vessel, for example, having a bycatch of, say, sablefish, was required by law to release the sablefish, because the vessel lacked quota to cover the sablefish. What was true for the groundfish trawl vessel was true for vessels operating in the sablefish or Pacific halibut fishery. Many of the released fish did not survive. The resultant fish mortality more often than not went unreported. The policy was destructive, both from an economic and a biological perspective.

The success of the multispecies groundfish trawl ITQ scheme pointed to a solution. After several years of preparation, the Department of Fisheries and Oceans integrated the groundfish ITQ systems. Return to the example of the groundfish trawl vessel and its sablefish bycatch. Post-2006, the groundfish trawl vessel would be required to (a) retain the sablefish, and (b) obtain quota to cover the retained sablefish. The vessel would meet its requirement by turning to the quota market.[42]

Let us now note the following. As we have stressed continuously, a fundamental characteristic of capture fisheries management is uncertainty. A standard method of mitigating risk and uncertainty in the world of finance is through portfolio diversification. Hilborn et al. (2001) addressed the issue of uncertainty in fisheries management and turned to the principle of portfolio diversification (Hilborn et al., 2001).

In their 2001 article, the authors observe that the then common limited entry programmes (with Olympics style TACs) compel fishermen to become specialists in one, or a small number, of fisheries (Hilborn et al., 2001, p. 9), which one can, in turn, see as reflecting a single species management approach. Needless to say, compelling fishermen to become specialists runs directly counter to the principle of portfolio diversification. Means should be found, they argue, to allow fishermen to diversify among fisheries, without undermining limited entry (Hilborn et al., 2001). The multiple species ITQ scheme, as described, does just that.[43]

We can re-express this in a slightly different way. In earlier chapters, we discussed non-malleability of produced and human capital in fisheries, and the consequences arising therefrom. Multiple species ITQ schemes have the effect of enhancing the malleability of produced and human capital with respect to specific fisheries.

Next, let us re-emphasize the fact, which should now be obvious, that multiple species ITQ schemes are inconceivable, unless the quotas are transferable, i.e., unless the IQs are ITQs. This points to the second 'other' argument for transferable IQs. Transferability of IQs makes multiple species IQ schemes feasible, which in turn serves to mitigate risk and uncertainty in fisheries management.

(g) IQs as cooperative games, and problems of surveillance and enforcement

It was argued at an earlier point that, if strategic interaction among fishermen continues to exist after the implementation of an IQ scheme, the scheme will be fully successful ultimately, only if it transforms a fishermen competitive game into a cooperative one.[44] We shall explore the theory of cooperative games in detail, when we discuss the management of internationally shared fishery resources. Suffice it to say at this stage, a fundamental condition that must be satisfied, if the solution to a cooperative game is to be stable, is referred to as the individual rationality constraint. Each and every 'player' must, at each and every point in time, believe that its payoff under cooperation is at last as great as would be its payoff under competition. If a 'player' believes that its payoff under competition would exceed its payoff under cooperation, then it would be irrational for the 'player' to cooperate.

In order for the individual rationality constraint to be satisfied, there must be some sort of mechanism for ensuring compliance among the 'players', and some means of guarding against free riding by non-'players' in the cooperative game. The reason is straightforward. Suppose that a 'player' in a fishery cooperative game believes that its allocation of the harvest is 'fair', but also believes that cheating by its fellow players and/or free riding by outsiders will go unchecked. The 'player' could then well calculate that its expected payoff from cooperation will be less than what it would enjoy under competition. What will be true of this 'player' will be true of its fellow 'players'. If this does indeed occur, then the cooperative game will disintegrate.

In the context of ITQ schemes, this means that surveillance and enforcement of regulations among ITQ holders must be effective. It means, as well, that poaching by non-ITQ holders must be firmly suppressed.

An example is provided by the most valuable shellfish fishery in Chile, the *loco* fishery,[45] which extends along the entire coast of Chile. After a pure open access induced management crisis resulted in the fishery being closed for three years, an IQ scheme was introduced into the fishery in 1993. Enforcement proved to be ineffective. After a few years, the IQ scheme collapsed in ruin (San Martín et al., 2010).

Prior to leaving the topic of ITQs, we point to one example of an ITQ scheme that had an immediate and dramatic impact upon implementation.

The case involves a fishery to which we have referred many times, the British Columbia Pacific halibut fishery. We used it as an example of the inexorable growth of fishing capacity under a limited entry with Olympics style TAC. To repeat, the maximum season length is approximately 240 days per year. By the end of the 1980s, the season length was down to six days (Munro et al., 2009). As remarked earlier, the short season costs in the fishery were extensive.

The IQ scheme was introduced into the fishery in 1991. Figure 6.2 shows the impact of the IQ scheme upon the season length in the fishery. The figure speaks for itself. The dramatic increase in the season length, from six days to the maximum of roughly 240 days, was not accompanied by an immediate reduction in fishing capacity in physical terms (Munro et al., 2009). It can be said that the introduction of the IQ scheme defanged the excess capacity.

This said, it has to be agreed that, due to enforcement and other problems, IQs/ITQs do not have universal application. Kurien and Willmann (2009), writing about developing fishing state small-scale fisheries, express serious doubt about the efficacy of IQs/ITQs in any of these fisheries. Other rights based harvesting schemes must be considered, the most prominent of which takes the form of community based management schemes—territorial use rights fisheries (TURFS).

6.7.2 TERRITORIAL USE RIGHTS FISHERIES (TURFs)

The first to introduce the concept of IQs was, as we observed earlier, Christy (1973). Christy was also the first to introduce the concept of TURFs (Christy, 1982). Christy (1982) recognized the infeasibility of ITQ schemes in most, if not all, artisanal developing coastal state fisheries. He was at the same time impressed by traditional community based fisheries management arrangements, in many developing states. He had also become aware of the centuries-old coastal based cooperative fisheries management arrangements in Japan. All of this led him to develop the concept of TURFs.

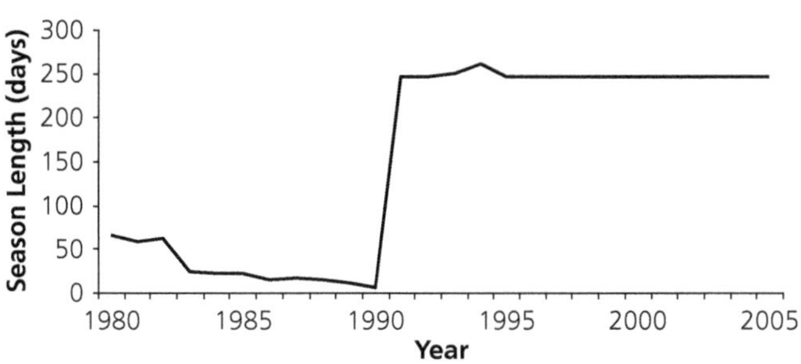

Figure 6.2. British Columbia Pacific halibut fishery season length (1980–2005)

Source: Munro et al. (2009, figure 1).

As the term implies, TURFs involve the granting of harvesting privileges to one, or a set of, fishing coastal communities. An example is provided by the Japanese Fishery Cooperative Associations (FCAs), which can trace their origins back to the sixteenth-century fishermen guilds. The FCAs focus on inshore and near shore coastal fishery resources. The Japanese federal government lays down broad resource management guidelines, and then devolves harvesting rights upon the FCAs (Cancino, Uchida, and Wilen, 2007).

The typical FCA may cover many diverse fishery resources. Within the FCA, fisheries management organizations (FMOs) emerged to focus on specific species and or gear types. Thus, the management unit, or TURF, consists of a FCA and a subset of FMOs (Cancino et al., 2007).

Another example is provided by Chile, in its response to the collapse of the *loco* IQ scheme. Harvesting rights were now given on a collective basis to groups of fishermen in coastal communities (Cancino et al., 2007).

A TURF is, almost by definition, a cooperative fishermen game. Christy (2000) sets out the problems to be addressed and solved, if the solution to the cooperative fishermen game that is the TURF is to prove to be stable through time. The problems to be faced will, with one possible exception, be found to be identical to those faced by ITQ based fishermen cooperative games.

First, the TURF must develop a mechanism for the sharing of the economic benefits from the fishery in an equitable manner. Secondly, the TURF must ensure compliance among TURF participants. Thirdly, the TURF must be protected against free riding by fishermen outside of the TURF (Christy, 2000).

Fourthly, the TURF must face a problem, for reasons to be explained, that does not confront an ITQ based fishermen cooperative game, the so-called 'new member' problem. This is a problem that will re-emerge and be discussed in detail under the heading of international fisheries management, when we come to discuss Regional Fisheries Management Organizations (RFMOs).

In any event, it is common that TURF 'charter' members are forbidden by national law from barring outright prospective new participants, expressing a 'real' interest in the relevant fisheries. Such, for example, is the case of both Chile and Japan (Cancino et al., 2007).

As we shall see, RFMOs are faced with the same sort of restriction emanating from international treaty law. Kaitala and Munro (1997), writing in the context of RFMOs, make the argument that the 'new member' problem could constitute a form of implicit free riding, which could in turn undermine the stability of the cooperative fishermen game. The argument basically runs as follows. If the cooperative game achieves success, it is certain to attract prospective 'new members'. If the 'new members', upon agreeing to abide by the management regime laid down by the 'charter' members (players), are given a pro rata share of the economic returns from the fishery free of charge, as it were, this constitutes a form of free riding, albeit implicit. Kaitala and

Munro (1997) demonstrate that, under these circumstances, it is quite possible that one, or more, of the 'charter' players will conclude that its (their) payoff(s) from the cooperative game will fall short of the payoff(s) it (they) would enjoy under cooperation. Ultimately the only solution is to have 'new members' buy their way in.

The 'new member' problem does not really arise for ITQ schemes. The reason is straightforward. The Kaitala–Munro solution is built in. Prospective 'new members' can enter only by buying or leasing quota from 'charter' ITQ holders.

The final problem identified by Christy (2000) (and one that really also applies to ITQ schemes, as well) is that conditions surrounding the TURFs may change over time and may do so in an unpredictable manner. For example, power relationships within the community, or communities, constituting the TURF scheme may shift through time. If the TURF scheme does not have sufficient resilience to absorb these shocks, the TURF scheme could well disintegrate (Christy, 2000). As described, this is the game theory version of the time consistency problem. It will be seen when we come to discuss international fisheries, that the time consistency problem is a major one confronting the cooperative fishermen games that are RFMOs.

An example, which illustrates the TURF problems to be solved, listed by Christy, is provided by a detailed case study of two fishing villages in the Indian southeastern state of Andhra Pradesh (Salagrama, 2005).[46] The fisheries off the two villages are complex multispecies ones. The fisheries have come under TURF management, not as a result of government decree, but as a result of history and custom. As such, they provide examples of traditional community based fisheries management.

Critical to the success of these fisheries cooperative games is the existence of fishermen castes and the system of governance arising under the caste system, the Caste Panchayats (Salagrama, 2005), which governs the entire village life. Non-members of the fishermen caste dominant in the village are effectively barred from the fishery. The system provides for the basis for sharing the benefits from the fisheries. The system does, as well, foster effective surveillance and enforcement. The ultimate punishment is to be banned from the caste, and hence excluded from the fishery and indeed from village life in general.[47]

Some of the exploited species are shared with neighbouring villages. Negotiations are undertaken with the neighbouring villages to ensure effective shared resource management and to guard against poaching. Resource conservation is maintained by virtue of the collective recognition of the importance of the fishery resources to the survivability of the villages, which implies in turn that the implicit discount rates being used by the villages are low (Salagrama, 2005).

The villages are acutely aware of the new member problem. New members are not barred outright, but gaining membership in the relevant caste, and hence fishery, can be difficult. The author reports that one of the villages had the practice of accepting new members equal to no more than the number of previous members who had left (Salagrama, 2005).

Finally the time consistency problem does arise. Changing conditions, social and economic, do stand as a threat to the stability of the TURFs. The author implies that one of the greatest threats arises from government, which tends to view the customary based resource management with some disdain, as being old fashioned (Salagrama, 2005).[48]

6.7.3 FISHERMEN COOPERATIVES

Fishermen cooperatives can be thought of as similar to TURFs, except that there is no land based geographical location. The United States provides an example of fishermen cooperatives, and their close relative, sector TAC allocations.

In 1996, a moratorium (subsequently to be lifted) in the United States was imposed upon IQ schemes. Harvester cooperatives were seen as an alternative, first for American West Coast fisheries (Holland, 2010). A prominent example is provided by the Pollock Conservation Cooperative (PCC), established in 1999, with the express approval and support of the American government.

The PCC consists of a group of companies operating catcher/processor vessels in the offshore pollock fishery in the Bering Sea. The PCC is allocated a quota, a segment of the overall pollock TAC, which the company members allocate among themselves. Thus, by definition, the PCC is a cooperative fishermen game. As such, it led to results similar to those in British Columbia groundfish ITQ fisheries. As a result of the establishment of the PCC season length was greatly increased;[49] and the product being produced by the vessels was substantially improved (Holland, 2010; At Sea Processors Association, 2011).

The harvester cooperatives on the American west coast came to serve as a model for harvester cooperatives in New England, referred to as sector allocations. Since 2004, self-selected groups of fishermen have been able to apply for groundfish allocations. The groups are then allowed to develop plans for managing their catches within the allocations (Holland, 2010). Once again, we are presented, by definition, with a cooperative fishery game.

The New England programme is recent. As a consequence, it is not really possible, at the time of writing, to offer an assessment.

6.7.4 A CONVERGENCE AMONG HARVESTING RIGHTS BASED SCHEMES

Until recently, the view was taken that IQ/ITQ schemes and TURF/fishermen cooperatives were separate and distinct. IQs are, after all, harvesting rights allocated to individual fishermen, while TURFs/fishermen cooperatives are based upon harvesting rights granted to groups of fishermen on a collective basis.[50]

We have now argued that, if strategic interaction among fishermen exists within an IQ/ITQ scheme, the scheme will evolve into a cooperative fishermen game, or its success will be limited, perhaps severely. If this argument is accepted, then it should now be clear that the distinction between IQ/ITQ schemes on the one hand, and TURF/fishermen cooperatives on the other, blurs. The first clear recognition of this blurring of distinctions, it might be added, came with the American National Research Council publication, *Sustaining Marine Fisheries*, which coined the term 'virtual communities' (National Research Council, 1999).

What follows from this is straightforward. If the Principal (coastal state) concludes that a harvesting right scheme is to be applied to a particular fishery, or groups of fisheries, then the Principal must look to its portfolio of harvest rights schemes (e.g., ITQs, TURFs, fishermen cooperatives) and choose the one, or the blend, that will minimize the *agency cost*. Thus, for example, in returning to the Andhra Pradesh case study, one has no difficulty in agreeing with the author of the case study that ITQs would be infeasible in the two villages that he is examining, and that TURFs are far more appropriate (Salagrama, 2005). On the other hand, one has to recognize that TURFs are likely to be limited to near shore fisheries.

6.8 **Marine protected areas (MPAs)**

An ancillary fisheries management tool consists of marine protected areas (MPAs), a tool which has been increasing in popularity. The National Research Council of the United States defines a MPA as follows:

> A marine protected area is . . . a spatially defined area in which all populations are free of exploitation . . . A primary purpose of such 'no take' zones has been to protect target species from overexploitation and to allow their populations to recover. (National Research Council, 1999, p. 84)

The National Research Council goes on to recommend MPAs encompassing as much of 20 per cent of the marine environment. The Council does, however, concede that the impact of MPAs upon the productivity of ocean fishery resources is very much open to debate (National Research Council, 1999, pp. 84–91).

From an economic standpoint, the strongest argument for MPAs is based, not upon the productivity of ocean fishery resources, but rather upon the issue of uncertainty, an issue that we have touched upon but lightly up to this point. The clearest version of this argument is put forward by Lauck et al. (1998).

Lauck et al.'s basic point is that fisheries managers are faced with *irreducible* uncertainty, which is to say uncertainty that cannot be eliminated by yet more research. To quote the authors: '. . . full understanding and predictability of anything as complex... as a marine ecosystem will forever remain a chimera' (Lauck et al., 1998, p. 874).

MPAs should be used, the argument continues, as a 'bet hedging' strategy to provide protection against inadvertent, and probably inescapable, resource management errors. The argument is really an application of the concept of portfolio diversification, which we encountered in our discussion of ITQs in the context of multiple species fisheries.

Think of a particular fishery resource, in a fishery with an MPA, as now being two interlinked assets. The segment of the fishery resource in the MPA is analogous to highly liquid (thus low risk) and low yielding, assets in a financial portfolio (e.g., government treasury bills).

The usual way of guarding against uncertainty in fisheries management, reducing the harvest rate, is an inferior approach, the authors argue. Obviously, if the harvest rate were to be set equal to zero, the resource would be fully protected. Once the harvest rate is positive, however, there will be risk of inadvertent resource depletion, even if the harvest rate appears to be 'low'. This arises by virtue of the fact that the 'low' harvest rate is applied to the entire resource. The 'low' harvest rate will be based on marine biologists' estimates, which may be seriously flawed, for example in terms of estimates of resource stock abundance (Lauck et al., 1998).[51]

The desirable characteristics of a programme of MPAs, the authors maintain, are as follows:

A. The MPA should be large enough to protect the resource in the event of inadvertent overfishing in the unprotected area. The National Research Council's proposed 20 per cent is to be seen as modest (and quite likely inadequate).
B. The MPA should serve as a 'source', capable of replenishing the exploited segment of the stock, in case of inadvertent overexploitation. Consequently, the MPA should protect spawning grounds and other areas critical to the viability of the population.[52] Think back to our simple sink-source model in chapter 5, in which both the sink and the source were commercially exploitable. Within the context of that model, the source would be an obvious candidate for a MPA.
C. The MPA must be thoroughly protected. Simply declaring an area to be a MPA is wholly inadequate. Obviously, a MPA will be an attractive target for poachers (Lauck et al., 1998, p. 875).

A few questions/comments are now in order. First, if the 'bet hedging' strategy through MPAs is adopted, can one be certain that the expected economic yield from the resource will be at least as great as the expected yield in the absence of such a strategy? The answer to the question is straightforward—No! Secondly, if the expected yield under the 'bet hedging' strategy is less than the expected economic yield without the strategy, should the bet hedging strategy then be ignored?

The answer to this second question rests upon the attitude towards risk of the resource managers. If the resource managers are strictly risk neutral, then the 'bet hedging' strategy should indeed be ignored. If the resource managers are risk averse, then the bet hedging strategy should be adopted.[53]

This follows from an application of elementary portfolio balance theory, normally cast in terms of financial portfolios. In the elementary version of the theory, the portfolio holder is seen as having to make a choice between a long-term, risky security, with a positive expected yield, and a riskless liquid security, with an expected yield of zero. If the portfolio holder allocates any part of his/her portfolio to the riskless security, then by definition the expected yield on the portfolio will be below the maximum. If the portfolio holder is risk averse, the portfolio holder can then be seen as facing a clear trade-off between expected yield on the portfolio and safety. Except under extreme circumstances, the risk averse portfolio holder will have a mixed portfolio. That is to say, a fraction of the portfolio will be allocated to the riskless securities, having an expected yield of zero.[54] Thus, to return to fisheries, the resource manager, if risk averse, can rationally opt for MPAs, even if there is clear evidence that by so doing, the resource manager will see the expected economic yield from the fishery reduced.

The second question is whether MPAs are compatible with rights based fisheries management. If the 'players' in a fishermen cooperative game are risk averse, then they should welcome MPAs, upon accepting the fact of irreducible uncertainties in fisheries management. It would be collectively irrational of the 'players' to oppose MPAs outright.

As a final comment, return to Lauck et al.'s second desirable characteristic of MPA programmes, that the MPA be a 'source', and return as well to our discussion of a simple sink–source model. From that discussion, it can now be argued that, before the second half of the twentieth century, there were for many fishery resources natural MPAs in the ocean. That is to say, there were offshore 'sources', which were uneconomic to exploit. The current move to MPAs can be seen as an attempt to replace, artificially, fish reserves that had once existed 'naturally', due to the then nature of the economics of the fishing industry.

6.9 **Subsidies**

We must not leave the topic of national/regional management of fisheries without referring to subsidies. While not the root cause of fishery mismanagement, they have, and do, serve to exacerbate the problem, often severely.

Subsidies, relevant to fisheries, can be defined as 'any government programme that potentially permits the firm [directly or indirectly] to increase its profits beyond what they would have been in absence of the programme' (Schrank and

Keithly, 1999, p. 156). That subsidies in world fisheries are massive has long been recognized. A recent study on subsidies in fisheries gives a mid-range estimate in the order of US $30+ billion per year (Sumaila et al., 2008).

Subsidies are deemed to be harmful, if they lead to excess fleet capacity, or in other ways encourage overexploitation of the fishery resources. Fuel subsidies, for example, which Sumaila et al. estimate to be in excess of US$6 billion per annum, lower the cost of harvesting and thus will, under conditions of actual, or near, pure open access, intensify the overexploitation of the fishery resources (Sumaila et al., 2008, p. 835).

Not all subsidies are harmful. Some subsidies may either have a neutral or beneficial impact upon fisheries management. Thus, for example, government expenditures on fisheries management, research, enforcement, and enhancement are deemed to constitute fisheries subsidies (OECD, 2000). No one would seriously contend that these expenditures are detrimental to fishery resource conservation—to the contrary.

The distinction between harmful and non-harmful fisheries subsidies is usually clear cut. There is one important form of subsidy, however, where the distinction is not at all clear cut. This takes the form of government subsidies used for vessel decommissioning/buyback purposes. We turn to this special class of subsidies at a later point.

With regards to subsidies that fall into the 'harmful' category, we have to make a distinction between subsidies used under conditions of pure or regulated open access, and those used in fisheries that are operating under well designed and managed harvesting rights schemes. In the case of the latter, one might maintain that the subsidies are no more than a transfer payment. One could argue against the subsidies on distributional grounds, but not on the grounds that they have negative resource management consequences, or so it would seem.

With regards to the use of 'harmful' subsidies under conditions of pure and regulated open access, we drive home the consequences by focusing on pure open access and taking what is admittedly an extreme example. We suppose that our Schaefer based dynamic economic of the fishery is relevant for the fishery at hand. Let us suppose further that the subsidies in question are fishing effort cost reducing subsidies.

We have now seen, over and over, that, in the absence of subsidies, we can, in our Schaefer based model, express unit harvest costs as:

$$c(x) = b/qx$$

where, as before b, a constant, denotes unit fishing effort costs, and q, the catchability coefficient. We have seen further that there is an economic brake on resource exploitation in that:

$$\lim_{x \to 0} c(x) = \infty$$

Now introduce fishing effort cost reducing subsidies, and suppose that these are fishing effort cost subsidies, *à outrance*, in that they reduce b, as seen by the fishing industry, to zero! Then the unit harvesting costs perceived by the fishing industry are:

$$c(x) = 0/qx = 0$$

The economic brake has been disabled, and the fishery resource, under open access conditions, will be sent careening towards extinction.

Munro and Sumaila (2002), who, to the best of our knowledge, first advanced this argument, concede that the assumption of subsidies reducing fishing effort costs, as perceived by the industry, to zero is an extreme one. They counter that the assumptions that the fishery resource is perfectly understood and that it can be perfectly modelled are also extreme. Fishing effort cost reducing subsidies, which do not go so far as to reduce the perceived b to zero, could, by loosening the economic brake, drive the fishery resource down to a level, which, after the fact, is realized to be dangerously low (Munro and Sumaila, 2002, p. 241). In any event, the subsidies will unquestionably make a bad situation worse.

Now turn to the case of a fishery operating under a well developed, effectively monitored, harvesting rights based scheme. Suppose, for the sake of argument, that the scheme is an ITQ scheme and that the solution to the fishermen cooperative game proves to be stable through time.

If the resource manager, constituting the Principal, exercises iron control over the harvest programme, and thus over the resource, through time, then it would be true that subsidies, deemed to be 'harmful' under conditions of pure open access, would amount to no more than transfer payments (Munro and Sumaila, 2002).

Suppose, however, that the Principal becomes the Principal in name only and that the fishery resource is effectively privatized, then subsidies could be damaging from a resource management perspective. Once again, following the lead of Munro and Sumaila (2002), we take the extreme example of a cost reducing subsidy that effectively reduces the fishing effort costs, as perceived by the ITQ holders, to zero.

We suppose, also as before, that the relevant dynamic economic model of the fishery is the Schaefer based model that we have used so many times. If perceived fishing effort costs reduced to zero, obviously the Marginal Stock Effect perceived by the private owners will equal to zero. If we return to chapter 3, it is clear that the resource investment decision rule confronting the private owners (acting collectively) is to be expressed as follows:

$$F'(x_p^*) = \delta_p \qquad (6.12)$$ [55]

where x_p^* denotes the optimal biomass level, as perceived by the private owners and where δ_p denotes their rate of discount.

Let us also recall from chapter 3 that: $\lim_{x \to 0} F'(x) = r$, where r is the intrinsic growth rate. From this it follows, that, if $r > \delta_p$, there will, in contrast to the pure open access case, be no risk of the resource being driven to extinction. We will be assured that: $x_p^* > 0$. We also know, however, from our discussion in chapter 3, about the threat of extinction, that if $r < \delta_p$, extinction may be a possibility.

What assurance is there that we will always have $r > \delta_p$? The answer is none whatsoever. There are some species for which r is very low. As examples, marine biologists estimate the intrinsic growth rates for two groundfish species, orange roughy and sablefish to be: $r = 0.025$, in each case (cited in Clark et al., 2010a, p. 217). Suppose now that the fishery resource effectively under private ownership has an intrinsic growth rate $r = 0.025$ and that $\delta_p = 0.05$ (i.e., 5 per cent), which can be viewed as a rather modest private sector rate of discount.

With $r < \delta_p$, the economic theory tells us the rational de facto private resource owners should drive the resource to extinction ($x^* = 0$) and invest the net economic proceeds from the fishery resource elsewhere in the economy. From the point of view of society, we would end up with an outcome that was worse than that of unsubsidized pure open access (Munro and Sumaila, 2002).

Once again, the assumption of subsidies reducing perceived fishing effort costs to zero is extreme. Also once again, however, less extreme subsidies could result in the fishery resource being driven down to a level that is recognized, after the fact, as being dangerously low.

We conclude with the contentious issue of vessel decommissioning/buy-back subsidies. This issue can be dealt with quickly, because we have addressed it already, at an earlier point in the chapter.

The argument for such subsidies is that they will serve to reduce excess fleet (and human) capital, seen as a threat to the security of the resource. What we now know is that such subsidies can be expected to achieve the desired result, if they are *unanticipated* by the fishing industry. If the subsidies are anticipated, then we now know further that we are confronted with a time consistency problem. The subsidies will exacerbate, not ameliorate, the threat of excess capacity.[56]

6.10 Uncertainty: a commentary

In chapter 5, we discussed uncertainty within the context of analytical economic models of the fishery. In so doing, we set forth categories of uncertainty, with by far the most significant category being irreducible uncertainty, i.e., uncertainty that cannot be eliminated by undertaking further research. Such uncertainty carries with it the threat of serious resource management error, and in the worst case, fish stock collapse. We concluded that incorporating uncertainty into analytical models carried us only a limited distance in dealing with irreducible uncertainty. This form of uncertainty is inherently a policy issue, and hence is properly addressed within this chapter. Such uncertainty has come up twice in this chapter, first in the discussion of harvest rights schemes and multiple species fisheries, and secondly in the discussion of marine protected areas (MPAs). Let us attempt to pull the strands together.

To begin, let us note that irreducible uncertainty is commonplace, occurring in many other aspects of the economy, for example finance. Those who believe that they can eliminate, or successfully ignore, irreducible uncertainty in finance and elsewhere, are more often than not called upon to pay a heavy price, as the Great Recession, commencing in 2008, has reminded us so painfully.

In terms of the fishery management regimes examined up to this point, what we seek, in the face of irreducible uncertainty, can be stated quite simply. We seek fishery management regimes that are characterized by *resiliency*, i.e., the ability to withstand major unpredictable shocks, without the risk of collapse of the fishery (Charles, 2001). The issue of resilience will arise again, when we come to talk about the economic management of international fisheries.

There are various techniques for attempting to ensure resiliency, for mitigating the effects of irreducible uncertainty, with one of the most common one being 'bet hedging', for example through portfolio diversification. We first, however, have to consider one aspect of fisheries that works against the resiliency of fisheries management regimes, namely non-malleable produced and human capital in the fishery, which we discussed at some length in chapter 5.

We illustrate by turning to Figure 6.3, which is almost identical for figure 5.1, except for one small change. Suppose that we commence at $t = t(0)$ with a virgin fishery. The social manager goes through the appropriate calculations, and determines the size of the fleet it will permit to be established. The social manager has complete control of the fishery. The fleet capital is quasi-malleable in that, while the rate of depreciation is positive, the re-sale value of the vessels is equal to zero. The allowed investment in fleet capital is such that the resource will be driven down to $\tilde{x}$ at $t = \tilde{t}$, or so it is believed. When that occurs, we find that $K = \tilde{K}$, meaning that there is no apparent temporarily redundant capital.

In contrast to chapter 5, however, we no longer suppose that we are living in a deterministic world. Uncertainty is a fact of life. The social manager, on the basis of imperfect stock assessment, thinks that at $t = \tilde{t}$, the resource has been stabilized at $x = \tilde{x}$. Further biological research now reveals, however, that the prior stock estimates were wrong, and that the true state of the world is almost certainly that: $x = x_{BE}$. The resource has been overestimated and we now find that it has been inadvertently overexploited. What is to be done? If the new estimates are correct, then Figure 6.3 seems to indicate that a temporary harvest moratorium should be declared. The vessels have no alternative use, however. If a harvest moratorium, even though temporary, is declared, and nothing further is done, many vessel owners are likely to be faced with bankruptcy, and fishing communities faced with closure, leading to vehement industry objections and resistance. The government could try to placate the industry by instituting an expensive decommissioning scheme, or subsidy programme, both of which may be politically unpalatable, and, in any case, would take time to implement.

There is no absolute certainty about the new research results. The easiest course of action for the social manager (government) is to hope that the new research results are wrong, and to allow the fleet to continue harvesting at the old level. If, indeed, the new research results are correct, then the easy course of action will lead to a difficult situation being transformed into a crisis. Basically, the fishery management programme lacks the resiliency to deal with the consequences of resource management error, brought about by uncertainty.

What does all of this have to do with harvest rights schemes and MPAs? First consider harvest rights schemes. What we found from our discussion of harvest rights schemes and multiple species fisheries, ITQs in particular, is that such schemes can be designed to allow fishermen to maintain diversified portfolios of harvesting rights. This has the consequence of making the vessel (and human) capital more malleable, with respect to specific fisheries. If a given fishery has to be subject to a temporary harvest moratorium, the industry will not be thrown into chaos. Thus fishermen diversified portfolios of harvesting rights add resiliency to fisheries management programmes.

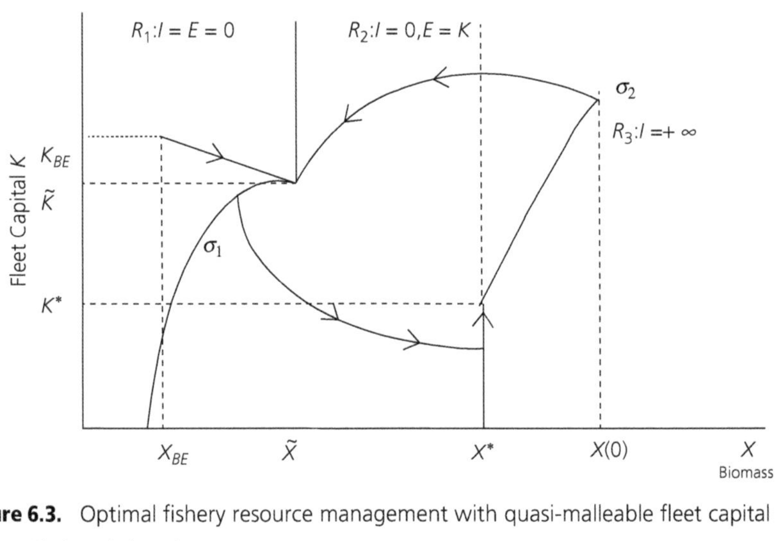

Figure 6.3. Optimal fishery resource management with quasi-malleable fleet capital

Source: Clark et al. (1979).

We characterized MPAs as another form of portfolio diversification. MPAs help to guard against stock collapse, and act as a buffer against management error arising from uncertain science. Thus, once again, it can be argued that a well designed MPA scheme serves to enhance the resiliency of the fisheries management programmes.

As a final question, we can ask, which of the two should be relied upon to enhance the resiliency of fisheries management programmes: fishermen diversified portfolios of harvesting rights, or MPA schemes? The answer is to rely upon both. There is no fear of the resource manager being faced with a surfeit of resiliency enhancing measures.

6.10.1 THE NORTHERN COD FISHERY OF ATLANTIC CANADA

The case of Canada's Northern Cod fishery illustrates the problems arising from uncertainty, combined with a lack of resiliency in fisheries management programmes. One can conjecture that the history of Northern Cod was a factor leading to the incorporation of the Precautionary Approach into international treaty law.

The Northern Cod resource is a cod stock complex extending from southern Labrador to southeast Newfoundland. The stock complex had been the mainstay of the Newfoundland fishing industry, going back for centuries, and had been of great importance to the fishing industries of other parts of Atlantic Canada (Munro, 1980; Hilborn et al., 2001).

The fishery had been divided into two broad components. The cod spawn offshore off southern Labrador in the early part of the year—January to April. When the resource was abundant, some portion of the resource would move inshore during the summer in search of food, for example capelin, giving rise to an inshore fishery. As we have recounted, the inshore fishery was the heart of the overall fishery until the late 1950s, when advances in fishing technology made it commercially feasible to prosecute an offshore fishery during the spawning season. As a consequence of the expansion of the offshore segment of the fishery, annual harvests of the resource rose from 300,000 tonnes in the mid-1950s to just under 800,000 tonnes in 1968 (Munro, 1980, p. 19). The massive increase was due almost entirely to distant water fishing state (DWFS) fleets. Canadian offshore harvests were negligible.

The fishery was not a pure open access one. The resource had been subject to management under the International Commission for the Northwest Atlantic Fisheries (ICNAF) since 1949. The harvests were, nonetheless, unsustainable. Moreover, the impact of the growth of the offshore harvesting upon the Newfoundland based inshore fishery was devastating. From the mid-1950s to the mid-1970s, inshore harvests declined by almost 80 per cent (Munro, 1980). The Northern Cod experience undermined Canadian confidence in ICNAF, and made the country a strong supporter of the move towards the coming EEZ regime.

Canada implemented its EEZ regime at the beginning of 1977. Its great prize under the new regime was Northern Cod, with some 95 per cent of the resource complex being encompassed by Canada's Atlantic EEZ. The Canadian government saw the Northern Cod resource having been seriously overexploited, while it had been under ICNAF governance. Its plan, consequently, was to engage in an extensive resource investment programme. The TAC would be set at an initial low of 170 thousand tonnes, not equal to zero to be sure, but well below F(x). The resource would steadily grow, and with it the TAC. The resource target goal was expected to be reached by the late 1980s. Once the resource target had been achieved, the TAC would be set at a conservative TAC of about 400 thousand tonnes, which implied a rate of fishing mortality of 0.2, i.e., about 20 per cent of the exploitable biomass. The TAC would remain at that level (allowing, of course, for resource fluctuations) thereafter. Prosperity would reign (Canada, 1982; National Research Council, 1999).

Within the Canadian EEZ, the DWFS fleets would be phased out, to be replaced by an all Canadian offshore fleet. Of the projected ongoing 400 thousand tonnes annual TAC, 95 per cent would be taken by Canadian inshore and offshore fleets. DWFS harvests would be restricted to the small amount of the resource outside of the Canadian EEZ (Canada, 1982).

Initially, the resource investment programme seemed to be proceeding according to plan. By the second half of the 1980s, however, there were signs that something was going wrong. The total harvests never exceeded 260 thousand tonnes per annum. While the offshore sector seemed to be doing well, the inshore sector was not prospering. After initially increasing, from 1977 onwards, inshore harvests went into a decline after 1982 (National Research Council, 1999). A Canadian government body, the Canadian Atlantic Fisheries Scientific Advisory Committee, was instructed to review stock assessment methods. Upon so doing, the Committee concluded that the Northern Cod resource had, in fact, been systematically overestimated since the implementation of the Canadian EEZ regime, and that, as a consequence, fishing mortality rates were at least double the target rate of fishing mortality. The Committee's conclusions were later confirmed by an independent Northern Cod Review Panel.

One reason that the resource managers had been misled is because the Northern Cod resource should have been one to which the Schaefer model was reasonably applicable, or so it was thought. From the Schaefer model, we have: $h = qEx$, which implies that the Catch Per Unit of Effort (CPUE) is: $h/E = qx$. This in turn implies that there is a linear relationship between the CPUE and x, implying further that CPUE is a good measure of stock abundance. There is clear evidence that marine biologists believed, at the time of the implementation of the Canadian EEZ regime, that there was a strong, although not perfectly linear, relationship between CPUE and stock abundance in the Northern Cod fishery (Munro, 1980, table 3–6, chapter 3; n. 62).

In the offshore sector, where reasonably accurate CPUE data were available, the CPUE appeared to be stable throughout the 1980s, thus leading to the comfortable belief that the resource was not under threat (Tsoa, 1996).

In response to the shocking revelations of the second half of the 1980s, the relevant scientists recommended a drastic reduction in the TAC, one which would have been sufficient to shut down the offshore sector (Roy, 1996). The Canadian produced and human capital operating in the Northern Cod fishery was highly non-malleable with respect to that fishery. The National Research Council states that the Canadian government was 'fearful of the economic disruption that such a draconian reduction in the TAC would entail' (National Research Council, 1999, p. 34).

Faced with the non-malleable produced and human capital in the fishery, and the resulting vehement industry resistance to draconian reductions in harvesting, the Canadian government adopted a compromise, which has been referred to as the 50 per cent rule. The Canadian government would, for a coming fishing season, look first at the most recent rate of fishing mortality, and then secondly at what the scientists were recommending as the 'safe' rate of fishing mortality. The government, the resource managers, would then take the mid-way point between the two as its target rate of fishing mortality for the coming season (Canada, 2002). As a 2001–2 Canadian government appointed independent panel on Atlantic fisheries states in its report, 'although the 50 per cent rule was not the only cause of the collapse [of Northern Cod], it was a significant factor' (Canada, 2002, p. 52).

While there were many factors leading to the Northern Cod collapse (Roy, 1996), management error, through systematic overestimates of stock size, played a significant role. Given the inherent and irreducible uncertainty in capture fisheries management, such management errors can never be eliminated. Having said this, resource management error was not—in, of, and by itself—sufficient to bring about the collapse of the resource. If the fisheries management regime had been resilient, if the resource managers had not been severely constrained by the presence of highly non-malleable produced and human capital in the fishery, disaster could have been averted. The drastic TAC reductions called for by the scientists could have been implemented.

As a final note, while the collapse of the Northern Cod fishery was particularly dramatic, the case was not unique. As Hilborn et al. (2001) inform us, there were several other capture fisheries, in which the management regime was devoid of resilience.

BIBLIOGRAPHY

There are many articles and books on domestic fisheries management. The reader should, in particular, look at Arnason (1990) on minimum information management in fisheries, Arnason (2007) on property rights and the efficiency of fisheries management regimes, Charles (2001, 2007) and Doyle et al. (2007) on sustainable fisheries management with stock uncertainty and costly capital adjustment, Dupont and Gordon (2007) on shadow prices for fishing quotas, Scott (2007) on the evolution of property rights in fisheries, Townsend et al. (2008) on Self Governance in Fisheries, Bastone and Sharp (2003) and Sharp and Batstone (2007) on minimum information management systems and ITQ fisheries and factor use and productivity change in rights based fisheries. A particularly good discussion of uncertainty and its impact upon fisheries management is to be found in Young et al. (1999).

NOTES

1. In passing, the management of EU fisheries can be regarded as the solution to a cooperative game played by all relevant member countries. There is a Common Fisheries Policy (CFP), which does not, however, encompass the Mediterranean Sea. The CFP is currently under review. In any event, we shall regard intra-EU fisheries as being essentially 'national' fisheries in nature.
2. Subject to uncertainty considerations. As we have already indicated, if the resource manager is risk averse, the manager could rationally opt for a management policy that would not maximize the *expected* economic returns from the fisheries.
3. We noted that the definition is akin to the economist's definition of general equilibrium—everything depends upon everything else.
4. An example is provided by the Arcto-Norwegian cod stock in the Barents Sea. The resource declined significantly in the mid-1990s. The resource managers realized after the fact that a significant cause of the decline was the decline in the low valued capelin resource. The cod prey upon capelin. The resource managers subsequently ensured that proper stock assessments of capelin were undertaken and that TACs for capelin were imposed, taking into account the consumption of capelin by cod (Armstrong et al., 2011).
5. Incentives that are perverse from society's point of view.
6. It is worth noting, in passing, that 'common pool' based perverse incentives are not unique to fisheries. They are to be found in many other 'common pool' situations, such as oil and gas, fresh water, air pollution, and radio spectrums (Libecap, 2007).
7. If the coastal state, as owner of the fishery resource, is able to implement its property rights effectively, then the 'common pool' nature of the fishery resource would vanish and the incentives of the fishermen would no longer be perverse. As we shall see at a later point, the state could, in theory at least, do just this through a judicious application of taxes. In practice, the coastal state property rights to capture fishery resources are seldom implemented effectively. Remnants of the 'common pool' nature of the fishery commonly linger on within the EEZ.
8. The vessel that wins the race gets the fish.
9. A major qualification is in order here. An implicit assumption that is being made here is that imposing the season by season TAC is costless. This is a heroic assumption. In reality, regulation of the fishery is a costly exercise, which can

affect the optimal stock level. Sutinen and Andersen (1985), applying the economics of crime to fisheries, argue that the greater the potential economic returns from the fishery, the greater will be the incentive of the fishermen to break the harvest regulation rules, and hence the greater will be the cost of enforcing the TAC. Thus, the optimal stock level will be below what it would be, if enforcing harvest regulations was a costless undertaking.

10. Fleet operating profits over the season are equal to: $(pq - b_{oper})KD$, but note that we have: $KD = Q/q$.
11. This is fully consistent with the simpler regulated open access model developed in chapter 2.
12. If $K_{ROA} = K_{opt}$, then there would be no deadweight loss. This exceptional case is that of a fishery just on the knife edge of economic feasibility.
13. An example is provided by the Pacific halibut fishery of British Columbia. In this fishery, D_{max} is approximately 240 days per year. By 1990, the actual D was equal to six days. The harvested fish was delivered to processors. Most of the fish then had to be frozen and processed over time, with predictable consequences for the quality, and thus the price received for the final product. The need to freeze the harvested fish reflected both the non-malleability of processing capital and the short run price inelasticity of harvested halibut (Casey et al., 1995). Note as well that what we are saying in summary is the following. Even if the cost of enforcing the TAC is zero, the resource rent from the fishery can be fully dissipated, due to the emergence of excess capacity and everything that follows from this.
    The discussion of the effects of a shortened fishing season does, however, raise another question. If shortened seasons lead to economic waste, then why not introduce regulations to spread out fishing effort over the maximum season length by, for example, by placing limits on the length of individual fishing trips—trip limits, a maximum catch quantity per trip, limiting the number of trips per month etc.? Such measures have in fact been introduced in several European and North American fisheries. The results have not been encouraging. A well documented example is the British Columbia groundfish trawl fishery, a complex multispecies fishery, to which reference will be made at a later point in the chapter. For the sake of the relevant fish processors, it was necessary to maintain a 12-month season. The resource managers introduced trip limits and monthly limits on the number of trips to achieve that goal. Fishing capacity nonetheless grew inexorably. Trip limits became shorter and shorter. The incentive to discard fish increased. Economic waste and the difficulties that the resources managers faced in enforcing the TACs increased steadily. The end result was that, although the 12-month season was maintained, the economic state of the fishery became, if anything, worse than that of the British Columbia Pacific halibut fishery, with its six-day seasons (Turris, 2000).
14. In other words, we now have to relax our assumption that enforcing the TAC is costless, and take into account the fact that the increasing capacity may drive the cost of enforcing the TAC to very high levels.
15. Unless the vessels have only a peripheral involvement in the fishery. Forcing such vessels to leave will do little to address the excess capacity problem.

16. Given the assumptions of the model, the entire loss due to excess capacity is captured by $\lambda(K_{ROA} - K_{opt})$. This loss is irreversible. Moreover, by assumption, there are no costs arising from the decreasing length of the fishing season. If we allow for positive short seasons costs, then the buybacks could indeed have positive benefits.
17. Berck and Perloff (1984) were the first to introduce the concept of rational expectations to fisheries. In their article they equate rational expectations with perfect foresight, for the sake of simplicity. We shall do the same.
18. Suppose, for the sake of argument, that the seasonal TAC = 10,000 tonnes, from which it follows that Y_T = 10,000 tonnes. Suppose further that the actual fleet capacity, expressed on a seasonal basis, is: Y_C = 15,000 tonnes. Then we would have $\theta = 0.5$, indicating that the actual fleet capacity exceeds the target fleet capacity by 50 per cent.
19. That is, $\theta_{2000} \approx \theta_{1970}$.
20. It will be recalled that the report, which gave rise to the first limited entry scheme in the British Columbia Pacific salmon fishery, was prepared by the Canadian economist Sol Sinclair. In the late 1970s, with the British Columbia Pacific salmon fishery limited entry scheme obviously encountering difficulties, Sinclair was invited by the Canadian government to prepare a review. In his second report, Sinclair recommended that the limited entry scheme be accompanied by a landings (harvest) tax (Sinclair, 1978). His recommendations, wise though they may have been, were ignored by the government.
21. Alternatively, one could impose a tax on effort. If set at the appropriate level, Maximum Economic Yield would be obtained.
22. In our discussion of extensions to the basic dynamic economic model of the fishery in chapter 5, we made a passing reference to the concept of 'marginal user cost'.
23. Suppose that the Marginal Stock Effect is zero and suppose that $x(t) < x^*$. The authorities would have to set the tax such that $\tau > p - c$. The instant that $x(t) = x^*$, τ would have to be adjusted so that $\tau = p - c$. If $x(t) > x^*$, the reverse would be called for. The administration of such a tax scheme would become impossible.

    There is one qualification, which we have to make. If unit harvesting costs are completely insensitive to the size of the biomass, but the fish is delivered to a processing sector which faces unit processing costs and/or a price for the finished product that is/are sensitive to the size of the biomass, the industry wide Marginal Stock Effect will be positive.
24. Could an alternative tax scheme, e.g., lump sum tax, be effective in these circumstances? There is no obvious answer to this question. Further investigation is required.
25. See Weitzman (2002).
26. While the SMCP demonstrated success, its success as a government monopsony proved to be short lived. In 1993–4, partly due to advice from the World Bank, the SMCP was privatized, with 65 per cent of the shares of the SMCP being sold to the industry and the national banks (Cunningham et al., 2005). The politics behind this move are not entirely clear, but Cunningham et al. (1995) and Cunningham

et al. (2005) imply that the fishing industry did not respond tamely to the resource rent extraction.
27. An example is provided by the British Columbia forest industry, in which some 95 per cent of the forest land is state owned. Pearse (see Commission on Pacific Fisheries Policy, 1982) had this example very much in mind.
28. Christy, for example, had serious doubts about the efficacy of IQs in developing coastal states. As will be seen, Christy was also an early proponent of TURFs, which he saw as being much more appropriate for developing coastal states than were IQs.
29. While the Third UN Conference on the Law of the Sea did not conclude its deliberations until 1982, the debate in the Conference on fisheries issues was effectively over by the mid-1970s. Several coastal states did, as a consequence, implemented their EEZ regimes well before 1982. Most European states and the three North American coastal states, for example, did so in 1977.
30. Ragnar Arnason, personal communication.
31. It is easy to gain the impression that IQs constitute a fisheries management instrument for developed coastal states only. While it is true that most of the states that have adopted IQs are developed, not all are. The developing coastal state, Namibia, for example, has a sophisticated ITQ scheme for its offshore fisheries (Manning, 2005). Other examples of developing coastal states employing IQs are Mexico, Morocco, and Mozambique (Arnason, 2007).
32. The theoretical foundation for this discussion rests upon a model developed by Kronbak and Lindroos (2006). Their game theoretic model can best be described as a multi-stage game: the fishermen playing either a competitive, or cooperative, game among themselves; the resource managers and fishermen playing a leader-follower game. We have talked in terms of PA analysis. As Mesterton-Gibbons (1993) has pointed out, the Principal-Agent relationship can be viewed essentially as a leader-follower game.
33. The existence of a TAC sharing mechanism can be thought of as a necessary, but not sufficient, condition for a stable cooperative fishermen game.
34. Specifically, the pieces of legislation are: Iceland, Fishery Management Act of 1988; New Zealand, Fisheries Amendment Act of 1986 (Connor and Shallard, 2010; Matthiasson and Agnarsson, 2010).
35. For NZ$80 million (Connor and Shallard, 2010).
36. See Connor and Shallard (2010) for the details of the industry reaction.
37. As such, they are similar to non-voting shares in a corporation.
38. Costello et al. (2008) report on an extensive empirical study, which they had undertaken, on fisheries under ITQ schemes versus those not under such schemes. According to the study, the former fisheries are much less prone to collapse than are the latter fisheries.
39. See Huppert (2007), for a discussion of IQ auctioning schemes in the American state of Washington. See, also Grafton (1995), for a discussion of rent capture in general in harvesting rights based fisheries.
40. Bruce Turris, personal communication.
41. See above, n. 13, para. 2.

42. Typically, the vessel owner would contact his/her quota broker by phone or radio, and find the quota awaiting him/her upon arrival in port (Bruce Turris, personal communication).
43. Return to our earlier discussion about an ecosystem approach to fisheries management. The multispecies ITQ schemes, just described, are fully in keeping with such an approach.
44. This is not to suggest that an ITQ scheme, in which the fishermen remain competitive, will have zero effect. If it possible to enforce the individual quotas, the scheme should indeed curb the race for the fish, which was seen, after all, as the original purpose of ITQs. The ongoing threat of the Prisoner's Dilemma will always be there, however. Having said this, are there any concrete examples of ITQ cooperative games to which one can point? The answer is yes. New Zealand with its quota owner associations is one example (Connor and Shallard, 2010). ITQ schemes off Canada's Pacific coast provide another example (Munro et al., 2009).
45. The *loco* species is closely related to abalone.
46. Uppada and Boddu Chinna Venkatya Palem.
47. Thus, the cooperative fishermen games are to be described as self-enforcing ones.
48. As described, the rights of the villages pertaining to the relevant fishery resources, come close to being de facto property rights, in contrast to what we said about IQs. Hence, the powers of the Principal, the state, are weak. One can conjecture that one of the motivations of the government is to assert property rights claims.
49. The season length went from 100 to 200 days (At Sea Processors Association, 2011).
50. See, for example, Cancino et al. (2007).
51. This is entirely in keeping with our earlier discussion of an ecosystem based approach to fisheries management. Recognition of the interaction between the relevant fish species and the aquatic environment leads to the recognition of irreducible uncertainty in fisheries management.
52. There is, in fact, nothing radical about this proposition. According to Walters and Martell (2004, p.181), 'the spatial organization [of fishery resources] can create a variety of powerful options for selectively protecting and exploiting fish, thus enhancing our ability to conserve stocks—such options have, in fact, long been a routine practice of fisheries management'.
53. One can argue that the Precautionary Approach, now embedded in international treaty law in the form of the 1995 UN Fish Stocks Agreement (UN, 1995, Article 6; Annex II), requires fishery managers to be risk averse.
54. If the portfolio holder is risk averse, the marginal utility arising from incurring risk is negative. Hence the portfolio holder will be willing to trade off some expected yield on the portfolio for greater safety. The reader is invited to turn to any basic textbook on money, banking, and the financial system.
55. See equation (3.20). This is assuming that the de facto private owners place an existence value on the resource equal to zero, which admittedly may not always be valid assumption.
56. The prominent American marine biologist Andrew Rosenberg once challenged one of the authors (Munro) to point to *any* instance after 1970 in which a decommissioning/buyback scheme was unanticipated by the industry. The author found himself incapable of rising to the challenge.

7 The Economic Management of Capture Fisheries at the International Level

In all of the analytical chapters up to this point, our focus has been exclusively upon fisheries at the domestic (intra-EEZ)/regional level. We assumed, moreover, that the fisheries within the EEZ of a coastal state were exploited exclusively by the fleet(s) of that coastal state. We must now recognize that many of the world's marine capture fisheries have significant international ramifications. Such fisheries present us with management issues that are considerably more complex than those we have encountered heretofore.

There are two critical facts of which we must be reminded. First, given the mobility of most capture fishery resources, the typical coastal state will find that some of the fishery resources encompassed by its EEZ cross the EEZ boundary, either into neighbouring EEZs, or into the adjacent high seas, or both. Secondly, the fishing fleets of coastal states do not all confine their operations to their respective coastal state's EEZ. In several instances, such fleets operate well beyond the waters of the home EEZs. The states having such fishing fleets have come to be referred to by the FAO and others as distant water fishing states (DWFSs).[1]

Under the heading of the management of international capture fisheries, we shall examine two issues, one major, the other, by comparison, minor. The major issue, not yet resolved at the time of writing, is the economic management of internationally shared fish stocks. An internationally shared fish stock is to be defined as a fish stock that is exploited by two or more states (or entities),[2] and which crosses an EEZ boundary, or is a discrete high seas stock, to be found exclusively in the high seas

The minor issue involves the economics of the relations between and among coastal states and DWFSs. The issue revolves around the fact that one, or more, DWFS(s) may seek access to the EEZ of a particular coastal state. Thus, for example, DWFSs operate extensively in the tuna-rich EEZs of the Pacific Island States of the western and central Pacific. The DWFSs do so under so-called access agreements. Are coastal states compelled to grant DWFSs access to their EEZs? If the answer is no, then what possible reason can a coastal state have for granting a DWFS access to its EEZ? These, and related questions are the ones that we shall be addressing under the heading coastal state–DWFS relations.

In the chapter, we shall do the following:

(i) Provide an overview of internationally shared fish stocks. We shall find that the two most important categories are transboundary and straddling fish stocks (7.1).
(ii) Examine the legal framework surrounding the management of these stocks (7.2).
(iii) Discuss the economic management of transboundary fish stocks (7.3–7.7).
(iv) Discuss the economic management of straddling fish stocks (7.8–7.10).
(v) Examine the economic relations between coastal states and distant water fishing states (7.11).
(vi) Conclude by discussing the links between fisheries management at the national level, and at the international level.

We turn first to an overview of internationally shared fish stocks.

7.1 Internationally shared fish stocks: an overview

In chapter 1, we referred briefly to the categories of internationally shared fish stocks. To be more specific and detailed, the FAO sets out four categories of internationally shared fish stocks, with the categories being:

(a) transboundary fish stocks—fish resources crossing the EEZ boundary of one coastal state into the EEZ(s) of one or more neighbouring coastal states;
(b) highly migratory fish stocks. The species of fish deemed to be highly migratory are set out in Annex 1 of the 1982 UN Convention (UN, 1982, Annex 1). They consist largely of tuna, and closely related species. Since the species are highly migratory they are to be found, both within the coastal state EEZ and in the adjacent high seas;
(c) straddling fish stocks—all capture fishery resources, other than highly migratory fish stocks, to be found both within the coastal state EEZ and in the adjacent high seas;[3]
(d) discrete high seas fish stocks—fish stocks to be found exclusively in the high seas (UN, 1982; Munro et al., 2004).

Two comments are in order. The first is that categories (b) plus (c) and category (a) are not mutually exclusive. There are numerous highly migratory and straddling fish stocks that are also transboundary in nature.

The second comment is that the distinction between highly migratory and straddling stocks reflects bargaining within the Third UN Conference on the Law of the Sea, 1973–82, rather than the world of marine biology and economics. The distinction is not based on biology. Marine biologists have pointed out that there are fish stocks, such as Chilean jack mackerel, which fall into the straddling fish stock category, that are as highly migratory as those listed in Annex 1 of the 1982 UN Convention (Hilborn and Sibert, 1988). The economics of the management of highly migratory species is identical to that of the management of straddling stocks.[4] We shall for our purposes, therefore, merge the two and talk in terms of straddling fish stocks broadly defined. We are thus left with, for our purposes, three categories of internationally shared fish stocks:

A. Transboundary fish stocks
B. Straddling fish stocks (broadly defined)
C. Discrete high seas fish stocks

as illustrated in Figure 7.1

The next question to be raised is how significant these resources are in terms of world capture fisheries. The answer is very. As mentioned in chapter 1, the FAO estimates that as much as one third of world marine capture fishery harvests are based upon internationally shared fish stocks (Munro et al., 2004). It is not for nothing that the FAO states the following: '... the effective management of [internationally] shared fish stocks stands out as one of the great challenges towards achieving long-term sustainable fisheries' (cited in Munro et al., 2004, p. iv).

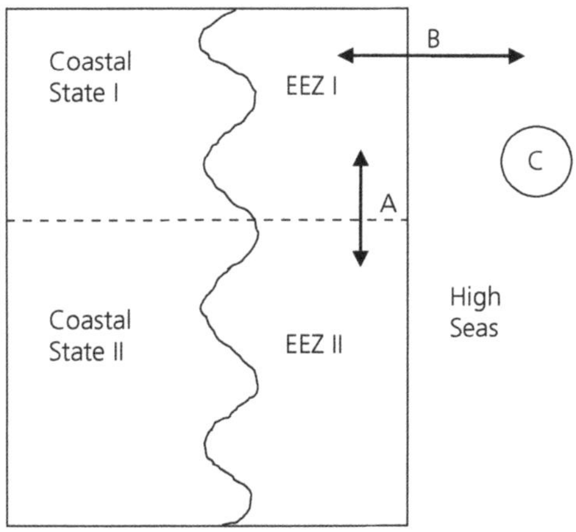

Figure 7.1. Internationally shared fish stocks
A. Transboundary fish stocks
B. Straddling fish stocks
C. Discrete high seas fish stocks

Prior to examining the economics of the management of internationally shared fish stocks, we must first look at the legal framework for the management of these resources established by international treaty law. If the legal framework is ignored, our economics will almost certainly prove to be wrong.

7.2 The economic management of internationally shared fish stocks: the legal framework

One can find examples of internationally shared fish stocks going back to the early history of ocean capture fisheries. The management of these resources did not become a major issue, however, to be studied by biologists, economists, and legal experts, until the coming of Extended Fisheries Jurisdiction (EFJ)—the advent of the Exclusive Economic Zone (EEZ) regime. There are, as a consequence, two post EFJ international treaties, and two only, that provide the basic legal framework for the management of internationally shared capture fishery resources. The treaties are: the 1982 UN Convention, and (as noted in chapter 1), the 1995 UN Fish Stocks Agreement[5] (Munro et al., 2004).

With respect to transboundary fish stocks (category A), the 1995 UN Fish Stocks Agreement is irrelevant. The 1982 UN Convention is relevant, but says little. Article 63(1) imposes a duty upon coastal states sharing a transboundary fish stock to come together and attempt to negotiate a cooperative resource management arrangement (UN, 1982, Article 63(1)). Importantly, however, the 1982 UN Convention does not impose a duty on these states to reach a successful conclusion to the negotiations. If the relevant coastal states do not reach a successful conclusion to their negotiations, then each coastal state is to manage the segment of the fish stock within its EEZ in accordance with the provisions laid down by the Convention pertaining to the rights and duties of a coastal state managing fishery resources within the EEZ (Munro et al., 2004, p. 9). We shall refer to this as the default option.

The legal framework underlying the management of straddling fish stocks is much more complex. To begin, there are two parts of the 1982 UN Convention relevant to these fish stocks: Part V, the Exclusive Economic Zone (Articles 55–75) and Part VII, the High Seas (Articles 86–120) (UN, 1982).

Straddling fish stocks are open to exploitation by both coastal states and by distant water fishing states. DWFSs cannot exploit the intra-EEZ segments of the stocks without coastal state permission, but they face no such restrictions in exploiting the high seas segments of the stocks.

The 1982 UN Convention, once again, admonishes fishing states exploiting this category of shared fish stocks to attempt to cooperate in the management of the resources. Beyond this, however, the drafters of the 1982 UN

Convention faced a difficult problem with respect to the segments of the straddling stocks in the adjacent high seas.

Under Article 56 of the 1982 UN Convention, the coastal states, as we have now noted several times, are given what amounts to property rights to the fishery resources within the EEZ. As far as straddling stocks are concerned, coastal states saw their intra-EEZ property rights to these resources being threatened by unrestrained exploitation of the high seas segments of the resources by DWFSs. DWFSs, on the other hand, feared 'creeping jurisdictionalism' on the part of the coastal states. Not content with their massive gains at their expense, maintained the DWFSs, the coastal states were ongoing in their attempt to extend coastal state jurisdiction, de facto, to the remaining high seas (Munro et al., 2004, p. 36, n. 19; UN, 1982).

The compromise reached is to be found in Part VII of the 1982 UN Convention, the High Seas. Part VII commences by proclaiming the freedom of the seas—the high seas are open to all states (certainly including DWFSs) (Article 87). The freedom granted to DWFSs to fish on the high seas is, however circumscribed by Articles 116–120. Article 117 states that all those fishing on the high seas shall cooperate with the other relevant fishing states for the purpose of conserving the fishery resources. Furthermore, those fishing on the high seas are to do so in recognition of the interests of affected coastal states (Article 116(b)) (UN, 1982; Munro et al., 2004).

The compromise proved to be ineffective. Part VII of the 1982 UN Convention leaves unclear the rights and duties of coastal states, as opposed to those of DWFSs, to the high seas segments of straddling stocks (Miles and Burke, 1989). At the close of the Third UN Conference on the Law of the Sea, in 1982, this weakness of Part VII of the 1982 UN Convention really did not seem to matter all that much. The expressed fears of the coastal state did not seem to be firmly based. It will be recalled from chapter 1 that, at the time, it was believed that only10 per cent of the commercially exploitable marine capture fishery resources were to be found in the remaining high seas (Alexander and Hodgson, 1975). The ensuing decade was to demonstrate that this sanguine view of the world of marine capture fisheries was wholly unfounded.

In the years following 1982, case after case of overexploitation of straddling stocks emerged. A well publicized example is that of Alaskan pollock, which historically had been the largest single species exploited in the North Pacific. Substantial concentrations of the species are to be found in the Bering Sea.

Following 1982, the two relevant coastal states, Russia and the United States, established EEZs on opposite sides of the Bering Sea. The two EEZs did not, however, encompass the entire Bering Sea. In the middle of the Bering Sea there remained a significant high seas enclave, to become known as the Doughnut Hole. Since the Alaskan pollock resources within the Doughnut Hole were subject to exploitation by DWFSs, Bering Sea Alaskan pollock was, and is, clearly a straddling stock.

Articles 116 and 117 of the 1982 UN Convention to one side, cooperative management of the Alaskan pollock resources within the Doughnut Hole was non-existent. In the words of the FAO, the pollock resource within the Doughnut Hole were not just overexploited, they were plundered (FAO, 1994). In 1992, Russia and the United States persuaded the four major DWFSs exploiting the Bering Sea pollock resources[6] to join them in declaring a 'temporary' harvest moratorium. The 'temporary' harvest moratorium remains in place at the time of writing (Munro et al., 2004).

The emerging capture fishery resource management disaster compelled the UN to take action, as discussed briefly in chapter 1. The result was the 1993–5 UN Conference on Straddling Fish Stocks[7] and Highly Migratory Fish Stocks, popularly referred to as the 1993–5 UN Fish Stocks Conference, which did in turn give rise to the 1995 UN Fish Stocks Agreement.[8] As emphasized in chapter 1, the purpose of the Agreement is not to supplant any part of the 1982 UN Convention, but is rather designed to buttress the 1982 UN Convention.[9]

The heart of the 1995 UN Fish Stocks Agreement lies in its call for straddling stocks (broadly defined) to be managed through Regional Fisheries Management Organizations (UN, 1995, Part III). A Regional Fisheries Management Organization (RFMO) can be defined as an 'intergovernmental fisheries organisation—that has competence to establish fisheries conservation and management measures' (Lodge et al., 2007, p. xviii). As such, RFMOs are to be distinguished from regional fisheries bodies, which generally have no more than a consultative or advisory role.

Several RFMOs were in fact already in existence in 1995. Many more have come into existence since the 1995 UN Fish Stocks Agreement was available for signing. For our purposes we can, at the time of writing, put the number at about 15.[10] Examples are provided by the Northwest Atlantic Fisheries Organization (NAFO), the Northeast Atlantic Fisheries Commission (NEAFC), the Inter-American Tropical Tuna Commission (IATTC), the International Convention for the Conservation of Atlantic Tunas (ICCAT), and the Western and Central Pacific Fisheries Commission (WCPFC).

A given RFMO is to bring together coastal states and DWFSs having a 'real' interest in the relevant fishery resource, or resources. Only those states that are members of the RFMO, or which agree to abide by the management regime of the RFMO, shall be allowed to fish in the area under the jurisdiction of the RFMO (UN, 1995, Article 8(4)).

Provision is then made for states that were not 'charter' members of an RFMO, but which at a later stage develop a 'real' interest in the fishery resources under RFMO jurisdiction, and wish to become a member of the RFMO. Under the 1995 UN Fish Stocks Agreement the 'charter' members of an RFMO must give consideration to such prospective 'new members'. New members cannot be automatically barred (UN, 1995, Article 8, Articles 10 and 11).

There are several legal issues that arise, pertaining to RFMOs, which are of direct relevance to the economics of the management of the fishery resources encompassed by the RFMOs. These are:

(i) Effective area of management. Article 3 of the 1995 UNFSA states that '... unless otherwise provided, this Agreement [and hence RFMO jurisdiction] applies to the conservation and management of straddling fish stocks and highly migratory fish stocks beyond areas of national jurisdiction [i.e., beyond the outer boundaries of the EEZs]' (UN, 1995, Article 3). In other words, RFMO jurisdiction is to be confined to the high seas—unless otherwise provided. Article 7 of the Agreement, however, states explicitly that the management of the portion of a straddling stock (broadly defined) in the high seas adjacent to the EEZ must be compatible with the management of the portion of the stock within the EEZ. While there is nothing in the 1995 Agreement explicitly permitting an RFMO to influence the intra-EEZ management of the stock, there is also nothing in the Agreement explicitly preventing such influence. Thus, depending upon relative bargaining strength, it is quite possible for DWFS members of an RFMO to influence the intra-EEZ management of the stock (Örebech, Sigurjonsson, and McDorman, 1998). For our purposes, when we talk of RFMO management of a straddling stock, we shall think in terms of RFMO management of the entire resource. The dominant influence over the management of the straddling stock may rest with the coastal state member(s) of the RFMO; with the DWFS members; or the power over management may be shared between the two groups.

(ii) Right to fish. As we have observed, under the terms of the 1995 UNFSA, which has come into force as international treaty law, only those states that are members of an RFMO, or agree to abide by the resource management provisions of the RFMO, are to be allowed to fish in the high seas under jurisdiction of the RFMO. A fundamental principle in international law is that a treaty is binding only on parties that have ratified the agreement,[11] unless the treaty has achieved the status of what legal authorities deem to be customary international law.[12] It is not at all clear that the 1995 UNFSA has achieved the status of customary international law (Munro, 2010a). Since states, which have yet to ratify the Agreement, are not clearly bound by the terms of the Agreement, this brings with it the threat of so-called unregulated fishing,[13] which we shall view as free riding. Free riding stands as a serious threat to the stability of the regime.[14]

(iii) 'Real' interest and new members. Two interrelated issues that are unresolved are (a) what, in fact, constitutes 'real' interest in fishery resources under RFMO jurisdiction; and (b) the terms and conditions under which prospective new members are to be admitted to the RFMO. It will be shown that (b) carries with it the threat of implied free riding. Particular attention will be devoted to (b) under the heading of the new member problem.

With the underlying legal framework in place, we at last turn to the economics of the management of internationally shared fish stocks. The three categories of such stocks are in ascending order of difficulty in terms of management. We commence with the easiest of the three, transboundary fish stocks.

7.3 The economics of the management of transboundary fish stocks: an introduction

We commence by observing that, with very few exceptions, there exists a strategic interaction between and among the coastal states sharing a transboundary fish stock. Suppose that the given transboundary fish stock is shared by but two states: coastal state I and coastal state II. The harvesting of the resource by I will have an impact upon the harvesting opportunities available to II and vice versa—hence the strategic interaction. As a consequence, economists have found themselves compelled to bring to bear the theory of strategic interaction, which we know more commonly as the theory of games. The result is that the economics of the management of all classes of internationally shared fish stocks is a blend of the dynamic economic model of the fishery, which we have encountered thus far, and the theory of games.[15]

We introduced the theory of games in a rather cursory manner in chapter 6, when discussing domestic fisheries management. In so doing, we made a distinction between competitive (also known as non-cooperative) games, and cooperative games. Competitive games prove to be straightforward; not so cooperative games. In cooperative games, the number of players matters a great deal. Once the number of players exceeds two, many complications arise.

We shall begin our examination of the management of transboundary stocks by assuming that the number of players is equal to only two.[16] Following this, we shall then ask what complications arise when the number of players is three or more.

We have now seen that the 1982 UN Convention calls upon the two states to negotiate on the cooperative management of the transboundary fish stock, but does not require the two coastal states to reach an agreement. Thus, the default option is that the two coastal states do not cooperate in the management of the resource. Rather, each coastal state commits itself to managing that segment of the stock within its EEZ to the best of its ability.

We are now faced with two questions:

(a) What are the consequences of the two coastal states adopting the default option and not cooperating in the management of the shared fishery resource?
(b) What conditions must be met, if a cooperative management arrangement between the two coastal states is to be stable through time?

Obviously, if the answer to question (a) is that the negative consequences of the two states not cooperating in the management of the resource are negligible, then question (b) becomes irrelevant.

In turning to question (a), we draw upon the theory of competitive (non-cooperative) games, with emphasis upon the model developed by Nash (1951).

7.4 Non-cooperative management of transboundary fish stocks

The analysis of non-cooperative management of transboundary fish stocks most relevant to our discussion is that developed, in two stages, by Clark (1980, 1985). In the first stage, Clark combines the Schaefer based dynamic economic model of the fishery with Nash's model of a two player competitive game (Nash, 1951; Clark, 1980). We look for the Nash equilibrium in the competitive game, continuing to suppose that the fishery resource is shared by coastal states I and II. A Nash equilibrium is achieved when player I (coastal state I) has no incentive to change its strategy, given the strategy adopted by II (coastal state II), and vice versa. In his 1980 model, Clark assumes that I is identical to II, except with respect to fishing effort costs. It is assumed that I has lower fishing effort costs than II. The players are thus asymmetric.[17]

Both players face a common perfectly elastic demand for harvested fish, and perfectly elastic supplies of the labour services, produced capital services, etc. comprising fishing effort. Perfect malleability of produced and human capital prevails. We can thus denote the unit fishing effort costs as b_{I} and b_{II} respectively. By assumption, we have $b_{\mathrm{I}} < b_{\mathrm{II}}$.

Let x_{I}^{*} and x_{II}^{*} denote the optimal biomass levels perceived by I and II respectively, and let x_{I}^{BE} and x_{II}^{BE} denote the Bionomic Equilibrium biomass levels perceived by I and II respectively. Since the now standard Schaefer based dynamic economic model of the fishery provides the foundation for the Clark (1980) model, the Marginal Stock Effect is positive in the two coastal state fisheries. It is thus easy to demonstrate that: $x_{\mathrm{I}}^{*} < x_{\mathrm{II}}^{*}$; $x_{\mathrm{I}}^{BE} < x_{\mathrm{II}}^{BE}$.

In the Clark (1980) model, the non-cooperative feedback strategies are expressed in terms of optimal rates of fishing effort. We use the superscript N to denote non-cooperative. For player I we have:

$$E_{\mathrm{I}}^{N}(x)=\begin{cases} E_{\mathrm{I}}^{\max} \text{ for } x > min(x_{\mathrm{I}}^{*}, x_{\mathrm{I}}^{BE}) \\ F(x)/x \text{ for } x = min(x_{\mathrm{I}}^{*}, x_{\mathrm{II}}^{BE}) \\ 0 \text{ for } x < min(x_{\mathrm{I}}^{*}, x_{\mathrm{II}}^{BE}) \end{cases} \tag{7.1a}$$

and for II we have:

$$E_{\mathrm{II}}^{N}(x)=\begin{cases} E_{II}^{max} \text{ for } x > x_{\mathrm{II}}^{BE} \\ 0 \quad \text{for } x \leq x_{\mathrm{II}}^{BE} \end{cases} \tag{7.1b}$$

The implication of (7.1a) and (7.1b) is that, if $x_{\mathrm{I}}^{*} < x_{\mathrm{II}}^{BE}$, the high cost player, II, will be hounded out of the fishery. We end up with the fish stock stabilized optimally at $x(t) = x_{\mathrm{I}}^{*}$. The fishery resource is then transboundary in name only. The default option carries with it no negative consequences, and cooperative resource management is beside the point.

If, on the other hand, $x_{\mathrm{I}}^{*} > x_{\mathrm{II}}^{BE}$, then the fish stock will be driven down to x_{II}^{BE}, a decidedly non-optimal, Prisoner's Dilemma, type of outcome. Clark then proves that, if the players are symmetric, i.e., $b_{\mathrm{I}} = b_{\mathrm{II}}$, the fishery resource will be driven down to common Bionomic Equilibrium level, $x_{\mathrm{I}}^{BE} = x_{\mathrm{II}}^{BE}$—a very unsatisfactory outcome indeed (Clark, 1980). We thus conclude from the Clark (1980) model that, while a Prisoner's Dilemma type of outcome is not guaranteed under the default option, such an outcome is definitely likely.

In the second stage of his analysis Clark (1985) alters a few of the assumptions used in his 1980 model. By so doing, he is able to present a lucid example of the Prisoner's Dilemma in the context of transboundary fisheries, complete with payoff matrices.

To begin, he assumes that the two players are symmetric, so that we have $b_{\mathrm{I}} = b_{\mathrm{II}}$. Importantly, he also assumes that harvesting costs are independent of the size of the biomass (given that $x > 0$). Thus the harvest production function for both I and II is: $h = qEx^0$ (Clark, 1985; Munro et al., 2004, appendix).[18]

As in his 1980 model, Clark assumes that the common demand for harvested fish is perfectly elastic, and that the supply of fishing effort in I is perfectly elastic, as it is in II. Produced capital and human capital are both perfectly malleable. The price of harvested fish is thus a constant, as are unit harvesting costs in I and II. We have:

$$c_{\mathrm{I}} = \frac{b_I}{qx^0} = c_{\mathrm{II}} = \frac{b_{\mathrm{II}}}{qx^0} = c.$$

As a final simplification, it is assumed that $p - c = 1$.

Given the symmetry of the players, the optimal biomass levels as perceived by I and II are: $x_{\mathrm{I}}^* = x_{\mathrm{II}}^* = x^*$. Suppose that x^* is positive and that we commence with $x(0) = x^*$. The perceived global economic return from the fishery, with $x(t) = x^*$, is simply: $F(x^*)/\delta$,[19] where δ is the common social rate of discount.

One option open to each of the two coastal states is to deplete the resource. Since harvesting costs are independent of the size of the resource, the resource could be reduced to zero. If we commence at $x(0) = x^*$, the global economic return from reducing the resource to zero, given our assumptions would be equal to: x^*. If x^* really is in fact positive, this implies that $x^* < F(x^*)/\delta$.

Coastal state I has two possible strategies: deplete the fishery resource or conserve it. If coastal state I adopts the deplete strategy, while coastal state II follows the conserve strategy, it is assumed that I can deplete the resource so quickly that II receives nothing. What holds true for I holds true for II, which faces the same set of strategies.

Next we assume that the two coastal states have equal bargaining strength and harvesting power. Hence, if the two follow the same strategies, they will share the economic returns from the fishery equally.

The payoff matrix can be expressed as follows (Clark, 1985; Munro et al., 2004):

Coastal state I/ Coastal state II	Conserve	Deplete
Conserve	$\frac{F(x^*)}{2\delta}, \frac{F(x^*)}{2\delta}$	$0, x^*$
Deplete	$x^*, 0$	$x^*/2, x^*/2$

If both I and II conserve, each will receive: $F(x^*)/2\delta$. If both I and II deplete, each will receive $x^*/2$. Since $\frac{x^*}{2} \leq F(x^*)/2\delta$, it follows that, if I and II could communicate and cooperate effectively, then the fishery resource would be conserved.

Now suppose that there is no communication, no cooperation, between I and II. Assume to begin with that $x^* > F(x^*)/2\delta$, and consider I. If II should follow the conserve strategy, I will receive $F(x^*)/2\delta$, if it conserves, and x^*, if it depletes. If II should follow the deplete strategy I would receive 0, if it follows the conserve strategy, and $x^*/2$ if it follows the deplete strategy. For I, the deplete strategy is obviously the dominant strategy. What holds true for I, holds true for II, and we end up with a deplete, deplete outcome. This is a perfect Prisoner's Dilemma case (Clark, 1985, pp. 151–3).

Next let us change the assumption and suppose to begin with that $x^* < F(x^*)/2\delta$. Coastal state I would be better off conserving, if coastal state II follows the conserve strategy. It is possible that we would end up with a conserve, conserve outcome (tacit cooperation). But, such an outcome is decidedly unstable. Suppose that I, guessing that II will conserve, adopts the conserve strategy, but is then proven wrong. II depletes, with the result that I is left with 0, and, to use an old expression, I ends up as the 'goat'.

There is, in the theory of games, a famous criterion for selecting strategies in non-cooperative games, which is particularly applicable when one's opponent is both aggressive and unpredictable. It is referred to as the maxmin criterion. The criterion states that one should look at the worst possible outcome from following each strategy, and then compare. Choose the strategy having the least worst outcome. In the case under discussion, $x^* < F(x^*)/2\delta$, the payoff matrix tells us that the worst outcome for I, if it follows the conserve strategy, is that it will receive 0. The worst outcome for I, if it follows the deplete strategy, is that it will receive $x^*/2$. An application of the maxmin criterion would lead I to choose the deplete strategy. If I and II each regard the other as aggressive and unpredictable, we can look forward to a deplete, deplete outcome. We might refer to this as the imperfect Prisoner's Dilemma case (Bacharach, 1976; Clark, 1985, Munro et al., 2004)

While the theory is straightforward, it does seem slightly unreasonable. If the coastal states do not cooperate, then surely the coastal states, if responsible, will manage their respective shares of the resource reasonably well, if not optimally. It is thus entirely appropriate to ask, if there is any evidence that the theory of competitive fishery games has predictive power. The answer is yes. Two examples will be put forward at this point. Others will emerge when we come to discuss the management of straddling stocks.

7.4.1 PACIFIC SALMON AND THE SOUTH TASMAN ORANGE ROUGHY FISHERY

The two cases selected involve Canada and the United States on the one hand, and Australia and New Zealand on the other. In both cases, the two relevant coastal states are developed and have substantial fisheries management expertise. Furthermore, in each of the two cases, the coastal states do not have significant cultural differences. Furthermore, there is both extensive Canadian–American cooperation, and extensive Australian–New Zealand cooperation in non-fisheries areas. Nonetheless, the Prisoner's Dilemma manifests itself in the two shared fisheries.

Turning first to Pacific salmon, to which we have referred in other chapters, the several Pacific salmon species together did, historically, constitute the most important fishery resource to the fishing industries of the American states of Washington, Oregon, and Alaska, and to the Canadian province of British Columbia. The Pacific salmon species, being anadromous, are produced in fresh water, make their way to the ocean, where they may live for up to four years, and then return to their fresh water habitats to spawn and die. The Pacific salmon are normally harvested as they approach the rivers and streams that will carry them to their fresh water habitats. See Figure 7.2.

It is inevitable that American salmon fishermen will catch ('intercept') some Canadian produced salmon; it is equally inevitable that Canadian salmon fishermen will catch some American produced salmon. Hence, the resource is unquestionably shared. It is also transboundary in nature.[20]

American–Canadian cooperation in the management of Pacific salmon commenced in the 1930s, focusing on the Canadian Fraser river, south of Vancouver. The Fraser river was then regarded as the most important salmon fishery in the Western hemisphere (Logan, 1974). It gradually became recognized, in the decades after the Second World War, that this narrowly focused cooperation was inadequate. In 1970, the two coastal states agreed to negotiate over the cooperative management of all Pacific salmon stocks from northern California to the Gulf of Alaska (Munro et al., 1998).

The negotiations proved to be long and arduous, with a treaty establishing the cooperative management regime not being ratified until 1985. What drove the negotiators on were manifestations of the Prisoner's Dilemma. The first

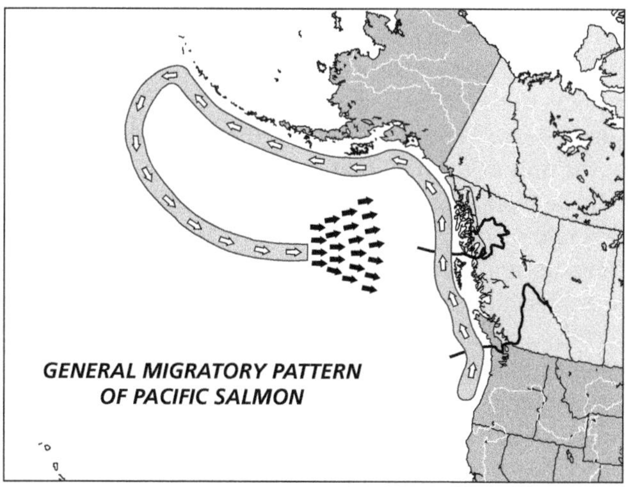

Figure 7.2. Pacific salmon—Northeast Pacific

Source: Canada 1997. Department of Fisheries and Oceans, 'Pacific Salmon Treaty: Moving Towards Equity and Conservation', paper prepared by Bud Graham, Director of Fisheries Management, Department of Fisheries and Oceans, Pacific Region, Vancouver.

manifestation pertained to enhancement projects. While the Pacific salmon stocks can be enhanced by reducing harvesting, they can also be built up through enhancement projects focusing on artificial improvements in salmon rivers, streams, and lakes designed to increase salmon survival, for example fish ladders and hatcheries. There was evidence at the time that there were opportunities for enhancement projects in both Canadian and American salmon rivers, which, if undertaken, would have proven to be mutually beneficial. Without a cooperative resource management regime in place, however, neither coastal state was willing to undertake enhancement projects for fear that the other would free ride on its efforts (Munro et al., 1998).

The second manifestation of the Prisoner's Dilemma took the form of outright 'fish wars', which have been described by a legal expert as the deliberate overexploitation of a fishery resource for the purpose of denying harvest opportunities to the other party or parties (coastal states, in this case) (Jensen, 1986).[21] When a treaty was finally ratified in 1985—the Canada–United States Pacific Salmon Treaty (Treaty, 1985)—the same legal expert commented that the Treaty could be best described as a 'peace treaty memorialising the end of the Pacific salmon war' (Jensen, 1986, p. 372).

As we shall see at a later point, the Treaty could have been better described as an armistice, rather than a peace treaty. The Treaty seized up in the early 1990s, with a reversion to competition and the re-emergence of Prisoner's Dilemma in the form of particularly virulent 'fish wars' (Miller et al., 2001).

The second case involves an orange roughy stock off Australia. Orange roughy is a deep water groundfish, which is normally harvested during its spawning phase when it is concentrated on seamounts.[22] Furthermore, the resource is very slow growing.[23] It is thus highly vulnerable to overexploitation.

The Australians became aware of the resource in 1997. Strictly speaking it is a straddling stock, in that it is to be found both in the Australian EEZ and the adjacent high seas (see Figure 7.3). At the time of discovery, the Australians believed that the only other state that would have an interest in exploiting the resource would be their neighbour, New Zealand. Hence, it was viewed as a de facto transboundary stock.

The Australians approached their New Zealand counterparts with the objective of establishing a cooperative fisheries management arrangement Given that both coastal states are developed, have exemplary records of domestic fisheries management, and have a common cultural background, establishing a cooperative fisheries management agreement should have been straightforward. An agreement for the cooperative management of the orange roughy resource was in fact duly signed in late 1997 (Munro, 2009).

The implementation of the cooperative fisheries management did, however, prove faulty. The cooperative fishery game degenerated into a competitive game, with recriminations on both sides. The Prisoner's Dilemma emerged in full force.

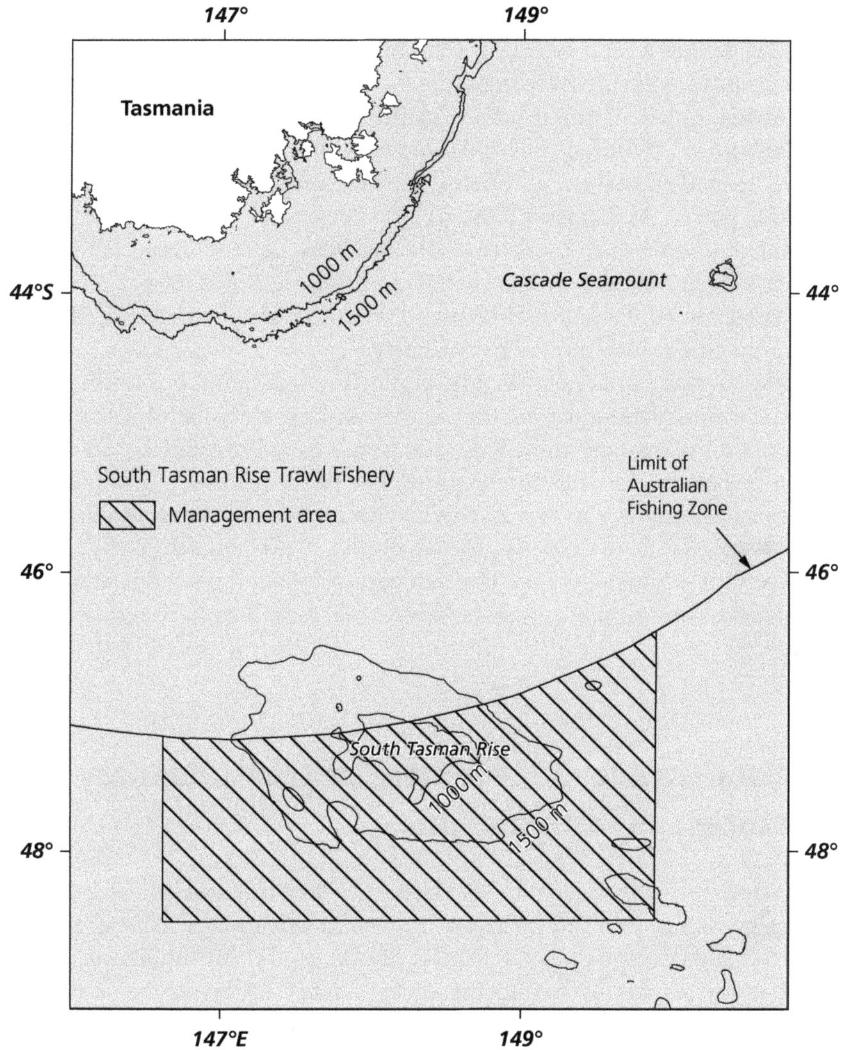

Figure 7.3. The orange roughy trawl fishery of the South Tasman Rise, Australia

Source: The Conservation and Management of Share Fish Stocks: Legal and Economic Aspects, FAO Fisheries Technical Paper 465, Rome 2004–Figure 3. p.52. Reproduced with kind permission of the Food and Agriculture Organization of the United Nations. Munro et al. (2004).

Ultimately cooperation was re-established in 2000. One can think of there having been a set of repeated Prisoner's Dilemma games leading ultimately to cooperation. The achievement of cooperation had, however, come too late. The orange roughy resource had been heavily overexploited during the competitive game period. The fact that the resource is exceptionally slow growing means that, at the time of writing, the recovery of the resource to commercially exploitable levels is very uncertain (Munro, 2009). The Clark (1985) model has a high degree of relevance to this case.

The point to be made in bringing forth these two brief case studies is not that Canada and the United States on the one hand, or that Australia and New Zealand on the other hand, should be held up for censure. Rather the point to be made is that the competitive fishery game model does indeed have predictive power. As Munro (2009, p. 19) states, with respect to the Tasman Rise orange roughy case, '... if Australia and New Zealand could fall into the Prisoner's Dilemma hole, then no set of countries sharing a fishery resource is safe'. Replace Australia and New Zealand with Canada and the United States, and the statement loses none of its validity.

Cooperation, then, does matter. In supporting this claim, it is not necessary for us to be able to demonstrate that every transboundary fish stock subject to non-cooperative management will, over time, be overexploited. All that we have to be able to do is to demonstrate that the risk of such an outcome is significantly high. It can be asserted, without fear of contradiction, that evidence of significant risk is overwhelming. Thus question (b), on the conditions that must be met if a cooperative fish resource management arrangement is to prove to be stable over time, must be addressed.

7.5 Cooperative management of transboundary fish stocks: an introduction

In the cooperative management of transboundary fish stocks, the cooperating coastal states start off with an important advantage. This is that the property rights to the relevant fishery resources are not ambiguous. If the fishery resources wholly within the coastal state EEZ are to be seen as the property of the coastal state, the fish stocks shared by neighbouring coastal states can be seen as being owned on a shared basis (McRae and Munro, 1989b). This stands in stark contrast to the cooperative management of straddling fish stocks, where the property rights to important segments of these stocks are nebulous. As we shall, see these nebulous property rights lead to chronic problems of free riding, i.e., the benefits of cooperation being enjoyed by players outside of the cooperative game. While in theory, free riding can occur in the cooperative management of transboundary fish stocks, in practice it is difficult to find significant evidence of such free riding (Munro et al., 2004).[24]

Having said this, we commence by retaining our assumption that the transboundary fish stock in question is shared by only two coastal states.

We then follow by relaxing this assumption and asking what the consequences are of the fishery resource being shared by three or more coastal states.

In any event, we now bring to bear the theory of cooperative games, relying most heavily upon Nash's (1953) theory of two-person cooperative games, because it is exceptionally useful for expositional purposes.[25] Nash's basic underlying assumption is that the players, in seeking to cooperate, are driven solely by self-interest. This gives rise to a fundamental condition that must be met, if the solution to the cooperative game (assuming that it exists) is to be stable through time.

This fundamental condition is most commonly referred to as the Individual Rationality Constraint. It states that each and every player, at each and every point in time, must expect a payoff from the cooperative game not less than what it could expect to receive under competition. Once stated, the condition seems almost stunningly obvious. Obvious or not, it is often ignored in the real world.

The theory of cooperative games is in essence a theory of bargaining. In the fisheries cooperative game, the players must obviously bargain over shares of the returns from the fishery. This raises an immediate question: what form should these shares take? It would seem obvious that the shares should take the form of shares of the TAC, or the equivalent, over time. This is indeed the form that the shares normally take in the real world of cooperative fisheries management.

We as economists, however, can argue that the focus should rather be on the sharing of the net economic returns from the fishery over time. Sharing the TAC (or equivalent) and sharing the net economic returns from the fishery are not necessarily the same thing. We shall refer to a famous, and very successful, cooperative fishery game, involving four players, in which the fleets of two of the players were allocated zero harvests, season after season. We shall in fact maintain, both on grounds of economics, but also on grounds of the stability of the cooperative resource management regime, that the focus should be on the sharing of the net economic returns from the fishery.

This raises the issue of what is called side payments, which can be seen as transfers between and among players that may be monetary, or non-monetary, in form. For our purposes, we shall define a transboundary fishery game *without* side payments as one in which the payoff to a given coastal state is determined exclusively by the harvest of the coastal state's fleet within the coastal state's EEZ. From this it follows that, if side payments are disallowed, the net economic returns from the fishery can be shared only through the sharing of the TAC. If side payments are allowed, this ceases to be the case. Let it be added that, if the players disallow side payments, they are, in effect, imposing a (possibly severe) constraint upon themselves.

The players, as well as having to bargain over the sharing of the net economic returns from the fishery, may have to bargain over the optimal resource management programme through time. The players, the coastal states, if they are asymmetric, can well have different resource management goals. Return to the Clark (1980) model, resting in turn upon the Schaefer based dynamic economic model of the fishery. In this model the players differ only in terms of their fishing effort costs (Clark, 1980). We assumed that coastal state I was the low cost player, and II the high cost player. We stated that it can be easily shown that: $x_{\mathrm{I}}^* < x_{\mathrm{II}}^*$. Thus, if the two coastal states agree to cooperate, player II will have a greater incentive to invest in the resource than will I.[26] The resource management goal differences will have to be resolved.

We now proceed by first assuming that side payments are disallowed in the cooperative fishery game. We also assume that, if the players achieve a cooperative agreement, the agreement will be binding through time. Both of these assumptions will be relaxed in due course.

Next assume that the two players are identical, except in terms of fishing effort costs and that, as before, $b_{\mathrm{I}} < b_{\mathrm{II}}$. The objective functional of the two coastal state resource managers can be expressed as follows:

$$max\ PV_{\mathrm{I}} = \int_0^\infty e^{-\delta t}(p - c_{\mathrm{I}}(x))\phi(t)h(t)dt \tag{7.2}$$

$$max\ PV_{\mathrm{II}} = \int_0^\infty e^{-\delta t}(p - c_{\mathrm{II}}(x))(1 - \phi(t)h(t))dt \tag{7.3}$$

where $\phi(t)$ is coastal state I's share of the harvest, $0 \leq \phi(t) \leq 1$.

In equations (7.2) and (7.3) we have allowed for the harvest shares to vary through time. Let us start out by simplifying further and suppose that the harvest shares are predetermined. In essence, the bargaining is a two-stage process. In the first stage there is bargaining over the harvest shares, while in the second stage there is bargaining over the resource management programme.

In the real world, predetermining the harvest shares is commonplace. What is usually done is to predetermine the harvest shares on the basis of a formula. Two such formulae are the historical catches of the two players from the transboundary fishery, and what is referred to as zonal attachment. Zonal attachment can be thought of as the amount of the transboundary resource that is to be found in I's EEZ, as opposed to that of II, per period of time (FAO, 2002). Commonly this is expressed in percentages, so that x% 'belongs' to the EEZ of I while $(1 - x)$% belongs to the EEZ of II.

We shall suppose that I and II agree to apply one of, or some combination of, the two formulae. The harvest shares, upon being determined, are to remain constant through time. The Individual Rationality Constraint demands that the agreed upon ϕ is: $0 < \phi < 1$.

A second fundamental condition that must be met, if the cooperative game is to be stable through time, is that the solution to the game be Pareto optimal.[27] There must not be an alternative solution that would make I (II) better off without harming II (I).[28]

A potential agreement is characterized as follows:

$$\max PV = \tau PV_{\mathrm{I}} + (1 - \tau)PV_{\mathrm{II}} \tag{7.4}$$

where $0 \leq \tau \leq 1$.[29] The coefficient τ is a bargaining parameter. If $\tau = 1$, the management preferences of I are wholly dominant; if $\tau = 0$, the management preferences of II are wholly dominant.

We take all values of τ between 0 and 1, and for each τ maximize PV in equation (7.4). By so doing, we will determine the Pareto frontier in the space of realized payoffs. The payoff to I (II), once again, is to be seen as the PV of the net economic benefits to I (II) arising from the fishery by following a given harvest policy (Munro, 1979).

The first fundamental condition for stability—satisfying the Individual Rationality Constraint—is dealt with through Nash's concept of the 'threat point', which consists of the payoffs that each player would receive under non-cooperation. Commonly the 'threat point' payoffs are thought of as those arising from solutions to a Nash non-cooperative game.

Let us denote the payoffs to I as θ and the payoffs to II as ρ. Denote the 'threat point' payoffs as θ^0 and ρ^0 respectively. We picture the Pareto frontier and the 'threat point' as follows in Figure 7.4. At any point on the Pareto frontier, it is not possible to make I better off except at the expense of II, and vice versa.[30] The 'threat point' indicates that there will be no solution to the game unless I receives a payoff at least equal to θ^0, and unless II receives a payoff at least equal to ρ^0.

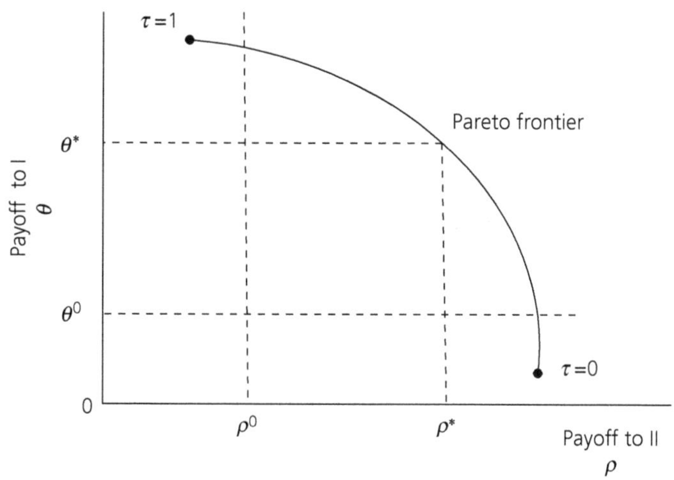

Figure 7.4. A cooperative transboundary fishery game

The dashed lines from the 'threat point' payoffs segment a section of the Pareto frontier. This segment of the Pareto frontier represents the 'core' of the cooperative game in that it shows all of the potential solutions to the game satisfying the two fundamental conditions. What assurance is there that the 'core' of the game will be other than empty? The answer is that there is no assurance, whatsoever.

Nash (1953) proves that, when the 'core' of the game is non-empty—there are potential solutions to the game—the solution will be determined by maximizing the following expression:

$$max(\theta^* - \theta^0)(\rho^* - \rho^0) \qquad (7.5).$$

Now let us relax the assumption that side payments are ignored. If side payments are allowed, then the world changes entirely. The harvest share parameter, ϕ, is unconstrained. It can vary over time,[31] anywhere between 0 and 1.

With side payments, the objective of the two players will be to maximize the PV of the net economic returns from the fishery from $t = 0$ to $t = \infty$. They will then bargain over the division of these fishery net economic returns.

The nature of the optimal resource management scheme is clear enough. Coastal state I is the low cost harvester, and as such places the greatest value on the resource. The harvest share parameter, $\phi(t)$, should be $\phi(t)=1$, for $0 \leq t \leq \infty$. Coastal state I should do all of the harvesting and should determine the harvest programme through time. Obviously, such an arrangement is feasible, if and only if, I makes side payments to II.

With side payments, the Pareto frontier will be a simple 45° line. The sum of the payoffs at any point on that frontier will be equal to the sum of the payoffs at any other point on the frontier. Consider Figure 7.5, in which the solution payoffs, given side payments, are denoted as θ^{**} and ρ^{**} respectively.

In Figure 7.5, it is assumed that, in the absence of side payments, ϕ will be predetermined and fixed through time. Thus, the without side payments Pareto frontier shown in Figure 7.5 is one of many possible such frontiers. Importantly, the Pareto frontier with side payments will lie everywhere above the Pareto frontier without side payments. The reason for this is straightforward. In the absence of side payments, we must have $\phi < 1$, otherwise II would receive nothing. With $\phi < 1$, the harvesting of the fish will be done at above minimum costs. Consequently, the global net economic returns from the fishery will be below the maximum.

Next let us introduce the concept of the cooperative surplus: $C(S)$. With side payments, the $C(S)$ can be expressed as follows:

$$C(S) = [\theta^{**} + \rho^{**}] - [\theta^0 + \rho^0] \qquad (7.6)$$

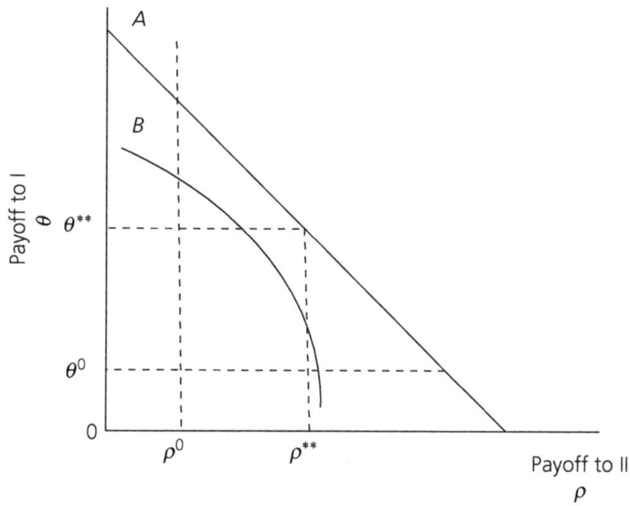

Figure 7.5. A cooperative fishery game with side payments
A Pareto frontier with side payments
B Pareto frontier without side payments

i.e., the $C(S)$ is simply the difference of the maximum global net economic returns from the fishery and the total net economic returns from the fishery under non-cooperation, both expressed in present value terms.

How then are the net economic returns from the fishery to be divided? It can be shown that the Nash model predicts that bargaining will lead to the cooperative surplus $C(S)$ being divided equally between the players, which is sometimes referred to as the egalitarian, or 'split the difference', rule (see, for example, Kaitala and Munro, 1997; Li, 1999). The intuition behind this outcome is that the two players make an equal contribution to cooperation. There can be no cooperation, unless I is prepared to cooperate; there can be no cooperation, unless II is prepared to cooperate. In any event, we shall have

$$\left.\begin{aligned} \theta^{**} &= \theta^0 + \frac{C(S)}{2} \\ \rho^{**} &= \rho^0 + \frac{C(S)}{2} \end{aligned}\right\} \tag{7.7}$$

Each player's solution payoff is equal to its threat point payoff plus one-half of the cooperative surplus.

In our example, a shift from a cooperative fishery game without side payments to one with side payments is clearly Pareto improving. Everybody stands to win. II obviously benefits from receiving side payments. I, the player making the side payments, also stands to gain. Without side payments, it would not be possible to maximize the global net economic returns from the fishery. I can anticipate enjoying a part of the enhanced economic pie.

Another way to view side payments is as a means of enhancing the scope for bargaining. A common-sense rule is that in order to achieve success in establishing a cooperative management regime, one should ensure that the scope for bargaining is as great as possible.

In the example discussed this far, a stable solution to the cooperative fishery game would be forthcoming in the absence of side payments. Side payments make a good situation better. It is possible that side payments will make the difference between a stable solution to the cooperative game and no solution at all.

Consider now the following. Let it be supposed that, in the absence of side payments, the two players find that the only acceptable means of determining ϕ is strictly on the basis of zonal attachment, which in turns means that there is but one Pareto frontier without side payments. We end up with the situation as pictured in Figure 7.6.

The cooperative game without side payments in Figure 7.6 looks exactly like the cooperative game without side payments in Figure 7.5, except that in Figure 7.6 the 'core' is empty. In the absence of side payments, there would be no solution to the cooperative game that would satisfy both of the two fundamental conditions. Attempts to manage the fishery resource cooperatively would founder and the two players would revert to competitive behaviour with all that that implies. The introduction side payments would 'save the day'.

In the examples of cooperative fishery games with side payments, discussed thus far, we find that coastal state I alone determines the optimal resource

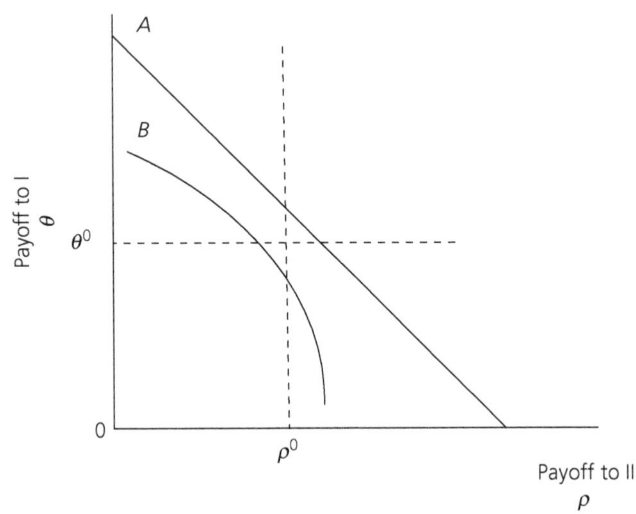

Figure 7.6. A cooperative fishery game with side payments: another view
A Pareto frontier with side payments
B Pareto frontier without side payments

management policy through time. These examples illustrate a general proposition that has come to be known as the *compensation principle.* The principle rests upon the fact that where differences in resource management objectives prevail between, or among, states sharing a fishery resource, it is almost invariably the case that one state places a higher valuation upon the resource than does the other(s). In our example, I, with the lower harvesting costs, will place a higher valuation on the fishery resource than II.

The *compensation principle* states that: '... the management preferences of that [state sharing the resource] placing the highest valuation upon the fishery [resources] should be given full reign. That [state] should, in turn, proceed to compensate [the other state, or states, sharing the resource] through the use of transfer payments [i.e., side payments]. The transfer payments may take any number of forms' (Munro, 1987, p. 282).[32]

The obvious objection to the examples we have discussed is that the purported optimal outcomes to the cooperative fishery games are politically infeasible—of theoretical interest only. What coastal state could possibly accept an agreement in which its fleet's harvests were reduced to zero? There are two responses to this objection. The first is that, if there are political constraints, one moves as close as possible to the *optimum optimorum* as the political constraints will allow.

The second response is that history provides us with a dramatic example of a cooperative fishery game in which the seemingly politically infeasible outcome did occur. This is the case of the North Pacific fur seal fishery.

During the second half of the nineteenth century, an active fur seal fishery developed in the North Pacific, which came in time to have four participants: Canada, Japan, Russia, and the United States. The management of the fishery was strictly non-cooperative. The resultant Prisoner's Dilemma played itself out, leading to resource overexploitation so severe that, by the early years of the twentieth century, there were fears that the seal resource was being driven to extinction.[33]

In response to these fears, the four participating fishing states came together in an attempt to transform the non-cooperative fishery game into cooperative one. The result was the North Pacific Fur Seal Treaty—the Convention for the Preservation of Fur Seals—signed in 1911 (Barrett, 2003). The players in the emerging cooperative game were distinctly asymmetric.

In effect, two sub-coalitions arose. The first consisted of Russia and the United States. The Russians and the Americans harvested the seals on land, on the Pribiloff Islands stretching from Alaska out towards eastern Russia. Canada and Japan constituted the second sub-coalition, with both states harvesting the seals at sea. The harvesting costs of the first sub-coalition were, to no surprise, significantly below those of the second sub-coalition. Furthermore, due to better handling of the seals, the Russians and the Americans received better prices for seal skins than did the Canadians and

Japanese. The *optimum optimorum* would have consisted of a cooperative arrangement in which the Russians and the Americans did all of the harvesting, and exercised full control of the management of the resource through time. The *optimum optimorum* was in fact achieved.

Under the terms of the Treaty, the harvest allocations for the Canadian and Japanese fleets were set equal to zero, year after year. The Canadians and the Japanese were compensated by receiving certain percentages of harvested seal skins each year. This was an example of pure side payments.

The Treaty remained in operation until the outbreak of war in the Pacific in 1941. It was resumed after the end of the Second World War, and remained in effect until 1984. The Treaty proved to be profitable to all four players and had very beneficial conservation effects. The FAO estimates that, during the first three decades of its existence—up to 1941—the seal herds increased 18 fold (FAO, 1992). In other words, the Treaty had given the four players a powerful incentive to invest in the natural capital that was the North Pacific seal herd.

An example of cooperative fisheries management involving side payments type of arrangements, somewhat less dramatic than the North Pacific fur seal fishery, but nonetheless highly instructive, involves fisheries shared by Norway and Russia in the Barents Sea. The example is of particular interest, because, instead of involving the cooperative management of a single species, it involves the cooperative management of several species simultaneously. It will be argued that the seemingly increased complexity may have actually facilitated, rather than hindered, the attempt at cooperative resource management.

In the mid-1970s, with the EEZ regime on the horizon, Norway and the then Soviet Union entered into an agreement to manage cooperatively several shared fish stocks, with the most important of these being groundfish stocks, and with the most important groundfish stock being Arcto-Norwegian cod (Armstrong and Flaaten, 1991; Stokke, 2003). The harvest shares were predetermined, and were generally set on a 50/50 basis (after about 15 per cent of the total allowable quota is set aside for third countries (Bjørndal and Lindroos, 2012)). Thus, if ϕ denotes Norway's share of the Arcto-Norwegian cod TAC, we have: $\phi = 0.5$.

There were no explicit side payment arrangements, but there were arrangements that had an unmistakably side payments flavour to them. Thus, for example, the Soviets/Russians were granted the right to take a large portion of groundfish allocations in the Norwegian EEZ. The groundfish are generally in their juvenile stage in the Russian EEZ and in their adult state in the Norwegian EEZ. Having the Russians take much of their allocations in the Norwegian EEZ serves to maximize the global economic returns from the fisheries (Stokke, 2003).

In addition, arrangements are made for quota swaps. If the Norwegian preference for cod is greater than that of the Russians (which has been the case), the Norwegians can acquire more than 50 per cent of the cod TAC by swapping quota of other species for cod—hence the argument that the fact that the cooperative fisheries management agreement involved several, rather than one, species facilitated cooperation[34] (Armstrong and Flaaten, 1991; Stokke, 2003).

7.6 Non-binding cooperative transboundary fishery management agreements

To this point, we have assumed that, if the coastal states sharing the transboundary fish stock successfully negotiate a cooperative fisheries management agreement, the agreement will be binding through time. No player will be able to deviate from the terms of the agreement. This is a truly heroic assumption.

Experience has taught us that cooperative agreements can become non-binding in fact, even if the agreements remain legally intact. An example is provided by the Canada–United States Pacific Salmon Treaty, which we discussed earlier. The Treaty effectively seized up in the early 1990s. There was no evidence of cheating on either side. It was rather a matter of a dissatisfied player 'throwing sand in the gears', and doing so very effectively. In any event, we must relax our heroic assumption and ask about the consequences of the cooperative agreement being less than fully binding through time.

There have been various attempts to address the problem of less than fully binding cooperative agreements, attempts which usually involve applying quite formidable game theory. The most effective of these attempts (in the opinion of the authors) has been undertaken by a group of Finnish economists and mathematicians, to whom we shall refer as the Helsinki group.

There are two problems that arise when the cooperative agreement is less than fully binding. The first is the obvious one, enforcement—the prevention of cheating. If, in a two-player cooperative game, one player assumes that the other player is likely to cheat, and can do so without fear of retribution, the player may well calculate that its payoff from cooperation will be less than its payoff under competition. What then is to be done? The question is addressed by two members of the Helsinki group, Kaitala (1985) and Kaitala and Pohjola (1988).

While the game theory applied by Kaitala (1985) and Kaitala and Pohjola (1988) is complex, the answer to the question is clear, almost disarmingly so, and carries with it a flavour of the economics of contract law. The authors talk in terms of an 'equilibrium agreement', which can be achieved by each player establishing a system of credible threats. Thus, for example, each player might threaten that it will revert to competitive behaviour, if the other players cheats and is found out.

In order for the threats to be credible, a player tempted to cheat must be led to believe that the costs of punishment will, if it cheats, exceed its ill-gotten gains. If the cheater is able to escape detection for an extended period of time, and if its social rate of discount is high, it may calculate that its gains from cheating will be greater than the cost of punishment. Ideally, each player will adopt a harvest and monitoring decision strategy that will bring instant punishment to the cheater—a 'trigger' strategy (Kaitala and Pohjola, 1988).

The second of the two problems to arise, when the cooperative agreement is less than fully binding, is less obvious than the first, but it is at least as important. Conditions will almost certainly change over time. A player satisfied with the cooperative agreement at $t= 0$, may not be satisfied at some later period of time because of changing conditions. The dissatisfied player may refuse to cooperate further. This leads to a third fundamental condition that must be met for solution to the cooperative game to be stable over time—time consistency.[35]

Kaitala and Pohjola (1988) illustrate the time consistency problem in the context of a deterministic model. We shall have something further to say about time consistency in a world of uncertainty under the heading of straddling fish stock management.

In any event, Kaitala and Pohjola take the two-player cooperative fishery game model that we discussed in section 7.5 and develop it further. We suppose as before that players I and II are identical except in terms of fishing effort costs, and that I is the low cost player. It is assumed further that $x_{\mathrm{I}}^{*} > x_{\mathrm{II}}^{BE}$.

It is next supposed that a non-cooperative game has played itself out, so that at $t=0$, we have: $x(0)=x_{\mathrm{II}}^{BE}$. I and II now contemplate entering into a cooperative resource management arrangement. The optimal arrangement is to have II cease harvesting, and to have I make side payments to II. I will have complete control over the resource management policy. Its resource management goal will be to rebuild the fishery resource up to the point that $x(t)=x_{\mathrm{I}}^{*}$.

Note that, if negotiations break down at $t=0$, and non-cooperation resumes, the payoff to II from the fishery, from $t=0$ onwards is zero. With a truly binding agreement, the side payments, which I would have to make to II, would be modest. Furthermore, it would matter nothing if the side payments were spread out over time, or whether they were made as a lump sum at $t=0$.

The cooperative agreement is not, however, truly binding. This means first that it would create an insupportable moral hazard problem to concentrate the side payments at $t=0$. The side payments must be spread out over time.

Next, coastal state I must take into account that, as the resource investment programme gets underway, II's threat point payoff will increase over time. Suppose that the resource achieves the goal of x_{I}^{*}. If the modest side payments agreed to initially at $t=0$, when $x(0)=x_{\mathrm{II}}^{BE}$, remain in place when the goal of x_I^* has been achieved, II could find it worth its while to breach the cooperative agreement, even if I's detection of the breach was instantaneous (Kaitala and Pohjola, 1988).

Kaitala and Pohjola use the term 'feasible' side payments, meaning side payments sufficient to keep II within the cooperative agreement. Denote the instantaneous side payment as T, and the set of side payments, commencing at $x(0)=x_{\mathrm{II}}^{BE}$ as μ. We thus have:

$$\mu(x_{\mathrm{II}}^{BE})=\{T : T \text{ feasible at each } x \in [x_{II}^{BE}, x_{\mathrm{I}}^*]\} \tag{7.8}$$

Kaitala and Pohjola (1988) then prove that

$$\mu(x_{\mathrm{II}}^{BE})=\{T : T \text{ feasible at } x_{\mathrm{I}}^*\} \tag{7.9}$$

The only set of feasible side payments through time are those equal to the side payments that will be feasible when the resource management goal is achieved, i.e., equal to the side payments needed to keep II from defecting, when II's incentive to defect will be the greatest. To rephrase Kaitala and Pohjola's conclusion, the only programme of side payments that is time consistent is a programme of side payments equal to the side payments, which prove to be feasible at $x(t)=x_{\mathrm{I}}^*$.

The Kaitala–Pohjola case is relatively simple, because the setting is a deterministic world. Since I knows the future with complete accuracy, it can plan accordingly. What if the future can only be guessed at? Then our concern with time consistency leads us into a concern about ensuring the resiliency of the cooperative resource management arrangement over time.

7.7 Cooperative transboundary fish stock games: three or more players and coalitions

All of our discussion of cooperative management of shared fish stocks up to this point has been within the context of two-player games.[36] Denote the number of players as n. Once we have $n>2$, then we must be prepared to think in terms of coalitions. Life is about to become considerably more complicated.

First some additional terminology is in order. Suppose that $n=3$, and that the players are I, II, and III. The coalition consisting of all players—I, II, and III—playing together cooperatively is referred to as the Grand Coalition. We can also, however, have sub-coalitions, for example: I and II, I and III, II and III. If I and II, for example, form a sub-coalition, the sub-coalition may play cooperatively with III, and thus form a Grand Coalition. On the other hand the I plus II sub-coalition may play competitively against III. Any player playing alone, and thus by definition competitively, is referred to as a singleton. One also has to ask, one should add, what payoffs I and II would enjoy, if they played as a sub-coalition against III.

The issue, which we now have to consider, is the 'fair' allocation of the net economic benefits from the cooperatively managed fishery in a world of coalitions, i.e., when $n > 2$. It will be found that the Nash 'split the difference' egalitarian rule is no longer adequate. The approach to the problem most commonly adopted is what is called a characteristic function game (c-game) approach in which the characteristic functions assign values to all possible coalitions. The value of a coalition is deemed to be equal to the cooperative surplus generated by the coalition.

Within the context of the characteristic function game, a widely accepted rule for a 'fair' allocation is one developed by Shapley (1953)—the Shapley value. Under this rule, a player's allocation is based upon the average of its contribution to all possible coalitions, in comparison with the average contributions of all other players.

To illustrate, we turn once again to an example developed by Kaitala and Lindroos (2007). The Kaitala–Lindroos example is basically an extension of the two-player cooperative game model we have discussed thus far. We now have three coastal states, I, II, and III, sharing a transboundary fish stock, rather than two. All of the linearity assumptions used in our $n = 2$ case are retained. It is furthermore assumed that the three states are identical in all respects, except for fishing effort costs. We have: $b_{I} < b_{II} < b_{III}$. It is then assumed that:

$$x_{I}^{*} > x_{III}^{BE} > x_{II}^{BE} > x_{I}^{BE} \tag{7.10}$$

Next denote the value of a possible coalition as $v(\{--\})$, where the value of the coalition is the cooperative surplus generated by that coalition. The smallest 'coalition' consists of single player—a singleton. For the sake of computational simplicity, it is supposed that:

$$v(\{I\}) = v(\{II\}) = v(\{III\}) = 0 \tag{7.11}$$

Thus, if there is all out competition, the sum total of the payoffs will equal zero.

We have, in addition to all out competition, the following sub-coalitions, and the Grand Coalition, denoted as follows:

(a) {I, II}
(b) {I, III}
(c) {II, III}
(d) {I, II, III}

The coalition (d) is, of course, the Grand Coalition. If (a) prevails, it is supposed that I and II will cooperate and will play competitively against III, and so on.

Given the assumption contained in expression (7.10), only two coalitions can have positive value (a) and (d). Suppose, for example, that I and III form a sub-coalition (b) and play competitively against II. Expression (7.10) tells us that $x_{II}^{BE} < x_{III}^{BE}$, which means that II can drive III out of the fishery, and thus destroy the I–III sub-coalition. All out competition will be the result. It is even more obvious that sub-coalition (c) will yield nothing.

We normalize with respect to the cooperative surplus arising from the Grand Coalition, and express it as 1. We have $0<v(\{I, II\})<1$. Denote the payoffs arising from a solution to the cooperative game as z_I, z_{II}, and z_{III} respectively. Kaitala and Lindroos demonstrate that the Shapley value payoffs arising from our three player game are given by the following pair of equations:

$$z_I = z_{II} = \frac{1}{3} + \frac{v(\{I, II\})}{6} \tag{7.12a}$$

$$z_{III} = \frac{1}{3} - \frac{v(\{I, II\})}{3} \tag{7.12b}$$

If the Nash 'split the difference' allocation rule had prevailed, we would have had the following (Kaitala and Lindroos, 2007):

$$z_I = z_{II} = z_{III} = \frac{1}{3} \tag{7.13}$$

The argument is that 'fairness' demands that I and II receive more than III, because I and II make a greater contribution towards cooperation than does III. While that may be so, it must be conceded that equations (7.12a) and (7.12b) are not intuitively obvious.

We can, however, make intuitive sense of the two equations in the Kaitala and Lindroos example, by focusing more on bargaining strength, rather than upon fairness. We shall also see the Nash 'split the difference' rule coming in through the back door.

Suppose now that I, II, and III attempt to establish a cooperative management arrangement. Let us ask what the consequences would be, if the attempt is unsuccessful. III would receive nothing; by assumption III's threat point payoff is equal to zero. I and II would also get nothing if wide open competition were to prevail which implies that I, II, and III all play as singletons. This is not necessarily the case, however. If I and II are able to maintain their sub-coalition, they will enjoy a positive return. Suppose, as is reasonable, that, if (b) prevails—I and II cooperate and play competitively against III—I and II would 'split the difference', i.e., I's payoff would equal: $v(\{I, II\})/2$. What is true of I would be true of II. We could then think of the 'true' threat point payoff of I and of II being equal to $v(\{I, II\})/2$. I and II are in a much stronger bargaining position than III.

If I and II do succeed in cooperating with III the total payoff pie would be increased. Using the 'split the difference' rule again, we could think of the increase in the payoff pie arising from I + II cooperating with III being split evenly three ways. The increase in the payoff pie is simply $[1 - v(\{I, II\})]$.

Now consider I's payoff from the {I, II, III} Grand Coalition. It would be equal to its 'true' threat point payoff plus one third of: $[1 - v(\{I, II\})]$. Thus one would have:

$$z_I = \frac{v(\{I, II\})}{2} + \frac{[1 - v(\{I, II\})]}{3} \tag{7.14}$$

What is true of I is true of II, i.e., $z_{II} = z_I$. Now look closely. Equation (7.14) reduces to equation (7.12*a*)![37]

To provide a numerical example, let us continue to suppose as before that $v(\{I\}) = v(\{II\}) = v(\{III\}) = 0$. We now suppose that: $v(\{I, II\}) = 60$ and that $v(\{I, II, III\}) = 120$. Full cooperation definitely pays. The net increase in the payoff pie arising from moving from the I, II sub-coalition playing competitively against III to the Grand Coalition, is equal to: $120 - 60 = 60$. I and II's payoffs from the Grand Coalition would be: $z_I = z_{II} = 60/2 + 60/3 = 50$. III's payoff would be: $z_{III} = 0 + 60/3 = 20$.[38]

7.8 The management of straddling fish stocks: an overview

We turn now to the management of straddling fish stocks, those fish stocks to be found both within the coastal state EEZ and the adjacent high seas.[39] It will be recalled that under the 1995 UN Fish Stocks Agreement such resources are to be managed through Regional Fisheries Management Organizations (RFMOs), which are to have both coastal states and DWFSs with a 'real' interest in the relevant fish stocks as members. We must now answer how the economics of the management of these resources differs from that of the management of transboundary fish stocks.

One part of this question, non-cooperative management of straddling fish stocks, can be answered quickly. The economics of non-cooperative management of these fish stocks differs not at all from the economics of the management of transboundary fish stocks. The Prisoner's Dilemma makes itself fully evident.

Indeed, it was the experience that the world had with the non-cooperative management of straddling fish stocks that compelled the UN to convene the 1993–5 UN Fish Stocks Conference. In our discussion of the legal framework for the management of these stocks, we stated that Part VII of the 1982 UN Convention on the Law of the Sea left the rights and duties of coastal states, as opposed to those of DWFSs, with respect to the high seas segments of these stocks in doubt. This in turn made it very difficult to establish stable cooperative management regimes for these resources. Attempts, more often than not, degenerated into competitive fishery games. The plundering of Alaskan pollock stocks in the Bering Sea Doughnut Hole was but one example of the consequences.[40] Additional examples will be provided in chapter 8.

By way of contrast, there are significant differences between the economics of the cooperative management of straddling fish stocks and that of the cooperative management of transboundary fish stocks. It is to the cooperative management of straddling fish stocks that we now turn.

7.9 The cooperative management of straddling fish stocks

To begin, while there are significant differences between the economics of the cooperative management of transboundary fish stocks and that of straddling fish stocks, it should be noted that the former provides the foundation for the economics of cooperative management of straddling stocks. Our economic model is, once again, basically a blend of the now standard dynamic economic model of an intra-EEZ fishery and the theory of cooperative games.

The first difference that we encounter between the economics of the cooperative management of the transboundary and straddling fish stocks is one of degree, rather than kind. It involves the number of players. In developing the economics of the cooperative management of transboundary fish stocks, we were able to make substantial progress by using $n=2$ cooperative games. We can make little progress in developing the economics of straddling fish stock management with such simple games. The typical RFMO has a 'large' number of players. The Northwest Atlantic provides an example. Canada and the United States share several transboundary fishery resources, e.g., haddock, in the Gulf of Maine region. Two-player cooperative games are entirely adequate for analysing the economics of the cooperative management of these resources. Off the Atlantic coasts of Canada and the United States is the RFMO, the Northwest Atlantic Fisheries Organization (NAFO). NAFO has 12 members, including members from Asia[41] (NAFO, 2011). Two-player cooperative games are wholly inadequate for analysing the economics of the cooperative management of straddling fish stocks under NAFO jurisdiction. The size of NAFO's membership, let it be added, is not unusually large for an RFMO. Another RFMO, the International Commission for the Conservation of Atlantic Tunas (ICCAT), which will be discussed in chapter 8, has 48 members or contracting parties.[42]

In analysing the economics of the cooperative management of straddling fish stocks, it is necessary that coalitional game theory be brought to bear right from the outset. The coalitions can be complex and subject to change over time. Furthermore, we need to keep in mind that, as a general rule of thumb, the larger the number of players, the more difficult it is to achieve a dynamically stable solution to the cooperative game.[43]

The second difference between the cooperative management of transboundary and straddling fish stocks is a difference in kind, and is potentially of far greater importance than the number of players. The difference relates to the property rights pertaining to the relevant fishery resources. We made the point that the property rights to fish stocks that are strictly transboundary in nature are unambiguous. By way of contrast, the property rights to the segments of straddling fish stocks to be found in the high seas are, at the time of writing, not at all free of ambiguity. The resource management problems arising therefrom we put under the headings of the threat of free riding and the new member problem.[44]

7.9.1 THE THREAT OF FREE RIDING

The FAO and others have laid great stress on the threat of IUU fishing, where IUU stands for: illegal, unregulated, and unreported (FAO, 2001a). While each is a cause for concern, the most serious is unregulated fishing. Return to our discussion of the 1995 UN Fish Stocks Agreement. Under the 1995 UN Fish Stocks Agreement, only states that are members of the RFMO, or which agree to abide by the management provisions of the RFMO, are to be permitted to engage in fishing activities in the high seas under the jurisdiction of the RFMO. This, however, does not and will not apply to those states that have yet to ratify the Agreement, until the Agreement achieves the status of customary international law—which it has yet to do. Vessels of such non-ratifying states, fishing in the high seas under RFMO jurisdiction in contravention of the RFMO management provisions, are seen, as we have argued, to be engaging in unregulated fishing.

Unregulated fishing is free riding, pure and simple, and clearly reflects a lack of clear property rights to the fishery resources under RFMO management. The potential effect of unrestrained free riding, with respect to the cooperative fisheries game, is essentially the same as unrestrained cheating. Suppose that a prospective player in the straddling fish stock cooperative game is convinced that its allocation is 'fair' and that cheating by its fellow members will be effectively controlled. On the other hand, the prospective player is also convinced that many of the economic benefits of the fishery cooperation will be skimmed off by free riders. The prospective player could well calculate that its payoff from the cooperative fishery game would be less than its payoff from a competitive fishery game. A recent international high level panel on best practices for RFMOs stated in its report that '. . . a core conclusion is that the success of international cooperation [pertaining to the management and conservation of straddling fish stocks] depends largely on the ability to deter free-riding' (Lodge et al., 2007, p. x).

The existing members of the RFMO could hope to be able to encourage (bribe) prospective free riders to join the club, by offering them legitimate shares of the economic returns from the cooperatively managed fishery, in exchange for good behaviour.[45] The other alternative is to bring some sort of legal sanctions against poachers. While accepting the *pacta tertiis* argument pertaining to the 1995 UN Fish Stocks Agreement, RFMO members could argue that free riders are in violation of Articles 116–18 of the 1982 UN Convention, which call upon all fishing states to cooperate for the purpose of conserving fishery resources on the high seas (UN, 1982). The articles are deemed to be a part of customary international law (Lodge et al., 2007, pp. 5–6).

Is it really necessary to look to legal sanctions? Once again we turn to game theory for guidance. While coalition game theory is required, the characteristic function game approach—*c*-game—discussed in the previous section is not sufficient for our purposes. The *c*-game approach does not address the issue of free riding. Instead, we turn to what is referred to as partition function games. In a partition function game, the payoff to a coalition depends, not only upon the players within the coalition, but also upon the way the other players are partitioned (Bailey et al., 2010). In the case of our straddling stock fishery game, the RFMO is the relevant coalition. Players outside the coalition (free riders) are seen as playing as singletons and thus playing against the RFMO coalition.

We turn to the Pintassilgo et al. (2010) partition function game analysis of the stability of RFMOs. Article 8 of the 1995 UNFSA to one side, it is assumed that each player has a choice of joining, or not joining, a given RFMO. It is assumed further that there are no legal sanctions on free riding.[46] A key question then becomes the externalities that are created by the establishment of an RFMO. Are non-members of the RFMO made better off (positive externality), or worse off (negative externality) by the formation of the RFMO? It is assumed, reasonably, that the externalities are positive. By forming the RFMO, the management of the fishery resource will be improved, to the benefit of outsiders, as well as those within the RFMO (Bailey et al., 2010).

Each player then must look to its potential payoff as a bona fide member of the RFMO and the payoff it would enjoy as an outsider, i.e., the payoff from free riding. Denote the (non-empty) set of players in the RFMO coalition as S. Denote the payoff to the ith player from participating in the RFMO as $V_i(S)$, and the same player's payoff by staying outside of the RFMO coalition and free riding, as $V_i(S/\{i\})$. Then clearly the ith player will willingly join the RFMO only if

$$V_i(S) \geq V_i(S/\{i\}) \tag{7.15}$$

This then gives rise to the concept of internal stability of the RFMO, also referred to as stand alone stability. Under the conditions described, the RFMO will be stable, if and only if:

$$\sum_{i\epsilon S} V_i(S) \geq \sum_{i\epsilon S} V_i(S/\{i\}) \tag{7.16}$$

(Pintassilgo et al., 2010, p. 382). If the total payoff from the RFMO Grand Coalition exceeds the sum of the payoffs to the players acting as singleton free riders, then a sharing rule can be devised that will give every player an incentive to remain within the RFMO.

If it should be the case that:

$$\sum_{i\epsilon S} V_i(S) < \sum_{i\epsilon S} V_i(S/\{i\}) \tag{7.17}$$

then no sharing rule exists that will ensure the stability of the RFMO. The incentive to free ride will be overwhelming.

Pintassilgo and Lindroos (2008) carry out a partition function game analysis, assuming that there are no effective legal sanctions against free riding, and assuming that the players are symmetric. Their pessimistic conclusion is that RFMOs with more than two players are inherently unstable (Pintassilgo and Lindroos, 2008).

Pintassilgo et al. (2010) extend the analysis by allowing for asymmetries, specifically with respect to fishing effort costs. The analysis becomes so complex that the authors have to turn to numerical analysis. Be that as it may, the authors first conclude that asymmetries enhance the scope for cooperation, rather than hinder it, by virtue of the fact that cooperation can eliminate cost inefficiencies (Pintassilgo et al., 2010).[47]

While asymmetry among the players enhances the scope for cooperation, the results are still not encouraging. Allowing for a reasonable degree of variation in harvesting costs, the authors conclude that the probability of an RFMO achieving stand alone stability, if $n > 5$, is virtually equal to zero. An RFMO, with five or fewer members, is a small RFMO indeed.

The authors conclude that '. . . if the international fishing community does prove to be incapable of suppressing unregulated fishing, the outlook for the emerging RFMO regime is bleak' (Pintassilgo et al., 2010, p. 346). It is quite simply not tolerable to allow unregulated fishing to go unchecked.

This conclusion has now been taken to heart among policy-makers. RFMOs are, in fact, beginning to take vigorous action against unregulated fishing. Vessels engaging in such activities are being blacklisted, with the result that they will be denied port facilities and their catches denied market access (Munro and Sumaila, 2011).[48] Furthermore, RFMOs are now beginning to work together to counter unregulated fishing. An example is provided by the two North Atlantic RFMOs, the Northwest Atlantic Fisheries Organisation (NAFO) and the Northeast Atlantic Fisheries Commission (NEAFC). If a vessel is blacklisted by NAFO, it is automatically blacklisted by NEAFC, and vice versa (Lodge et al., 2007).

If unregulated fishing is effectively curbed, there yet remains the destabilizing threat to RFMOs of new members, which we shall see reflects, in large part, the nebulous nature of property rights to the high seas fishery resources under RFMO jurisdiction. It is to the so-called new member problem that we now turn.

7.9.2 THE NEW MEMBER PROBLEM

It might seem reasonable to suppose that all coastal states and DWFSs having a 'real' interest in a straddling stock, or set of straddling stocks, would come together to establish an RFMO, and that would be that. The distinct possibility exists, however, that a state, not a part of the 'charter' group, will, after a time, develop a 'real' interest in the relevant fishery(ies) and request membership in the RFMO.

Most commonly, these prospective new members of an RFMO will be DWFSs. There is the possibility, however, that a prospective new member could be a coastal state. We shall commence by assuming that all prospective new members are DWFSs. At a later point, we shall enquire into what the consequences would be, if the prospective new member is a coastal state.

An example of an RFMO being confronted by prospective DWFS new members is provided by the Western Central Pacific Fisheries Commission (WCPFC) (see Figure 7.7). When established, the WCPFC consisted largely of the member states of the South Pacific Forum Fisheries Agency,[49] a few neighbouring Southeast Asian coastal states, and DWFSs that had had a history of operating in the region, e.g., Japan and the United States. In recent years, Latin American DWFSs, e.g., Ecuador and Mexico, have expressed a 'real' interest in the tuna resources under WCPFC jurisdiction (South Pacific Forum Fisheries Agency, 2011; WCFPC, 2011).

This example, it might be added, provides one reason why new members may appear on the scene. The aforementioned Latin American DWFSs, with

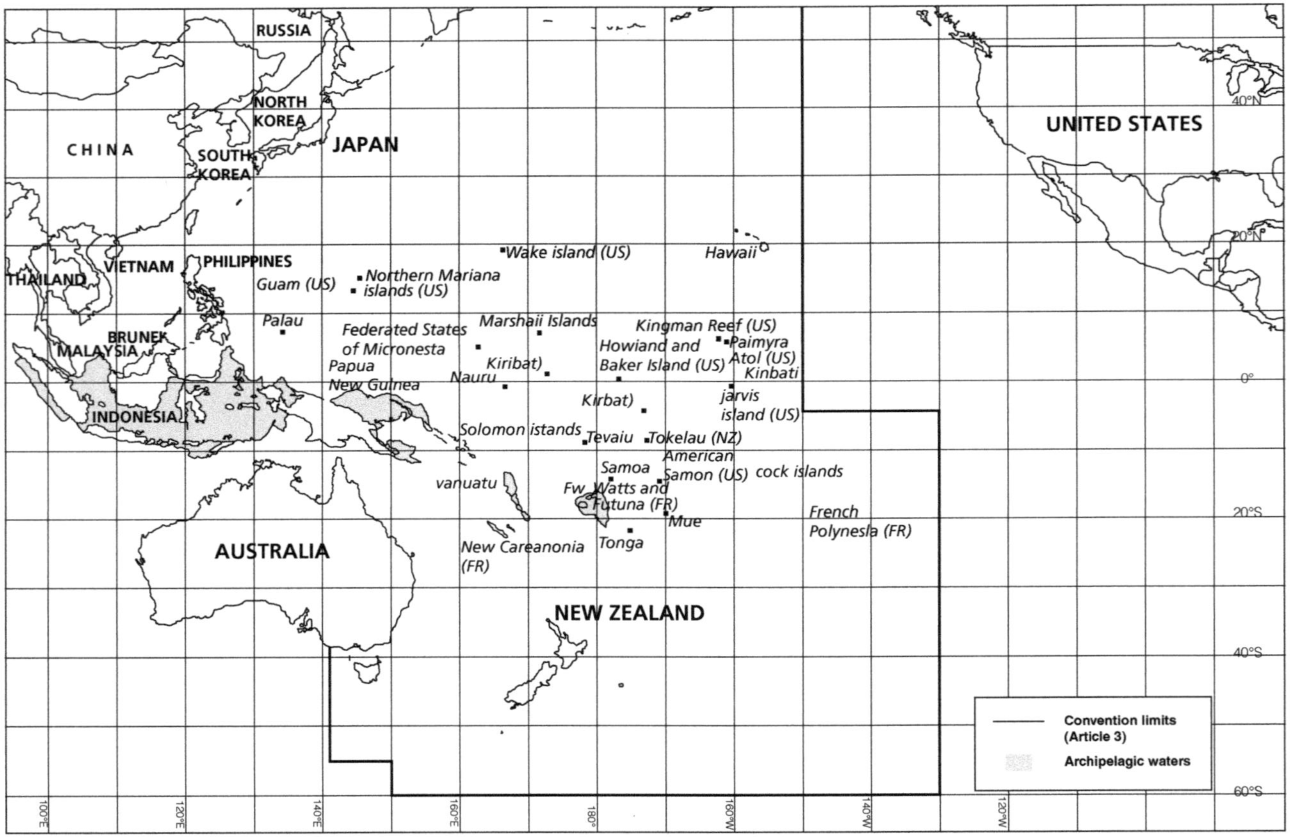

Figure 7.7. The Western and Central Pacific Fisheries Commission Convention Area

Source: WCPFC. This map is illustrative only and is without prejudice to maritime claims and maritime limits of members of the Commission.

an interest in tuna, do not have a long history as DWFSs. Their emergence as DWFSs can be ascribed to shifting comparative advantage in tuna harvesting.

In any event, Article 8 of the 1995 Fish Stocks Agreement (1995 UNFSA) which maintains that all states having a 'real' interest in the relevant fisheries may seek membership in the RFMO, means that the 'charter' members of the RFMO cannot bar would-be new members without just cause (Lodge et al., 2007, p. 35; UN, 1995).[50] This then raises the question of what the consequences might be of the RFMO being approached by a flood of prospective new members, promising to refrain from explicit free riding, and agreeing to conform to the resource management provisions of the RFMO. Return to the example of the Western and Central Pacific Fisheries Commission.

Legal experts Örebech, Sigurjonsson, and McDorman, (1998, p. 123) have argued that Article 8 of the 1995 UNFSA requires that prospective new members, seen to be acting in good faith, 'must be offered *just and reasonable* shares of the TAC available under a [RFMO] management plan'. If just and reasonable is taken to mean that new members should be given pro rata shares of the TAC (or the net economic benefits arising from the fishery(ies)), then it can be shown that this just and reasonable scheme could seriously undermine the stability of the RFMO. The argument is that, given a large enough number of new members, the estimated payoffs to the 'charter' members could be driven below their threat point payoffs, and the RFMO would be stillborn (Kaitala and Munro, 1997).

What then is to be done? One possible solution that has been put forward is to have the new members subject to a waiting period, during which they would receive none of the net economic benefits from the fishery. It is, in fact, now common for would-be new members to be first granted the status of cooperating non-members. This can be seen as a transitional, 'second class' citizen status (Lodge et al., 2007). In any event, the hope is that this would ensure that the 'charter' members' expected payoffs would not be driven down below their threat point payoffs. The waiting period solution has led to some controversy (Pintassilgo and Costa-Duarte, 2000). Doubt has been cast on the efficacy of the solution by Kaitala and Munro (1997), who demonstrate that the solution may do nothing to enhance the stability of the RFMO.

Kaitala and Munro suppose the following. A coastal state, C, and a DWFS, D_1, establish an RFMO. Immediately after the establishment of the RFMO, at $t=0$, a new member, D_2 appears, is accepted, but is informed that it will receive no economic benefits from the fishery until $t=t_{NM}>0$. Assume that side payments are feasible, that C, D_1, and D_2 are identical, and that sub-coalitions are infeasible. We now have a simple three-player game, in which the Nash 'split the difference' rule can be invoked, so that each player's payoff will be equal to its threat point payoff plus one third of the cooperative surplus. Denote D_2's share of the net economic return from the fishery, evaluated at $t=T_{NM}$, as $w_{D_2}^{T_{NM}}(x(T_{NM}),T_{NM})$, which is to be seen as a share of the PV of the net economic returns from the fishery evaluated at $t = T_{NM}$. Denote D_2's threat point payoff as $\sigma^0 \geq 0$. In any event, D_2's payoff, evaluated at $t = 0$ is:

$$e^{-\delta T_{NM}} w_{D_2}^{T_{NM}}(x(T_{NM}), T_{NM}) = \sigma^0 + \frac{C(S)}{3} \tag{7.18}$$

where $C(S)$ is the cooperative surplus.[51] The implication of equation (7.18) is that the payoff to D_2, which is expressed in *PV* terms, is independent of the length of the waiting period, T_{NM}. Essentially, nothing has been gained, in terms of ensuring the stability of the RFMO, by introducing a waiting period (Kaitala and Munro, 1997).

Kaitala and Munro (1997) put forward an alternative solution, which they adopt from a draft Agreement presented for consideration by a group of 'like minded' coastal states during the UN Fish Stocks Conference (UN, 1993; Kaitala and Munro, 1997).[52] The 'like minded' coastal states put forward, as one way of dealing with the problem, what can be termed as transferable membership. The fishery resources under the jurisdiction of the RFMO could be declared to be fully exploited. A new member could enter the RFMO, given that a 'charter' member was willing to transfer its membership to the new member. Kaitala and Munro then translate this into a 'charter' member transferring to the new member its share of the TAC (or equivalent) to the new member, in all or in part.

Kaitala and Munro then go on to say that it is unreasonable to suppose that the 'charter' member would transfer its share of the TAC—its national harvest quota—to a new member free of charge. It is, however, quite reasonable to suppose that the 'charter' member might be prepared to lease or sell its national quota in part, or in all. Thus, new members would effectively buy their way into the RFMO. The risk that the 'charter' members would see their expected payoffs from the cooperative management regime—RFMO—driven below their threat point payoffs would be eliminated.[53]

The proposal is of academic interest only, if it proves in fact to be contrary to international law. A specialist in international law, Serdy (2007), argues that the proposal is definitely not contrary to international law. On the contrary, he continues, the transferability of quota, between 'charter' members and prospective new members, would serve to hasten the 1995 UN Fish Stocks Agreement's achievement of the status of customary international law.[54]

At the time of writing, there is no specific example to which the authors can point in which new members have effectively bought their way into an RFMO. The concept has, however, gone beyond the realm of academia and has found its way into policy circles. Thus, for example, the Royal Institute of International Affairs (Chatham House) based Independent Panel on RFMO Governance has the following among its summary recommended best practices for RFMOs:

'In each RFMO, members should seek means of accommodating new members that will not undermine the long-term stability of the RFMO, such as by allowing new members to purchase or lease fishing opportunities from existing RFMO members' (Lodge et al., 2007, p. 177).

More recently, the solution has come up for discussion within the OECD. A document published by the OECD Committee on Fisheries on strengthening RFMOs contains several case studies, one of which is on the RFMO, the Commission for the Conservation of Southern Bluefin Tuna (CCSBT) (OECD, 2009). The bluefin tuna resource is to be found off Australia and New Zealand and in the Pacific between New Zealand and South Africa. The resource had been heavily depleted in the past. The economic case for a major resource investment programme in the fish stock is all but overwhelming (Bjørndal and Martin, 2007).

Attempts at effective management of the resource have been hindered by the new member problem. In the case study on the CCSBT it is stated that:

an issue... is what role... quota trading might play in... attracting new members... given the nature of the fishery... quota trading may occur in *the not too distant future* [italics added]. In fact... it makes sense where the stock is severely depleted and being rebuilt, and where there is little or no scope to accommodate an increase in the global TAC for states wishing to enter the fishery. (OECD, 2009, p. 38)

If but one RFMO adopts the transferable membership (quota trading) solution, others are certain to follow.

The case of the CCSBT brings up another aspect of the new member problem referred to by Kaitala and Munro (1997) as implicit free riding. The CCSBT, like many other RFMOs, is undertaking a resource investment programme. If new members were to appear after the resource investment programme has been completed and receive shares of the net economic returns from the fishery free of charge, they would be engaging in implicit free riding. They would enjoy a share of the returns of the resource investment, while having incurred none of the investment costs. Indeed, the OECD case study on the CCSBT reveals just this concern. The authors of the case study talk about 'the extent to which these hard fought gains [from resource investment] could be dissipated by new fishing states entering the fishery' (OECD, 2009, p. 36).

Suppose now that the 1995 UN Fish Stocks Agreement achieves the status of customary international law and that a system of international tradable harvest quotas is implemented among the RFMOs. The 'charter' members would then have collective property rights to the fishery resources under RFMO management, de facto, if not de jure, a fact recognized by the FAO a decade ago (Munro et al., 2004). This would, of course, go to the very heart of the most serious RFMO resource management problem.[55]

To this point, we have restricted ourselves to the case of DWFSs as RFMO new members. It is possible that a new member will be a coastal state. As an example, some RFMOs, which were established before the 1995 UN Fish Stocks Agreement, did not include all relevant coastal states as 'charter' members.[56] These states can be expected to demand membership. Another example is provided by Northeast Atlantic bluefin tuna, for which the International Commission for the Conservation of Atlantic Tuna (ICCAT) is the relevant RFMO. As will be discussed in chapter 8, the distribution pattern of this tuna resource has been severely constricted as a consequence of stock depletion. If the stock were to recover to historical high levels, the distribution pattern would be expected to return to what had been observed in the past. This would imply that tuna would migrate through the EEZs of a number of states, where no Northeast Atlantic bluefin tuna is currently to be found.[57]

These coastal state new members, with segments of the relevant resources in their respective EEZS, do, or would, have a proprietorial claim to share in the resource. It makes no sense to talk about these states having to 'buy' their way into the RFMO. What then is to done? At this stage, all that we can say is that it is an issue that has received little, or no, attention in the literature, and which cries out for research, both economic and legal.

7.10 **Time consistency in cooperative international fisheries arrangements and uncertainty**

In section 7.6, on non-binding cooperative transboundary fishery management agreements, we introduced the concept of time consistency. A time consistent cooperative fisheries management agreement is one that is both stable at the commencement of the management programme, and remains stable through time. The Kaitala–Pohjola (1988) model that we presented at an earlier point[58] to illustrate the problem is deterministic. We remarked that, if the world is deterministic, plans can readily be made to deal with the changes occurring through time, and thus maintain time consistency.

The world, of course, is not deterministic. The real problem is that most of the economic, political, and environmental changes, which can threaten the future stability of a cooperative fishery management agreement, cannot be accurately predicted. They fall within the realm of irreducible uncertainty. Such time consistency problems under uncertainty stand as a threat to the cooperative management of all categories of internationally shared fishery resources. We illustrate, by looking at two international fisheries, to which we have referred several times before, one a transboundary fish stock—Pacific salmon—and the other a straddling fish stock—Norwegian spring spawning herring.

We make the following two points, before looking at the two examples. The first is that the game theoretic analysis designed to address effectively the fisheries time consistency problem under uncertainty awaits future development. The second is that, while most future changes threatening the stability of cooperative fisheries management agreements cannot be predicted with accuracy, many can be anticipated. Miller and Munro (2004) use the analogy of earthquake prone zones. Those living in such zones cannot predict with any degree of accuracy the timing or severity of future earthquakes. They can, however, anticipate future earthquakes and take preventive measures.

7.10.1 PACIFIC SALMON

We referred earlier to the fact that the Canada–US Pacific Salmon Treaty, 1985, seized up in the early 1990s. We gave no reasons. The time has come to do so.

The Pacific salmon fisheries off Pacific North America are affected by climate regime shifts, in particular the Pacific decadal oscillation (PDO), which goes through alternating warm and cool phases. The warm phase of the PDO is beneficial to salmon stocks off Alaska, and detrimental to Pacific salmon stocks off southern British Columbia and the states of Washington and Oregon. The cool phase of the PDO has the reverse effects (Miller and Munro, 2004). The shift from cool to warm and vice versa can now be anticipated, but it cannot be accurately predicted.

In the cooperative game that is the management of Pacific salmon under the Canada–US Pacific Salmon Treaty, Canada can be viewed as a single player. The United States must be viewed a sub-coalition, consisting of the states of Washington and Oregon together, the state of Alaska, the Treaty Indian Nations, and the American federal government[59] (Miller and Munro, 2004). In the negotiations leading up to 1985 Treaty, Alaska had been a reluctant player, but cooperated in the end, due in considerable part to pressure brought to bear by Washington/Oregon and the Treaty Indian Nations (Munro et al., 1998). In any event, when the Treaty was signed in 1985, the Pacific salmon produced by the Fraser River and Columbia river systems were seen as being of paramount importance (Miller and Munro, 2004). The Pacific salmon resources off Alaska and northern British Columbia were seen as a 'side show'.

What the negotiators were unaware of in 1985 was the PDO shift from a cool to a warm phase, which had been under way from the late 1970s. The shift was to have a dramatic impact upon Pacific salmon productivity in Alaska. Alaskan harvest of all species of Pacific salmon experienced a ten-fold increase between the mid-1970s and 1995 (Miller and Munro, 2004, p. 384). The impact was particularly great on salmon stocks in southeastern Alaska, where the Alaskan stocks are intermingled with Canadian produced salmon. The PDO shift had a decidedly negative impact upon Columbia river system produced salmon.

As we remarked earlier, from 1985 until the end of the 1990s, there was a complete absence of side payments in the Pacific salmon cooperative game. In the early 1990s, the then Canadian Minster of Fisheries and Oceans was quoted as saying that the Treaty was designed to give each nation the 'opportunity to harvest the fish produced in its own rivers, or failing that, to harvest an equal amount of the other nations' fish' (Huppert, 1995, p. 12). This was popularly known as the 'fish for fish' rule.

In the years immediately following the signing of the Treaty, the 'fish for fish' rule was more or less met by the American harvest of Fraser river system salmon being matched by the Canadian harvest of Columbia river system produced salmon (see Figure 7.2). With the dramatic increase in Alaskan salmon production, the Alaskan fishermen fished harder, thereby increasing their harvests of salmon produced in northern British Columbia.[60] The 'fish for fish' rule was violated. The Americans were in no position to compensate the Canadians by allowing the Canadians to harvest more Columbia river system produced salmon, because these stocks, thanks to the PDO, were in sharp decline. They did offer to reduce their harvests of Fraser river system salmon, but this was not enough. The Alaskans, called upon to reduce their harvests, looked upon this call as a demand that they bear an uncompensated cost. By 1993, the players in the Pacific salmon game were at an impasse. The Treaty remained officially intact, but essentially the cooperative fishery game degenerated into a competitive game, with an unstable sub-coalition, consisting of Canada, Washington/Oregon, and the Treaty Indian Nations, playing competitively against a now powerful player, Alaska.

With regard to the Alaskans, think of their threat point payoff having been substantially increased, due to the PDO. The consequence was that the Alaskan Individual Rationality condition was no longer being met (Miller and Munro, 2004).

The Pacific salmon cooperative agreement lacked time consistency. To put it another way, the cooperative agreement lacked resiliency, in that it was unable to absorb an unexpected shock, in this case arising from environmental conditions. The consequence was the re-emergence of fish wars. The Canadians, for example, began aggressive fishing of both Fraser and Columbia river salmon stocks, a strategy which one could characterize as a 'punish your friends' strategy (Miller and Munro, 2004). Eventually alarm in both Canada and the United States about the state of several key salmon stocks drove the two sides to work out a 'patch up' agreement. It took six dangerous years to bring this agreement to fruition.

As we noted earlier, one consequence of the 1999 Agreement was the quiet introduction of side payments. There was a realization that the former 'fish for fish' rule provided too narrow a scope for bargaining (Miller and Munro, 2004). As we also noted, the Treaty was, several years later, formally renewed, and now appears to be on reasonably firm footing.

7.10.2 NORWEGIAN SPRING SPAWNING HERRING

The case of Norwegian spring spawning herring is of particular interest, because it shows that even a cooperative fishery game with side payment like arrangements can prove to lack time consistency. It will be recalled that the resource, which historically was among the largest fishery resource(s) in the North Atlantic, had a near death experience in the late 1960s–early 1970s. The resource recovered and then came under cooperative management, as a straddling stock, by Norway, Iceland, Russia, the Faroe Islands, and the EU, during the second half of the 1990s (Bjørndal, 2009). For a time, Norwegian spring spawning herring was seen as model of cooperative fishery management (Munro, 2000).

The allocation of the TAC is an example of a predetermined allocation largely based upon zonal attachment. The zonal attachment formula, needless to say, depends upon accurate measures of the resource's migratory pattern, which is, of course, subject to change. All went well until 2002, when Norway, the largest player, claimed that the true zonal attachment was significantly different from what it had been thought to be, when the cooperative fishery management arrangement was entered into. Under the true zonal attachment, Norway argued, it should be given a significantly larger share of the TAC than it was then receiving. Other players disagreed; an impasse was reached. The cooperative fishery management arrangement effectively seized up, commencing in 2003. The cooperative fishery game degenerated into a competitive one.

The Prisoner's Dilemma exhibited itself in at least two ways. First, under cooperation, the agreed upon TAC was, and is, based upon independent advice from the International Commission for the Exploration of the Seas (ICES). Commencing in 2003, each player began announcing its own quota, with the sum of the autonomous quotas, as of 2006, exceeding by a significant margin the global quota implied by ICES advice (Bjørndal, 2009). We shall expand upon this point in chapter 8.

Secondly, Norway had had bilateral agreements with its fellow players, allowing its fellow players to take part of their harvest quotas within Norwegian waters. By so doing, the fellow players were permitted to harvest herring when the fish were at their most valuable state. Furthermore, Norway's fellow players were allowed to land their harvests in Norway, thereby reducing the transportation that the herring would have to undergo before processing, which in turn led to a superior final product. The bilateral agreements were manifestly Pareto improving. Commencing in 2003, Norway announced that, with the exception of Russia, the bilateral agreements were to be suspended, thereby reducing the global economic returns from the fishery—one of the clearest examples of the Prisoner's Dilemma that one is likely to find (Bjørndal, 2009).

Like the Pacific salmon cooperative game, the Norwegian spring spawning herring lacked time consistency. It was unable to adjust to the shock of shifting migratory patterns of the resource. Eventually, a renewed cooperative fishery management agreement was entered into, coming into force in 2007—four dangerous years after the breakdown. In the end the Norwegian share of the TAC was increased, but only by a very modest amount[61] (Bjørndal, 2009).

To return to Miller and Munro (2004), the moral of these two stories is that cooperative fishery management arrangements are inherently fragile. If they lack time consistency, they are at risk of foundering at some time in the future. The players in such a game cannot, as was said earlier, predict future shocks with any accuracy. What they can do, and indeed must do, however, is to do their utmost to anticipate the possibility of such shocks and plan accordingly (FAO, 2002; Miller and Munro, 2004). It is also obvious that time consistency, resiliency, demands that the scope for bargaining be made as broad as possible—hence the importance of side payments, or the equivalent.

7.11 Coastal state–distant water fishing state relations

We next turn to a lesser, but still important, issue in international fisheries management, namely the relationship between coastal states and DWFSs within the coastal state EEZ. We have maintained that fishery resources within the EEZ do, to all intents and purposes, constitute coastal state property. The implication is that a DWFS will have access to fishery resources within an EEZ, only if the coastal state is willing to grant such access.[62]

This immediately raises the question of why a coastal state would ever grant a DWFS fleet access to its EEZ. Surely the coastal state would wish to retain its EEZ fisheries for exploitation by its own fleets alone. The answer lies more in international economics than it does in renewable resource economics.

The coastal state will, if rational, not grant a DWFS access to its EEZ unless it is in the economic interest of the coastal state to do so. The law of comparative advantage now comes into play. A coastal state granting EEZ access to a DWFS fleet can be thought of as 'importing' DWFS harvesting services, and 'importing' processing services, if the DWFS vessels have processing capacity. The DWFS fleets may have a comparative advantage in harvesting/processing services with respect to coastal state fleets and/or processing plants. In such a case, it is to the economic benefit of the coastal state to import such services (Chan, 1978; Munro, 1983).

In principle, the 'importation' of the DWFS fleet services is no different from a state exporting goods into the world market 'importing' the services of foreign merchant vessels to ship its goods abroad, rather than relying upon its

The ultimate goal of the coastal state, as principal, is to stabilize the resource at $x(t) = x_p^*$, as given by

$$F'(x_p^*) - \frac{c'(x_p^*)F(x_p^*)}{p - c(x_p^*)} = \delta_p \tag{7.23}$$

where $c(x_P) = b/qx$. Recall that we are assuming that $q = 1$, so that we can rewrite $c(x_P) = b/x$.

In a first-best situation, the coastal state would dictate the $E(t)$ profile to ensure that x_P^* was achieved. It would also ensure that $R_A = L$. It would be a matter of indifference whether the coastal state implemented t_h or t_e. It would be pointless to use both taxes. The tax system has but one role, namely that of a resource rent extraction device.

If we are in a second-best situation, then our first conclusion is that using one tax alone will be inadequate. Agency costs to the principal then become inescapable. This can be illustrated as follows. Suppose that the one tax to be used is t_h and suppose further that $\delta_A = \delta_P$. Since the DWFS, as agent, has freedom of choice, we have to think of the agent as having a view on both the optimal fishing effort profile, $E(t)$, and the optimal biomass level which we shall denote as x_A^*. We have:

$$F'(x_A^*) - \frac{c'(x_A^*)F(x_A^*)}{(p - t_h) - c(x_A^*)} = \delta_A \tag{7.24}$$

where $c(x_A) = b/x$.

Given that $t_e = 0$ and $\delta_A = \delta_P$, we will find that for $t_h > 0$, we will have $x_A^* > x_P^*$. The agent will be 'excessively' conservationist. The gap between x_A^* and x_P^* is a measure of the agency cost. The more that the coastal state attempts to extract resource rent through the use of t_h, the further will the DWFS be driven from what the coastal state deems to be the optimal fishing effort profile, $E_P^*(t)$. This is but an example of the rule that one cannot achieve two goals simultaneously by using only one instrument.

Clarke and Munro (1987) demonstrate that, if $\delta_A = \delta_P$, it will be possible to achieve a first-best outcome through the use of a judicious combination of t_h and t_e.[72] The assumption that $\delta_A = \delta_P$ is an extreme one, so extreme as to make the Clarke and Munro (1987) conclusion of very limited interest. Let us now consider the much more realistic case in which $\delta_A \neq \delta_P$.

We commence by assuming that $\delta_A > \delta_P$, and continue to assume a second-best world. We shall see that a consequence of the difference in discount rates is that principal and agent will value future payments to the agent differently. Clarke and Munro (1991) proceed by employing the Lagrangian multiplier technique.

The Lagrangian for the problem of maximizing R_P, subject to the constraint $R_A \geq L$, and continuing to assume that $q=1$, is given by the following equation:

$$\mathcal{L} = \mathrm{R_P} + \lambda \mathrm{R_A} \tag{7.25}$$

where λ now denotes the Lagrangian multiplier. We then have the Lagrange multiplier rule:

$$\partial \mathcal{L} / \partial_{t_h} = 0 \tag{7.26}$$

$$\partial \mathcal{L} / \partial_{t_e} = 0 \tag{7.27}$$

$$R_A \geq L \text{ with } \begin{cases} \lambda = 0 \text{ if } R_A > L \\ \lambda \geq 0 \text{ otherwise} \end{cases} \tag{7.28}$$

Given that both t_h and t_e are employed, the authors prove that the following optimality conditions (equations 7.29–7.31) hold:[73]

$$R_A([p - t_h], [b+t_e]) = L \tag{7.29}$$

$$\frac{1}{\delta_A}\frac{d}{dx}\left\{\left((p - t_h) - \frac{(b+t_e)}{x_A^*}\right)F(x_A^*)\right\} = p - \frac{(b+t_e)}{x_A^*} \tag{7.30}$$

$$\begin{aligned} &\frac{1}{\delta_P}\frac{d}{dx}\left\{\left(\left(p - \frac{b}{x_A^*}\right) - \left([p - t_h] - \frac{[b+t_e]}{x_A^*}\right)\right)F(x_A^*)\right\} \\ &= \left(p - \frac{b}{x_A^*}\right) - \left([p - t_h] - \frac{[b+t_e]}{x_A^*}\right) \end{aligned} \tag{7.31}$$

Equation (7.29) states that the agent is to receive no more than the minimum, while equation (7.30) states that x must be stabilized at what the agent perceives to be the optimal level, given the taxes, t_h and t_e, to which it is subject.

The seemingly complicated equation (7.31) can be seen as the principal's investment decision rule. The principal will wish investment in the resource to take place up to the point that the PV of the incremental 'net' addition to the resource rent, as perceived by itself, is equal to the 'net' cost of the incremental resource investment, also as perceived by itself.

This sounds almost identical to the argument used in chapter 3, but there is a difference. Look at the L.H.S. of equation (7.31). The 'net' addition to sustainable resource rent, as seen from the principal's perspective, is equal to the addition to global sustainable resource rent minus the ongoing 'premium' payments to the DWFS fleet. The R.H.S. of equation (7.31) tells us that the cost of the marginal resource investment, to the principal, is equal to the marginal forgone current resource rent minus the marginal current 'premium' payments to the DWFS fleet.

Suppose that $\delta_A = \delta_P$. If the second optimality condition holds, it can be seen that equation (7.31) will reduce to:

$$\frac{1}{\delta_P}\frac{d}{dx}\left\{\left(p-\frac{b}{x_A^*}\right)F(x_A^*)\right\}=\left(p-\frac{b}{x_A^*}\right) \quad (7.32)$$

This resource investment decision rule is not influenced by the principal's 'premium' payments to the DWFS fleet, the implication being that $x_A^* = x_P^*$. If the first optimality condition is satisfied, then we have a first-best situation (Clarke and Munro, 1987). The principal will manipulate t_h and t_e in such a manner as to stabilize the resource at $x(t) = x_P^*$.[74]

Now return to our assumption that $\delta_A > \delta_P$, and return as well to equation (7.31). If the second optimality condition is satisfied, note that we will then have:

$$\frac{1}{\delta_P}\frac{d}{dx}\left([p-t_h]-\frac{[b+t_e]}{x_A^*}\right)F(x_A^*)>[p-t_h]-\frac{[b+t_e]}{x_A^*} \quad (7.33)$$

simply by virtue of the fact that $\delta_P < \delta_A$. The consequence is that equation (7.31), the third optimality condition can hold only if:

$$\frac{1}{\delta_P}\frac{d}{dx}\left\{\left(p-\frac{b}{x_A^*}\right)F(x_A^*)\right\}>p-\frac{b}{x_A^*} \quad (7.34)$$

This implies, in turn, that $x_A^* < x_P^*$. Optimality leads the principal to establish a tax incentive scheme that will cause the fishery resource to be 'overexploited', as seen from the principal's perspective. Even though the agent receives no more than the minimum, the principal is forced to accede to the agent's resource management preferences, at least to a degree.

This requires some further explanation. While the outcome is independent of whether the principal deems the fishery resource to be overexploited, or underexploited, at $t=0$, an intuitive understanding is most readily obtained, if we take the case in which the principal perceives the fishery resource to be overexploited at $t=0$.

Commence by noting that the first optimality condition, $R_A = L$, states that the agent is to receive the minimum stream of 'premium' payments, measured in *PV* terms at $t=0$, employing the discount rate δ_A. This same stream of 'premium' payments, in *PV* terms, will be seen by the principal as greater than L, because the principal employs a lower rate of discount than does the agent.

Next, assuming that the relevant produced and human capital are perfectly malleable, the optimal resource investment policy commencing at $t=0$ is to declare a harvest moratorium until the resource investment goal, whatever that might be, is achieved. This means that the agent will receive no 'premium' payments until some time in the future. Denote the time at which the 'premium' payments to the agent commence as: $t = t^{\dagger}$. The farther out in the future is $t^{\dagger}$, the greater will be the gap between L, and the PV of the 'premium' payments to the agent as perceived by the principal.[75] Thus, commencing at $t=0$, the greater will be the principal's cost of 'importing' the agent's services, the more the principal invests in the biomass x. It is for this reason that it is not worth the principal's while to invest in the resource all the way up to $x(t)=x_P^*$.

In any event, the difference between x_A^* and x_P^* is a measure of the agency cost. As a general rule, the larger the extent to which the ratio δ_A/δ_P exceeds one, and the larger the minimum 'premium' payments to the agent, L, the greater will be the agency cost, i.e., the greater will be the extent to which x_A^* falls below x_P^*[76] (Clarke and Munro, 1991).

While $\delta_A > \delta_P$ is an obvious possibility, it is conceivable that we could have: $\delta_A < \delta_P$. If the DWFS agent is a capital rich developed state, and the coastal state principal is a capital poor developing state, then it is indeed possible that $\delta_A < \delta_P$. The analysis would then go into reverse. With the principal placing less value on future 'premium' payments to the agent than does the agent, we will end up with the result that $x_A^* > x_P^*$ —indeed 'overinvestment' in the fishery resource. Suppose, for the sake of argument that we were to commence at $t=0$, with the principal taking the view that the resource had been overexploited in the past. It then follows that, the farther out in the future is $t = t^{\dagger}$ when the 'premium' payments to the agent commence, the lower will be the costs of 'importing' the agent's harvesting services, as perceived by the principal. If one has a conservationist frame of mind, one can refer to the resultant agency cost as a 'virtuous' agency cost (Munro, 1994).

As a final comment, agency costs in the coastal state–DWFS principal-agent framework are seen to arise, even though it is assumed that we are living in a deterministic world. This reinforces the point made by Clarke and Munro (1987, p. 85) that the heart of the agency cost problem does not lie in uncertainty, but rather it lies in monitoring and enforcement imperfection. Having said this, however, the real world of coastal state–DWFS economic relations is a world of uncertainty and asymmetry of information, all of which will undoubtedly serve to intensify the agency cost problem.

7.12 Links between the economic management of capture fisheries at the national/regional level and at the international level

We conclude this chapter with a brief commentary on the linkages between the economic management of capture fisheries at the national/regional level and at the international level. In the introduction to chapter 6, it was stated that the investigation of the economic management of capture fisheries at the two levels would be done in isolation from one another. The reader was assured, however, that there are linkages between the two. We now make good on that assurance.

Let us deal first with coastal state–DWFS relations. Here the link is obvious. If the coastal state is to attempt to maximize the economic returns from the coastal state fisheries over time, it must decide on the extent to which the harvesting and processing is to be done by domestic fleets and processors, as opposed to foreign ones. This is a part of the coastal state's overall principal-agent problem. We have seen that, if the coastal state opts for domestic fleets and processors, it does not escape the principal-agent problem. Rather the problem simply changes form.

Beyond this, much of the relevant economics comes from international economics, rather than renewable resource economics. For example, several coastal states (particularly developed coastal states), upon establishing their EEZs, opted for exclusively domestic fleets and processors, or as close as they feasibly could come to that goal. It has been pointed out that, in many instances, the arguments for so doing are largely arguments for protection, familiar to every student of international economics, with one of the most prominent arguments employed being the infant industry argument. Arguments for not banning foreigners from the EEZ are variants of the arguments for free trade (Munro, 1992). Economists are fond of pointing out that, while a few arguments for protection are legitimate, many are spurious. They point out further that spurious arguments for protection often have great political appeal. If coastal states are beguiled by such political appeal, they are all but guaranteed of dissipating some portion of their fisheries resource rent.[77]

With regards to internationally shared fish stocks, the linkage is also clear. Commence with transboundary fish stocks. If two coastal states share a transboundary fish stock and have domestic fisheries management regimes guaranteed to dissipate fully all resource rent, then cooperation is futile. On the other hand, if the two coastal states have exemplary domestic fisheries management, but do not cooperate in the management of the transboundary resource, then we now know what the consequences will be. The exemplary fisheries management applied by each coastal state to its share of the resource will be at serious risk of being undermined.

We have talked about side payments, and side payments-like arrangements, between the two coastal states. We can see this possibly intersecting with domestic management. One such side payments-like arrangement involved quota swaps when several transboundary stocks are being managed. If the two coastal states have ITQs their fisheries, there is no obvious reason why the quota swaps could not take place through private quota markets, given that this is done with inter-government oversight (Munro, 2007). In the case of single species management, if one state should be encouraged to refrain from harvesting, and both have ITQ schemes in place, the desired outcome could be effected by the fishing firms of one state leasing or selling their quota to fishing firms of the other state, once again with inter-government oversight (Munro, 2007).

In turning to straddling fish stocks, we have similar issues. If the high seas portion of a straddling stock is subject to ineffective management, the intra-EEZ management of the resource will be undermined, regardless of how sophisticated the intra-EEZ management scheme may be. With regards to quotas, if the 'charter' members' national shares of the overall quota are divided up into ITQs, then there is no obvious reason why the purchasing, or leasing of quota, by a new member could not be done, in part at least through the ITQ market, with the purchasing/leasing of quota being done by new member fishing firms.

In any event, we arrive at a fundamental conclusion. Domestic/regional and international management of marine capture fishery resources are linked inextricably.

BIBLIOGRAPHY

The interest in international fisheries, from a theoretical and policy perspective, has grown substantially over the past two decades. Empirical applications will be discussed in the chapter to follow. Along with the many articles cited in this chapter, the interested reader is encouraged to turn to the articles by Hannesson (1997) on fishing as a supergame and the incentive compatibility of fish-sharing agreements (2006), the article by Lindroos et al. (2007) on coalition games in fisheries, the article by Kobayashi (2007) on dynamic games and Markov perfect equilibrium, Miller (2007) on the impact of climate regime shifts and international fisheries, and McKelvey et al. (2007) on gauntlet international fisheries.

NOTES

1. Thus, by definition, a DWFS is also a coastal state. The state may have clear coastal state interest in some fisheries, and DWFS interests in others. The United States, for example, did, and does, have very clear coastal state interests in New England and Alaskan fisheries. It has, at the same time, had strong DWFS interests in tropical tuna fisheries, although that interest is fading somewhat.
   It used to be thought that DWFSs were all developed fishing states. Prominent examples were: Japan, the Soviet Union/Russia, Spain, and the United States. This is no longer the case. There are now several fishing states that fall into the developing fishing state category, for example: Ecuador, Mexico, and Thailand.
2. We exclude from this definition fisheries within the waters of European Union states. These fisheries we regard as regional fisheries, to be regarded as essentially 'domestic' in nature. There is for these fisheries an overriding EU Common Fisheries Policy. See chapter 6, n.1.
3. Excluding anadromous and catadromous fish stocks.
4. We shall see later that the number of members of a Regional Fisheries Management Organization (RFMO) is highly relevant for resource management. It should be pointed out that the largest RFMOs are tuna RFMOs.
5. To repeat from chapter 1, the 1995 UN Fish Stocks Agreement is the name popularly given to this piece of international treaty law. The legal title is: Agreement for the Implementation of the Provisions of the United Nations Convention on the Law of the Sea of 10 December 1982 Relating to the Conservation and Management of Straddling Fish Stocks and Highly Migratory Fish Stocks (UN, 1995).
6. People's Republic of China, Japan, Republic of Korea, and Poland.
7. Narrowly defined.
8. See n. 5.
9. In particular Part VII of the 1982 UN Convention.
10. What is, and what is not, to be considered an RFMO is somewhat uncertain. Some listings, for example, include the International Pacific Halibut Commission (IPHC) even though the fishery resource in question is strictly transboundary in nature. We would choose to exclude the IPHC.
11. *Pacta tertiis nec nocent nec prosunt.*
12. Customary international law arises out of state practice, a conviction, by major powers as a minimum, that states are obliged to abide by the rules laid down by the law. If a state is to disregard such customary international law, it must explicitly object to the law, which is, of course, often costly politically (Buergenthal and Murphy, 2002, pp. 22–3).
13. In contrast to illegal fishing. If illegal fishing takes place, the affected state can take direct action. Unregulated fishing is a much less certain concept. Such fishing may be seen as morally reprehensible, but it is not entirely clear what can be done about it.
14. For future reference, Part VII of the 1982 UN Convention is deemed to be a part of customary international law.

15. Economists studying the economics of other transboundary resources, e.g., the atmosphere (global warming) and water, find themselves having to make extensive use of game theory.
16. The assumption of only two players is not unreasonable when analysing transboundary fish stock management. One can find several cases in which only two states are involved. An example was provided in chapter 4, namely, North Sea herring, which is managed cooperatively by the European Union and Norway.
17. The further implicit assumption is that there are sufficient barriers to the international flows of factors of production to maintain the fishing effort differential. We should also add that asymmetry between and among players in real world fishery games is the norm.
18. This is a case of a pure schooling fishery. See chapter 3.
19. Recall the simplifying assumption that $p - c = 1$.
20. It is true that many salmon go well beyond the boundaries of the American and Canadian EEZs into the adjacent high seas. However, an article in the 1982 UN Convention expressly prohibits the direct harvesting of the salmon while they are in the high seas (UN, 1982, Article 66).
21. Canadians and Americans continued to cooperate over the management of Fraser river Pacific salmon stocks under a Convention ratified in the late 1930s. The Convention set out a specific area within which the cooperation was to occur. Pacific salmon to be found outside of the Convention area were not protected and became the focus of the 'fish war' (Munro et al., 1998).
22. This means that a fundamental assumption underlying the Schaefer model, namely the assumption that the resources are always uniformly spread throughout the aquatic environment, is invalid for this fishery. Consequently, one cannot safely assume that harvesting costs will steadily rise as the resource is depleted.
23. Marine biologists commonly use the intrinsic growth rate, r (maximum rate of growth), as a means of classifying species with respect to relative growth rates. The estimated r for orange roughy (in decimal terms) is $r = 0.025$ (Clark, et al., 2010a, p. 217). This can be contrasted with another groundfish resource, Western Channel sole, with an r of 0.398 (chapter 4).
24. An apparent exception is provided by mackerel in the North Atlantic. This has traditionally been managed by an internationally set TAC shared by the EU, Russia, and Norway; however, in many years the TAC has been exceeded by significant amounts (Bjørndal, 2009). The situation is currently even more difficult. Due to changes in migratory patterns, mackerel is now being found in the Icelandic zone, where it is being exploited outside of the terms of the cooperative management arrangement. Having said this, Iceland does now have a proprietorial claim to share in the resource, since the resource has entered the Icelandic EEZ. The cooperative resource management arrangement will have to modified in such a manner as to include Iceland, the new member of the resource (Bjørndal, 2009).
25. Having said this, there are several entirely acceptable alternatives. See Armstrong (1994).
26. Think of the high cost player placing greater value on the harvesting cost reduction consequences of investing in the resource than its low cost partner.

27. An outcome is Pareto optimal, if it is not possible to make one player better off except at the expense of the other player.
28. This is sometimes referred to as collective rationality. If the solution is not Pareto optimal, it is collectively irrational.
29. See Munro (1979).
30. Given the possibility of using correlated mixed strategies, we are assured that the feasible set of outcomes will be convex (Roth, 1979; Thomson, 1989; Munro, 1990).

    It must be realized that there is not just one possible Pareto frontier, but rather a family of such frontiers, one for each ϕ, $0<\phi>1$.
31. Would it be possible to allow $\phi(t)$ to vary over time, in the absence of side payments? The answer is yes, but the forthcoming solution is so complex and cumbersome as to be of no more than theoretical interest. See Munro (1979).
32. The *compensation principle* was soon to find its way into policy documents and articles. See for example, FAO (1992); Caddy (1997).
33. This was one of the few exceptions to the view still held in the early part of the twentieth century that ocean fishery resources were free from the threat of serious overexploitation.
34. Why? The answer is straightforward. Having several rather than one species to be managed cooperatively broadens the scope for bargaining.

    Armstrong and Flaaten (1991) carried out an empirical study of Arcto-Norwegian cod, fitting the two-player cooperative fishermen game model that we have discussed. They assume rigid harvest shares and ignore possible quota swaps. They attempt to estimate the cooperative surplus. While they concede that the cooperative management arrangement is far from the *optimum optimorum* and concede that their revenue and cost data are subject to uncertainties, their estimate of the cooperative surplus is so massive that it is impossible to reject the conclusion that the cooperative resource management arrangement was highly profitable to both players in the early 1990s. There is no reason to believe that these conclusions are now less true than they were in 1991 (Ola Flaaten, personal communication).
35. If the cooperative agreement is in fact fully binding over time, then time consistency is not a concern. Such an agreement is close to being a theoretical extreme.
36. This is true even of the discussion of the North Pacific fur seal case. Four players were involved. The two sub-coalitions, Russia–USA and Canada–Japan, were each so tight that it was perfectly reasonable to treat the case as a two-player cooperative game.
37. That is, $z_{\mathrm{I}} = 1/3 + v(\{\mathrm{I,II}\})/2 - v(\{\mathrm{I,II}\})/3 = 1/3 + 3v(\{\mathrm{I,II}\})/6 - 2v(\mathrm{I,II})/6$ which is equation (7.12a). One final comment remains to be made. If $n = 2$, the Shapley value would give us the Nash 'split the difference' rule (Kaitala and Lindroos, 2007).
38. Hence, we have $z_{\mathrm{I}}+z_{\mathrm{II}}+z_{\mathrm{III}}=50+50+20=120$.
39. The reader is reminded that the UN talks in terms of highly migratory stocks (tuna primarily) and all other fish stocks found within the EEZ and the adjacent highs seas—straddling stocks narrowly defined. We are using a broad definition of straddling fish stocks, which includes both of the above sets of stocks.

40. Munro, Van Houtte, and Willmann (2004, p. 45) state that '. . . the overexploitation of straddling fish stocks [broadly defined], which provided the rationale for the UN Fish Stocks Conference, bears powerful testimony to the predictive power of the economic analysis of the non-cooperative management of such resources'.
41. Japan and the Republic of Korea.
42. Source: <www.iccat.es> (accessed 30 June 2011).
43. For example, the larger the number of players, the more difficult it is to achieve compliance through time, due to monitoring problems, if nothing else.
44. We shall at this point attempt to distinguish between free riding and non-compliance. Non-compliance essentially refers to cheating by members of the cooperative resource management arrangement. Free riding can be thought of as the skimming off of the economic benefits of cooperative resource management by non-members of the cooperative resource management arrangement.
45. This is a marine version of the ancient practice of turning poachers into game keepers.
46. Pintassilgo et al. (2010) are thus abstracting from the difference between those states that have ratified the 1995 UN Fish Stocks Agreement, and those which have not. If the legal sanctions against free riding are non-existent, or wholly ineffective, then the Agreement Article 8 requirement that no state ratifying the Agreement is to exploit fisheries under the jurisdiction of an RFMO, unless it abides by the RFMO management regime, may be unenforceable—what legal experts would refer to as 'paper law'.
47. This conclusion is consistent with our discussion of the simpler transboundary fish stock case. Return to the case of the North Pacific fur seal fishery. There were marked asymmetries between Russia and the United States on the one hand, and Canada and Japan on the other. It is reasonable to conjecture that a stable cooperative management arrangement would have been more difficult to achieve if the players had been symmetric.
48. RFMO members are well aware that attempts are made to 'launder' such catches. Elaborate measures are being taken to counter fish 'laundering' (Munro and Sumaila, 2011).
49. The South Pacific Forum Fisheries Agency (FFA) members consist of Australia, New Zealand, and 15 Pacific Island developing states, such as Fiji, the Marshall Islands, Papua New Guinea, Samoa, and the Solomon Islands (South Pacific Forum Fisheries Agency, 2011).
50. With just cause being a refusal to abide by the resource management regime set forth by the RFMO. There could be cases of new members being more efficient than the 'charter' members. If the number of new members was small enough, the 'charter' members could actually benefit from the entry of the new members into the RFMO.
51. Recall that the cooperative surplus is the difference between the sum of the solution payoffs and the sum of the Threat Point payoffs.
52. The 'like minded' coastal states were Argentina, Canada, Chile, Iceland, and New Zealand. The reason that Kaitala and Munro draw upon the draft Agreement is that it gives them some assurance that the proposed solution was within the realm

of legal and political feasibility. We shall comment further on the legal feasibility of the proposed solution.

53. This is, of course, very similar to new entrants buying their way into fisheries with ITQs and TURFs in domestic fisheries. It can also be argued that the solution results in the 'charter' members of the RFMO being granted de facto collective property rights to fishery resources under RFMO jurisdiction. This should not come as a surprise, since the lack of clear property rights to the high seas segments of straddling fish stocks helped to foster the straddling fish stocks management problem.
54. One should add that, if there is to be transferability of quota between 'charter' and new members, it would be nonsensical not to allow transferability of quota among 'charter' members.
55. It would mean as well that, from a fishery perspective, the high seas under the jurisdiction of an RFMO would be high seas in name only.
56. An example is provided by two tuna RFMOs, the International Commission for the Conservation of Atlantic Tuna (ICCAT) and the Inter-American Tropical Tuna Commission (IATTC). Both left out certain relevant developing coastal states at the time that they were founded (Lodge et al., 2007).
57. Even transboundary stocks may be confronted with this form of new member problem. An example is provided by Northeast Atlantic mackerel. Let us be reminded that due to changes in its migratory pattern, be it for climatic or other reasons, mackerel is now found in the EEZ of Iceland (Bjørndal, 2009).
58. See Section 7.6.
59. The Treaty Indian Nations refers to 24 Native American tribes of the states of Washington, Oregon, and Idaho, granted fishing privileges under the 1854 Oregon Treaty. The American federal government has weak powers within the US sub-coalition. The players having critical power within the US sub-coalition are Alaska, the Treaty Indian Nations, and Washington/Oregon (Munro et al., 1998).
60. Recall that Alaskan salmon, off southeastern Alaska, are intermingled with salmon produced in British Columbia. If Alaskan fishermen fished harder off southeastern Alaska, in order to realize their bounty, it would be impossible for them not to catch more British Columbia produced salmon.
61. What drove the players to return to cooperation? We have no definitive answer. We can conjecture, however, that having the players take a good hard look into the abyss was sufficient. The players knew from past history that competitive fishing could lead to the resource crashing again. They knew, as well, that the recovery of the resource, after the late 1960s–early 1970s crash, was due, in large part, to good fortune. If the resource were to crash a second time, it might never recover.
62. Occasionally, coastal states may choose to give DWFSs harvesting allocations within their EEZs free of charge. Thus, for example, when Norway and the Soviet Union/Russia established their EEZs in the late 1970s, and established the Barents Sea cooperative fisheries management arrangement, they agreed to set aside 15 per cent of the agreed upon TACs for DWFSs that had traditionally fished in the area. Other coastal states have not shown the same generosity.
63. Consider the following simple example. A coastal state has an offshore fishery, within its EEZ, in which the most economically efficient manner to exploit the

fishery is to process the fish on board the vessel, rather than on land. The finished product is sold into the world market, with the current price being €1000 per tonne of finished product. If the fish is harvested by a domestic fleet, the combined harvesting and processing costs will be €900 per tonne of finished product. Assume that these costs reflect true opportunity costs. Thus, the net economic return to the coastal state would be €100 per tonne of finished product, if the domestic fleet were employed.

The coastal state has the alternative of granting access to a DWFS fleet to exploit the fishery. The DWFS fleet will be granted access only if it agrees to abide by the coastal state's resource management regime, and only if it is willing to pay a fee for the right to harvest the resource and process the catch. In no sense is the coastal state obligated to grant the DWFS fleet access to its EEZ free of charge.

Now let it be supposed that the negotiated fee that would be imposed on the DWFS fleet is €300. The coastal state is in effect paying the DWFS fleet €700 per tonne of finished product for harvesting and processing services. The €700 must obviously be great enough to cover the DWFS fleet's harvesting and processing costs, but may include a premium over and above that. Suppose that the true harvesting and processing costs of the DWFS fleet are €600 per tonne of finished product. The fleet does, however, possess other fishing opportunities outside the coastal state EEZ. In its next best fishing opportunity the fleet could expect to earn a profit, which would be the equivalent of €100 per tonne of finished product from the coastal state EEZ fishery. This €100 is an opportunity cost from the perspective of the DWFS fleet. It obviously will not operate in the coastal state EEZ fishery unless it retains €700 of the gross revenue per tonne of finished product.

In any event, from the perspective of the coastal state, €300 per tonne is better than €100 per tonne. The DWFS fleet alternative is the more attractive of the two. The attractiveness of the DWFS fleet alternative rests ultimately upon the fact that the DWFS fleet has lower harvesting and/or harvesting costs than does the domestic fleet.

There is, however, an obvious political objection to this policy, namely opposition from domestic fishing companies. New Zealand demonstrated how one can avoid this objection. When New Zealand established its EEZ, its EEZ was found to encompass some potentially valuable offshore fisheries. Upon setting TACs for these fisheries, New Zealand fishing companies were granted the equivalent of ITQs to exploit the fisheries. The fishing companies were then given the choice of using their own vessels in the fisheries, or of chartering foreign vessels. Any New Zealand fishing company that chose to ignore the dictates of comparative advantage was subject to swift and certain financial punishment (Munro, 1992).

64. An important point needs to be made at this juncture. Comparative advantage is, as every textbook in international economics informs us, dynamic, i.e., subject to continuous change through time. Thus, it is possible that the DWFSs in the western and central Pacific tuna fisheries could lose their comparative advantage over time. If that were to be the case, it would cease to be to the economic advantage of the coastal state to grant EEZ access to the DWFSs.

65. This should come as no surprise. After all, in chapter 6 we stated that the management of domestic fisheries, with domestic fishers only, should be thought of in principal-agent terms.
66. The Clarke and Munro (1987) article has been cited many times, e.g., Vestergaard (2010); the 1991 article by the two authors has been cited only seldom. This is unfortunate, because the 1991 article represents a marked advance over the 1987 article.
67. A profits tax is not considered. In many real world situations, it is far too easy for the agents to disguise profits.
68. This is not entirely unrealistic. We have referred now many times to the tuna fisheries of the western and central Pacific, where the bulk of the harvesting is done by DWFS fleets. The Pacific Island coastal states impose fees upon the DWFS fleets in the form of taxes. According to Doulman (1987, p. 19), formerly of the South Pacific Forum Fisheries Agency, in addition to having the purpose of extracting resource rent, the tax system has the purpose of 'fostering operational efficiency in the use of the resource; and providing an instrument for [coastal state] government to regulate, develop, conserve and generally manage the fishery'.
69. It is also assumed that the DWFS fleet and human capital are perfectly malleable with respect to the fishery—a reasonable assumption in this case.
70. In the Clarke–Munro model processing is ignored, so that we are to regard the services of the DWFS as being strictly harvesting services. The model could be easily modified to include processing costs as well, if one so desired.
    It is assumed that R_P exceeds the PV of the net economic returns from the fishery that the coastal state would enjoy if a domestic, rather than a DWFS, fleet was used to exploit the fishery. If the assumption were to be invalid, then it would be pointless for the coastal state to grant the DWFS access to its EEZ.
71. Why might L exceed 0? The reason is straightforward enough. The DWFS may have other harvesting opportunities, and thus face an opportunity cost in fishing in the coastal state's EEZ.
72. Where t_e would have to be negative, i.e., a subsidy (Clarke and Munro, 1987).
73. See Clarke and Munro (1991, appendix) for the details of the proof.
74. We would get the same result if $L = 0$, i.e., the DWFS fleet would ask nothing more than having its fishing effort costs covered, which in turn would imply that its opportunity cost, with respect to fisheries outside the coastal state EEZ, is zero. This deserves to be placed in the category of totally unreasonable assumptions.
75. If the problem arises from the fact that the agent places a lower value on future 'premium' payments to itself than does the principal, why not have these 'premium' payments to the agent be made as a lump sum at $t = 0$? The reason is that with the principal exercising imperfect control over the agent, such a policy would lead to an impossible moral hazard problem. The principal is subject to what might be called a 'continued motivation' constraint (Munro, 1994). Given this 'continued motivation' constraint, there will still be an advantage to the principal of having the 'premium' payments to the agent shifted towards the present and near future.
76. Let x_A^{**} denote the agent's perception of the optimal biomass level in a tax free world—$t_h = 0$; $t_e = 0$. Is it conceivable that in a non-tax free world—$t_h \neq 0$; $t_e \neq 0$–

and given that $\delta_A > \delta_P$, we could ever have $x_A^* = x_A^{**}$? We might refer to this as the triumph of the agent's will. Clarke and Munro (1991) prove that the answer is no, the agent will never achieve the ultimate triumph. Rather, we shall have: $x_A^{**} < x_A^* < x_P^*$.

77. One aspect of the issue, not considered in Munro (1992), is that the foreign fleet capital may well be more malleable from the coastal state perspective than domestic fleet capital. DWFS fleets are nothing if not mobile between and among fisheries of the world.

8 Empirical Applications of International Fisheries

In chapter 7, it was seen that, among internationally shared fish stocks, straddling fish stocks (broadly defined) pose particularly formidable management problems. In this chapter, we shall focus on the modelling of these stocks.

The United Nations, as we saw (chapter 7.1), breaks down straddling fish stocks (as we have used the term) into two categories, highly migratory fish stocks, and straddling fish stocks, narrowly defined. Highly migratory fish stocks, let us be reminded, are those so defined in Annex I of the 1982 UN Convention, and consist primarily of tuna stocks (UN, 1982, Annex I). Straddling fish stocks, narrowly defined, is a catchall term for all other fish stocks to be found both within the EEZ and the adjacent high seas.

We shall, in this chapter, turn our attention to the study of one straddling stock (narrowly defined), Norwegian spring spawning herring (8.1), and one highly migratory stock, East Atlantic bluefin tuna (8.2). Both stocks are under RFMO management. One provides an example of a striking resource management success; the other an example of quite the reverse.

8.1 International management strategies for Norwegian spring spawning herring

The Norwegian spring spawning herring (*Clupea harengus*), or Atlanto-Scandian herring, is a resource that came up for brief discussion in chapters 3 and 7. We now examine the case more fully.

The resource is, as stated, a straddling stock (narrowly defined). Let us recall that the resource is distributed throughout large parts of the North-East Atlantic during its lifespan. The fishery is important for employment and revenue in many countries, including Norway, which records the largest annual harvest, Iceland, Russia, Faeroe Islands, and some member countries of the EU. The fishery for Norwegian spring spawning herring follows the migration of the stock closely as it moves from the wintering and spawning grounds along the Norwegian coast to the summer feeding grounds in the Faeroese, Icelandic, Jan Mayen, and international areas.

During the early 1950s, the total biomass of the resource ranged between 15 and 20 million tonnes, while the spawning stock biomass (SSB) averaged 10 million tonnes (Figure 8.1). Average fishing mortality was generally less than 0.1, but it produced annual catches in excess of 1 million tonnes, peaking at 2 million tonnes in 1966. In the late 1960s, as we noted in chapters 3 and 7, the stock collapsed, due to the introduction of new technologies leading to overfishing (cf. chapter 4, North Sea herring) but also, to some degree, as a result of worsening environmental conditions.

A moratorium on fishing was imposed in 1969, but the stock continued to decline, with the spawning stock biomass (SSB) reaching a nadir of 7,000 tonnes in 1973. Let us further recall that the International Commission for the Exploration of the Seas (ICES) declared the minimum safe SSB level for the resource to be 2.5 million tonnes.

The fishing moratorium was imposed for the purpose of rejuvenating and restoring the stock. The restoration was not completed until the mid-1990s. The spawning stock biomass is currently at levels not seen since the 1950s, standing in 2010 at almost 9 million tonnes. Landings were recorded at close to 1.7 million tonnes in 2009.

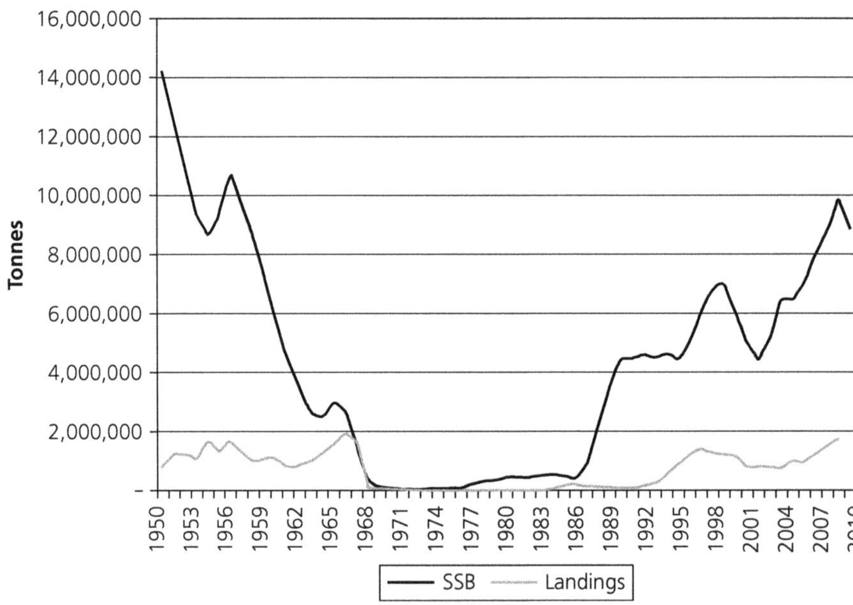

Figure 8.1. Spawning stock biomass (SSB) and landings of Norwegian spring-spawning herring (1980–2004; mill. tonnes)

Sources: ICES Report of the ICES Advisory Committee, 2008. ICES Advice, 2008. Book 9. 345 pp. (table 9.4.5.3 for the years 1950–87). Available at <http://www.ices.dk/products/icesadvice/2008/ICES20ADVICE20200820Book209.pdf>. ICES Report of the ICES Advisory Committee, 2010. ICES Advice, 2010. Book 9. 299 pp. (table 9.4.5.3 for the years 1988–2010). Available at <http://www.ices.dk/products/icesadvice/2010/ICES20ADVICE20201020Book209.pdf>.

Herring spawns off the coast of Norway, then passes through its EEZ into international waters to feed. The stock forms dense schools during this feeding period, making it particularly vulnerable to fishing. If the stock of fish is abundant, an international migratory cycle is maintained, which extends from Norwegian coastal waters through the EEZs of the EU, the Faeroe Islands, Iceland, and the high seas enclave called the 'Banana Hole' to the summer feeding area near Jan Mayen Island. However, when the stock becomes depleted, as it did in the 1960s, the species remains solely in Norwegian coastal waters, under Norwegian fisheries management jurisdiction. Consequently, Norway plays a pivotal role in deciding fisheries management policies for spring spawning stock, and may maintain this role by keeping the stock in a non-migratory state.

Management of the herring fishery is complicated by its multi-nation exploitation, as the straddling nature of the species causes it to move between several coastal state EEZs and the high seas ('Banana Hole'). The authority for fisheries management lies with the individual countries when the stocks are in their coastal waters, but on the high seas the stock is in principle open for harvesting by many fishing nations. As we now know from our discussion in chapter 7, non-cooperative management of such a resource would be fraught with danger.

Following the stock collapse, what little fishing that was allowed was almost entirely confined to Norwegian coastal waters. During the summer of 1994, however, catches in the offshore areas of the Norwegian Sea were recorded for the first time in 26 years, due to the herring resuming its traditional migratory pattern.

In 1995, the Advisory Committee on Fishery Management (ACFM) of ICES recommended a total allowable catch (TAC) of 513, 000 tonnes. The participating states did, however, choose to ignore the recommendation. The collective harvests of Norway, Russia, Iceland, Faeroe Island, and the EU exceeded 900,000 tonnes, almost twice the quantity recommended by ACFM. The fishery expanded further the subsequent year (Table 8.1). It is entirely possible that the five were setting high unilateral TACs in the hope of boosting their bargaining power when a cooperative management regime came to be negotiated. Nonetheless, with catches exceeding the TAC recommended by ICES by 80 per cent, what was emerging was a clear example of the Prisoner's Dilemma, and the dangers associated therewith.

In 1996, Norway, Russia, Iceland, and the Faeroe Islands moved away from competition, reaching an agreement for cooperative management of the resource. The EU was initially not included, and continued to fish at full capacity. In game theory parlance, we were faced with a cooperative sub-coalition playing competitively against a singleton, the EU.

With the 1995 UN Fish Stocks Agreement providing the framework, the EU, the singleton, was later in the year brought into the cooperative management arrangement. The Grand Coalition of the EU, the Faeroe Islands,

Table 8.1. ICES advice about total allowable catch quotas (TACs), actual TACs, and catch quantity per year (1995–2010; '000 tonnes)

Year	ICES advice	TAC	Catch	Comment
1995	513	900[a]	906	Non-cooperative exploitation
1996[c]	–	1,425[b]	1,220	Norway, Russia, Iceland, and the Faroe Islands reached agreement for total TAC. The EU was fishing at full capacity and set its own quota.
1997[c]	–	1,500	1,427	Cooperation
1998[c]	–	1,300	1,223	Cooperation
1999	1,263	1,300	1,235	Cooperation
2000	Max 1,500	1,250	1,207	Cooperation
2001	753	850	766	Cooperation
2002	853	850	808	Cooperation
2003	710	711[d]	790	Breakdown of cooperation
2004	825	825[d]	794	Breakdown of cooperation
2005	890	1,000[d]	1,003	Breakdown of cooperation
2006	732	967[d]	969	Breakdown of cooperation
2007	1,280	1,280	1,267	Renewed cooperation
2008	1,518	1,518	1,546	Renewed cooperation
2009	1,643	1,642	1,687	Renewed cooperation
2010	1,483	1,483	n.a.	Renewed cooperation

a Autonomous TACs.
b Autonomous TACs were set by April 1996.
c For 1996 and 1997, ICES advice was 'keep SSB over 2.5 mill tonnes'; for 1998, it was 'do not exceed the harvest control rule'. For these three years, the advice was not quantified in tonnes.
d The number is the sum of autonomous quotas for the individual parties.
Sources: Bjørndal (2009); ICES Advice 2010, book 9.

Iceland, Norway, and Russia agreed to implement a long-term management plan for the Norwegian spring-spawning herring.[1] The management plan was part of the international agreement on total quota setting and sharing of the quota during the years 1997–2002. The parties agreed to maintain a level of spawning stock biomass (SSB) greater than the critical level (Blim) of 2.5 million tonnes, and to restrict their fishing on the basis of a TAC consistent with a fishing mortality rate of less than 0.125 for appropriate age groups as defined by ICES for the year 2001 and subsequent years.

To repeat what was said in chapter 7, there were, in addition, a number of bilateral agreements among the countries involved. Fishermen from other countries were allowed to harvest part of their quota in the Norwegian EEZ and the control zone around Jan Mayen, which is under Norwegian jurisdiction, thus enabling them to harvest at a time of year when the herring contain more fat and thus are more valuable. Moreover, fishermen from other countries are allowed to land their harvests in Norway, which would tend to reduce transportation distances and thus increase the prices they would fetch. This policy also benefits the Norwegian fish processing industry. The herring spawn in Norwegian waters while, in the juvenile state, they are in the Russian EEZ. To compensate Russia for not harvesting juvenile herring, which would imply growth overfishing, Russia is given a quota in the Norwegian EEZ. While not strictly side payments, these arrangements can be seen as side payment-like arrangements, leading both to the enhancement of the global economic returns from the fishery, and to the broadening of the scope for bargaining.

In the years 1997–2002, the partners agreed on the setting of the annual TAC and the shares for each country. Moreover, catches never exceeded the TAC. As was discussed in chapter 7, the agreement broke down in 2002–3 over Norwegian demands for a higher share of the TAC. It will be recalled that these claims were based on the zonal attachment principle or the concept of 'biomass by time' within the zones (stock size within a zone multiplied with the duration of the stay). It turned out that the herring spent more time in the Norwegian EEZ than expected when the first agreement was reached—hence the Norwegian claim for a greater share of the quota.

It was further noted in chapter 7 that, as a result of the dispute, the management plans and coastal state agreements were suspended for four years between 2003 and 2006. The consequences of this suspension can now be seen in Table 8.1. The countries in question set quotas unilaterally. For 2003 and 2004 the sum of autonomous quotas did not exceed the ICES advice. This changed in 2005, when the sum of quotas was one million tonnes compared to a recommended quota by ICES of 890,000 tonnes. In 2006, the total TAC was 967,000 tonnes compared to ICES advice of 732,000 tonnes, i.e., the difference between the sum of autonomous quotas and the ICES advice increased both absolutely and relatively. Although the stock was in a healthy state (Figure 8.1), this development could have been very detrimental, if it persisted.

Let us recall that during this period, the bilateral agreements between Norway and other countries were suspended, except for the one between Norway and Russia regarding juvenile herring. The suspensions, along with the collective unilateral TACs exceeding the ICES advised levels by increasingly wide margins, provide a clear example of the Prisoner's Dilemma at work.

The dispute was ultimately resolved. In January 2007, the EU, the Faeroe Islands, Iceland, Norway, and the Russian Federation signed an agreement on the management of this stock for 2007. The parties agreed on a TAC for the Norwegian spring spawning herring of 1,518 million tonnes in 2008.[2] The allocation of the quotas is as follows: European Community 98,822 tonnes (6.51%); Faeroe Islands 78,329 tonnes (5.16%); Iceland 220,262 tonnes (14.51%); Norway 925,980 tonnes (61.0%); and Russian Federation 194,607 tonnes (12.82%). The relative shares have remained constant ever since. The agreed TACs, compared to the actual catch and ICES advice over time, are shown in Table 8.1.[3]

The Norwegian spring spawning herring fishery provides a very interesting example with regard to the management of straddling fish stocks. As mentioned, when the stock was in a depressed state, it stayed fully in the Norwegian EEZ.

The recovery of the Norwegian spring-spawning stock offers the opportunity for substantial annual harvests on a sustainable basis for the benefit of all nations involved. The fact that the SSB is back at levels not seen since the 1950s is an indication of the remarkable success of this cooperative straddling stock management scheme.[4]

The history of the previous decade, does, however, leave us with the warning that, if the current cooperative arrangement (cooperative game) among the five were to fail, the return to a competitive game type of situation would be very damaging biologically, economically, and politically.

ICES classifies the current status of the stock as having full reproductive capacity and being harvested sustainably.

8.1.1 BIOECONOMIC MULTI-AGENT SIMULATION MODELS

In turning to the bioeconomic modelling of the fishery, in this section, we shall examine in detail the model developed by Bjørndal et al. (2004). This will be followed by a description of an earlier model developed by Arnason et al. (2000). We will then go on to compare the results of these two models.

Bjørndal et al. (2004) develop a bioeconomic model consisting of stock dynamics, described by a discrete-time age-structured population model, and an economic component, representing the different national fleets. The bioeconomic model characterizes the harvest strategy for national fleets that are differentiated by costs and efficiency, to evaluate alternative management strategies.

Based on historical coalitions in the fishery, three players are specified:

1. Norway (including Russia).
2. Iceland (including the Faeroe Islands).
3. The European Union.

As noted above, Norway has traditionally cooperated with Russia so that these two countries can be considered a sub-coalition. The same applies to Iceland and the Faeroe Islands. In this way, the number of players is reduced from five to three.

For all three countries, the purse seine technology is predominant. Regardless of the similar technology used, the cost of harvesting may still vary across the national fleets. One factor explaining this is that the distance to the fishing grounds varies. For example, Norway is able to harvest herring on its coast during the spawning season while others may harvest on the high seas. For this reason Norway is assumed to be the most efficient harvester, followed by Iceland and then the EU. The price may also vary, among other things, because the quality of the herring varies over the year.

The model of population dynamics

A very complex model of population dynamics, based on the characteristics of the herring, is developed. While in chapter 4 aggregate biomass models were considered, we now turn to an age class model. The herring biomass consists of 17 age classes, i.e., new recruits to the stock (age class 0) and 16 older age classes (1–16). For each age class, mortality and weight need to be specified.

Harvesting by the three countries is determined by the *fishing mortality* they inflict on the stock, where fishing mortality is defined as the rate at which fish are removed from the biomass over the year due to fishing.

A Ricker discrete time age-structured model is used to define the population dynamics for the herring stock:

$$N_{0,t}=R_t \tag{8.1a}$$

$$N_{a+1,t+1}=N_{a,t}e^{-m_a-\sum_{j=1}^{3} S_{a,t,j}f_{t,j}} \tag{8.1b}$$

where

$N_{a,t}$ = the number of fish of each age class a in year t, a = 0, 1, . . . , 16;

$R_t = N_{0,t}$ = recruitment (age class zero) in year t;

m_a = natural mortality for age class a;

$f_{t,j}$ = fishing mortality which is related to the effort applied by the different national fleets (j = 1, 2, 3) to the stock each year; and

$S_{a,t,j}$ = selectivity rate per fleet which sets the vulnerability of each age class and year.

In the empirical analysis to follow, the initial estimate of $N_{a,0}$ is based on historical data. A Beverton–Holt stock-recruitment relationship is used in linking the number of recruits, R, to the spawning stock biomass.

Thus, according to equation (8.1a), a certain number of fish are recruited (age class zero) in year t. During the year, the number of fish in each age class is reduced due to natural mortality (m_a), which varies with age class and fishing (see equation 8.1b). The number of fish harvested from a year class is determined by fishing mortality ($f_{t,j}$) and gear selectivity $S_{a,t,j}$. It should be noted that fishing mortality varies with country (j) and year (t), while selectivity also depends on the age class (a).

The total population biomass in year t (B_t) is simply the number of fish in each age class multiplied by individual weight. This is expressed by

$$B_t = \sum_{a=0}^{16} SW_a N_{a,t} \tag{8.2}$$

where

SW_a = individual weight at age in stock.

The annual spawning stock biomass (SSB_t) is given by

$$SSB_t = \sum_{a=0}^{16} MO_a SW_a N_{a,t} \tag{8.3}$$

where MO_a is the proportion of mature individuals among the age class as constant averages, for each age class. It is assumed that only the older component of the population (from age class 7 on) is fully mature, i.e., $MO_a = 1$, whereas the younger age classes (age class 0–3) do not spawn so that $MO_a = 0$. The intermediate age classes are only partially mature, giving an MO_a between zero and one.

The catch for fleet j of an age class in numbers in year t is defined in terms of the fleet's share of total catch and natural mortality of the age class:

$$C_{a,t,j} = \frac{S_{a,t,j} f_{t,j}}{m_a + \sum_{j=1}^{3} S_{a,t,j} f_{t,j}} (N_{a,t} - N_{a+1,t+1}). \tag{8.4}$$

$N_{a,t} - N_{a+1,t+1}$ is the change in the number of fish in age class a from year t to year $t + 1$. This can be attributed to the combined natural and fishing mortality affecting the age class, as represented by the denominator. $S_{a,t,j} f_{t,j}$ if the fishing mortality of country j; dividing this by total mortality gives the fraction of total mortality that is attributed to country j's fishing. When this is multiplied by the change in the number of fish, it will give the catch in numbers from this age class.

To find total catch for each country in year t, we apply equation (8.4) to each age class and multiply by weight. Combining equations (8.1) and (8.4), the annual total yield $H_{t,j}$ for fleet j is obtained:

$$H_{t,j} = \sum_{a=0}^{16} CW_a N_{a,t} \frac{S_{a,t,j} f_{t,j}}{m_a + \sum_{j=1}^{3} S_{a,t,j} f_{t,j}} \left(1 - e^{-m_a - \sum_{j=1}^{3} S_{a,t,j} f_{t,j}}\right). \tag{8.5}$$

In this equation CW_a is the catch weights of each age class. The catch weights at each age may differ from the stock weights in equation (8.3) and are estimated as from historical data.

The rate of catch for each fleet j is controlled via two parameters, fishing mortality, $f_{t,j}$ and selectivity, $S_{a,t,j}$. Selectivity is essentially a technological parameter, expressing how selective the fishing gear is to the different age classes.

Fishing mortality is related to the effort applied by fishermen on the stock and is considered a control term. In the applied analysis of Bjørndal et al. (2004), fishing mortality is set at 1.0, 0.9, and 0.6 for Norway, Iceland, and the EU, respectively.

In the model, natural mortality is set to 0.9 for age classes 0–2, while it is set to 0.15 for age classes 3–16.

Selectivity depends on the interaction between fish and vessel technology. The selection parameter is defined as $S_{a,t,j}=0$, for age classes that are not harvested by fleet *j* and $S_{a,t,j}=1$, for all age classes that are harvested by fleet *j*. The first fishing age is for age class 4.

The bioeconomic model

For each fleet *j* a cost function is specified. For SSB more than 5 million tonnes, the unit cost of harvesting per kg herring is constant. Unit cost is exponentially increasing for stock levels less than five million tonnes. Norway is the low cost harvester followed by Iceland and the EU.

The economic objective is to maximize rent over the planning horizon. As the planning horizon extends for decades, net profit comparisons for the three national fleets for alternative management strategies are based on the present discounted value of profit or

$$\pi_j=\sum_{t=1}^{T}(\frac{1}{1+i})^t(p_jH_{t,j}-\text{Cost}_{t,j}) \tag{8.6}$$

where Π_j is net present value for country *j*, *i* is the discount rate, p_j is the herring price in country *j*, and $\text{Cost}_{t,j}$ are harvesting costs for country *j* in year *t*.

Optimal harvest strategy contains optimal effort and the possibility to select the optimal catch.

Fleet adjustment, also in open access, occurs with a lag effect and is constrained due to investments in vessel and gear. The fleets are able to react more rapidly when it comes to increasing effort (due to positive profits), while reducing effort (due to negative profits) tends to be much slower.[5]

Management strategies

Based on the history of this fishery, five strategies are evaluated for managing the herring fishery:

1. Open access is the base case.
2. Open access with fishery closure. This case is defined by imposing a simple regulatory scheme of complete fishery closure on the open access case as

the stock reaches critically low levels in order to ensure sustainability. B_{lim} = 2.5 million tonnes as noted above.

3. Non-migration: the stock is restricted to a depleted non-migratory state, remaining only in Norwegian waters and solely under Norwegian fishery jurisdiction. Stock level of non-migration is set at 0.5 million tonnes.[6]
4. Monopoly: monopoly control for Norway over an abundant migratory fish stock. In this case, side payments (rather than catch levels) are allocated to other states allowing Norway to maintain monopoly control.
5. Cartel: an abundant migratory fish stock under international fisheries management, where each participating state receives a share of the total catch (= today's situation).

The first management strategy is the open access case, showing the cycle of stock collapse and recovery. The second case is defined by imposing a simple regulatory scheme of complete fishery closure on the open-access case, as the stock reaches critically low levels. This has substantial positive benefits for both the stock level and potential net profits for all participants in the fishery. The third strategy restricts the stock to a depleted non-migratory state remaining only in Norwegian waters, solely under Norwegian jurisdiction. The fourth case investigates the consequences of sharing among the three agents the rents obtained from monopoly harvesting over an abundant migratory fish stock, and the fifth case envisages sharing the total catch among the three agents for an abundant migratory fish stock under international fisheries management (cartel solution).

The bioeconomic model is simulated for a 60-year period. The discount rate is set to 4 per cent. Results are presented in Table 8.2.

The base case—open access—is seen to give rise to a net present value of NOK 10.69 billion. This result may seem a bit counterintuitive. It may be explained by the fact that the initial stock level is very high. Moreover, although profits will lead to entry to the fishery, in the model there is a constraint on how quickly effort can increase. Thus, it will take some time to deplete the stock. In this period, rents are earned.

Table 8.2. Net present value of profit, billion NOK, 4% discount rate

Management regime	Norway	Iceland	EU	Total
Open access	5.49	3.89	1.31	10.69
Open access with fishery closure	13.72	8.04	1.88	23.64
Non-migration	8.85	–	–	8.85
Cartel	19.39	8.26	1.92	29.57
Monopoly	38.04	–	–	38.04

Monopoly harvesting by Norway gives maximum profits of NOK 38.04 billion. This is more than that of a cartel, which gives a total net present value of NOK 29.57 billion. Thus, the monopoly case can be seen as the *optimum optimorum*. In theory, it might be possible for Norway to buy out the other countries. For political reasons, this does not appear to be very likely.

Two alternatives involving cooperation are analysed (monopoly and cartel), while three alternatives with non-cooperation are considered (open access, open access with fishery closure, and non-migration). It is seen that cooperation gives a higher net present value than all three cases of non-cooperation, not only in total, but also for all three countries involved. It is of particular interest to note that the strategy of restricting the stock to a depleted non-migratory state shows minimal benefits in terms of net profits to the Norwegian fishery when compared with the competitive open-access fishery with closure.

Management under either a monopoly or a cartel solution such as these can bring significant benefits to all participants in the industry. Under a monopoly, the largest potential profits are earned in the fishery, but international agreement is required to share the rent among the non-participants who would allow the monopoly to exist. Under a cartel, potential profits are smaller than under a monopoly, but larger than under open access with fishery closure, although this requires international agreement to share the monopoly harvest level.

Model simulations show that the benefits of international cooperation far exceed the returns of a competitive open-access fishery on a timescale of several decades. Whether international agreement allocates shares of rent or catch levels, management cooperation in the spring spawning herring fishery can achieve substantial economic benefits for all participants to the fishery under sustainable stock levels.

We now consider the Arnason et al. (2000) model. This is a migratory model, based on the life-cycle of herring. Exploitation by different nations is represented by a game theoretic model. A great number of game simulations are run using different specifications regarding biomass growth, migrations, harvesting costs, number of players and coalitions.

The results derived from general experiments lead to the following conclusions.

- Cooperation always leads to more profit than competition.
- The presence of many exploiters under competition often leads to the virtual (or complete) depletion of the stock after a short period of time, provided the stock has initially spent a good deal of time in the high seas.
- If a particular exploiter has a relatively small zone to fish from, it will fish very aggressively whenever the stock wanders into his possible zone of fishing.
- The higher the rate of discount, the more aggressive the fishing.

- A high-cost exploiter (one that is located far from the area where the fish roam) generally does not have a great effect on the behaviour of the other exploiters.

A model is employed to determine the possible outcomes of the harvesting game for the Norwegian spring spawning herring fishery, simulating the exploitation by the five players, Norway, Iceland, Russia, the Faeroe Islands, and the EU. It assumes, in each of the simulations, that the players begin with a virgin stock.

Results suggest that player cooperation is needed to save the stock from (near) extinction, since the competitive game always resulted in the stock being (almost) fully depleted. The cooperative game also offers substantially higher profits than the competitive situation, and the higher the level of cooperation, the greater the profits. However, despite its suboptimal nature, the competitive game yields substantial present value of profits to the players, because of the assumption that the stock is a virgin stock at the start of the game. Large profits can be made from the initial large catches, as the stock is being depleted.

Full cooperation generates considerably higher net value of profits than competition, but even the fully cooperative solution involves intense harvesting at the outset, with a quick reduction to the long run optimal level of about 4.4 million tonnes. After that, the harvesting continues at approximately the level needed to keep the stock unchanged, with some variability as a result of migration. The initial stock reduction is rapid in both the competitive and the cooperative scenarios, requiring a great deal of harvesting and processing capacity, which may well be in excess of what is actually available, as aggregate capacity constraints were not included in the model.

There is a substantial difference between the long run levels of quarterly catch and biomass under the two game solutions; in particular, the stock becomes extinct quite rapidly when no cooperation is assumed. Nevertheless, harvest rates and stock levels fluctuate when cooperation is assumed, even in the long run. This is caused by migrations that move the stock between exclusive fishing zones and affect the profitability of fishing. Therefore, since the migrations are stochastic, the fluctuations are also stochastic.

Despite varying details, the same broad characteristics are exhibited in each case. The more extensive the cooperation between the players, the greater the aggregate present value of profits and the equilibrium stock level. The higher the rate of discount, the less benefit to coalitions, as the initial high state of the stock implies that the competitive and optimal harvesting paths are initially quite similar. Therefore, the great benefits from cooperation are primarily realized later, when the optimal and competitive harvesting paths diverge.

Without side payments few coalitions seem to be stable, in the sense that each player benefits from participating, which is not the case in a competitive situation. Thus Norway cannot be induced to enter into a coalition without side payments, as it will find it beneficial to break the agreement. However, with side payments the coalition seems very stable, yielding almost three times more aggregate benefits than the competitive path. Thus, in the Norwegian spring spawning herring case, stable coalitions are only likely to be achievable if side payments or other ways of inducing players to participate in coalitions are possible.

8.1.2 DISCUSSION

In important areas, these two papers give qualitatively similar results. Cooperation is seen to yield significant benefits in terms of rent generation, as compared with open access. Also, the more extensive the cooperation, the higher the net present value of rents.

When the TACs could be substantially increased in the mid-1990s and traditional migration to the Norwegian Sea resumed, the stock was at a very high level. Therefore, transitory rents under open access were found to be substantial. This cannot be considered a general result applicable to other fisheries as it very much depends on the high initial stock level combined with delays in entry to the fishery in response to positive profitability.

While cooperation is seen to give rise to an optimal stock at a very high and sustainable level, open access is seen to lead to (near) extinction of the stock. This latter prediction is in line with what was observed in the fishery in the 1960s.

The stability of coalitions is analysed by Arnason et al. (2000), who find that in many situations side payments are necessary to achieve this objective. Side payments are also part of the actual management arrangements for the stock, as described above. Owing to a different modelling approach, Bjørndal et al. (2004) do not explicitly consider stability of coalitions. It is, however, shown that no player will have an incentive to break out of the cartel solution. In other words, in the cartel solution case, the Grand Coalition achieves internal stability.

8.2 The East Atlantic bluefin tuna fisheries: stock collapse recovery?[7]

We next turn our attention to the fisheries of the East Atlantic bluefin tuna, a highly migratory stock. The fisheries stand in contrast to the highly successful Norwegian spring spawning herring fishery.

The Northern Atlantic and Mediterranean bluefin tuna (*Thunnus Thynnus*) is a large oceanic pelagic fish and is also the largest of the tunas. It can grow to a length of over two metres, weigh more than 500 kg and live up to 25 years. In 1982, the International Commission for the Conservation of Atlantic Tunas (ICCAT) established a dividing line between the East and West Atlantic, separating the stocks in order to facilitate stock assessment. The eastern stock is distributed from the east of the Canary Islands to Norway, in the North Sea, in Ireland, in the whole of the Mediterranean and in the south of the Black Sea. Occasionally, it migrates to Iceland and Murmansk (Figure 8.2). Spawning is located in the warm waters of the Mediterranean around the Balearic Islands and in the south of the Tyrrhenian Sea, starting in June and continuing until July.

Bluefin tuna is the most valuable fish in the ocean. High quality tuna fetches a price premium in the Japanese sashimi market, where a single fish can command a price of up to US$250,000. Moreover, the price has been increasing in recent years.

The bluefin tuna fisheries are characterized by a variety of technologies:

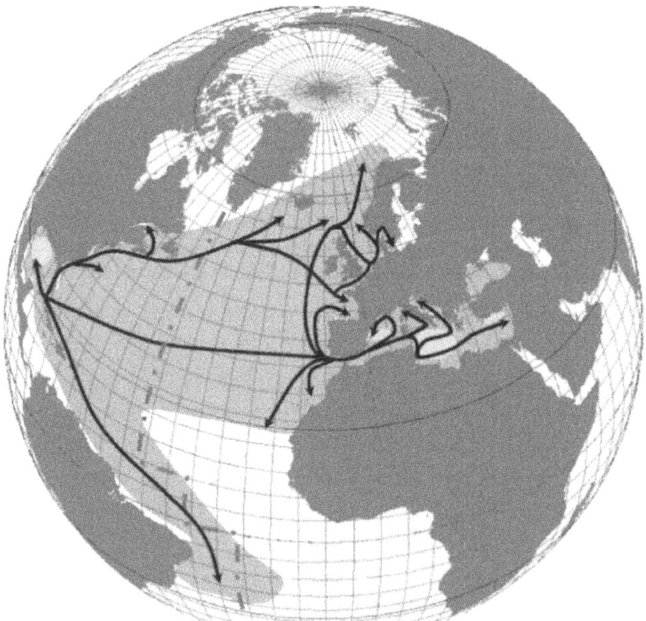

Figure 8.2. Map of the spatial distribution of Atlantic bluefin tuna (dark grey), main migration routes (black arrows) and main spawning grounds (pale grey areas) deduced from current and historical fisheries data as well as traditional and electronic tagging information. The vertical dashed line depicts the stock delimitation between the two current ICCAT management units.

Source: ICCAT Manual. Available at <http://www.iccat.int/en/ICCATManual.asp?mId = 5> (accessed 26 August 2011).

- The *purse seine* is a huge net that is cast into the sea, gathering fish in its sweep. Generally, the fish caught are of medium size and weigh about 150 kg.
- The *long line* consists of a cable to which smaller independent cables are attached at intervals of several metres.
- The *trap* is a kind of labyrinth created in the sea that leads the fish to an area where they remain until they are taken at convenience.
- The *bait boat* catches the fish using live bait and fishing rods. The fish caught are smaller since the strength of the fishermen is required to land the catch.
- *Remainder:* A number of other minor gears participate in the fishery.

Prices vary substantially with different gear types due to differences in quality and size of fish harvested.

The spatial distribution of the different gears has changed through the years. The most important change in this respect has been the relocation of the long line fishery to latitudes above 40° and longitudes between 20° and 50° west, i.e., to fishing grounds on the high seas outside coastal state 200-mile EEZs.

The management of the Northern Atlantic bluefin tuna is the responsibility of the ICCAT. ICCAT, which has 48 contracting parties,[8] has two main functions:

(i) to provide scientific assessments of Atlantic tunas and tuna-like fish, and
(ii) to give management recommendations that will permit a sustainable fishery.

ICCAT was established many years before the 1995 UN Fish Stocks Agreement came into force. Having said this, ICCAT is now an RFMO in the sense of the 1995 UN Fish Stocks Agreement.

Official fishing regulations were implemented by ICCAT in 1975, but these were disregarded by fleets and therefore had little impact on fishing effort. Present regulations include catch limits, prohibition of juvenile landings, and closed seasons, but all have proved ineffective owing to the inability of ICCAT to monitor and enforce them. This is compounded by the large number of participants in the fishery, some of which are members and some of which are not members of ICCAT.

Historically, more than 50 countries have participated in the fishery for East Atlantic bluefin tuna; currently 25–30 participate, including distant water fishing states (DWFSs) such as Japan and Korea, employing long line. While the number of participants in the fishery has been reduced, 25–30 is still a large number.

The reduced number of participants in the fishery reflects the low stock levels and the reduced distribution of the stock. As a result, countries such as Norway, Iceland, and Russia are not active in the fishery. The profitability of the stock is such that if it were to recover and extend its distribution pattern, there would be many potential new entrants to the fishery, partly as coastal states.

The RFMO governing the East Atlantic bluefin tuna thus suffers from having a very large number of players, and from free riders, i.e., participants in the fishery that are not members of ICCAT. Illegal, unreported, and unregulated (IUU) fishing is in many years believed to amount to up to 50 per cent of legal catches.[9]

From everything that we learned in chapter 7, it should come as no surprise that ICCAT has been ineffective in its management. The fishery has been described by Bjørndal and Brasão (2006) as bordering on Pure Open Access. Severe overexploitation has been the consequence.

The situation has been aggravated by tuna farming, which has become fairly widespread in the Mediterranean. This involves catching juvenile tunas which are 'fattened' in captivity for subsequent live export to Japan. While aquaculture often reduces the pressure on wild stocks of the same species (Asche and Bjørndal, 2011), this does not appear to be so for bluefin tuna. It is found that farming increases, rather than decreases, the pressure on the stock.

The situation is very grave. If the current trend is maintained, a complete stock collapse may occur. Already the trap fishery off Sicily, a mainstay of the island's economy since the Middle Ages, is facing extinction.

The bioeconomic analysis to follow rests upon a model developed by Bjørndal and Brasão (2006). We first note, by way of background, that in the 1970s, annual catches varied between 10,700 tonnes in 1970 and 22,200 tonnes in 1976 (Figure 8.3). Catches then increased and reached a maximum of 50,800 tonnes in 1996. Thereafter, there has been a decrease to 33,800

Figure 8.3. Annual catches of bluefin tuna (1970–2009; tonnes)

Source: Table 8.4.

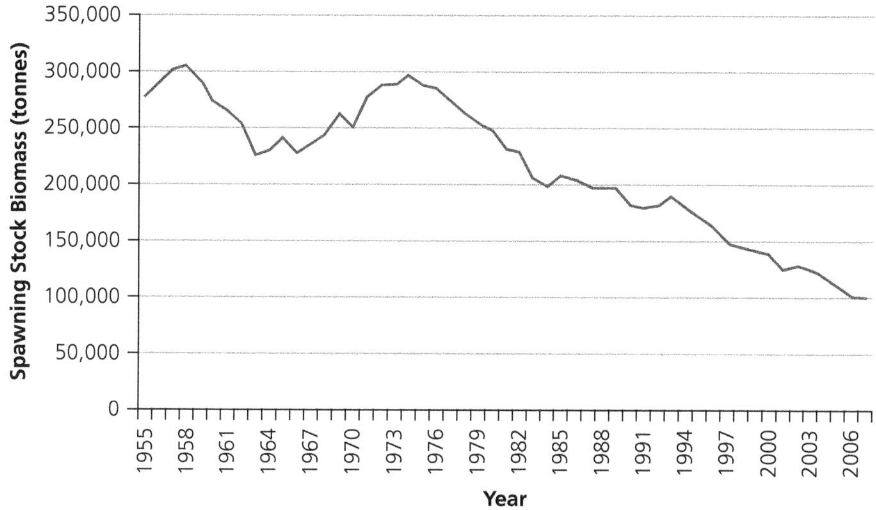

Figure 8.4. Spawning stock biomass Northeast Atlantic and Mediterranean bluefin tuna (1955–2007; tonnes)

Source: ICCAT Report of the 2008 Atlantic bluefin tuna stock assessment session, Madrid, Spain, 23 June–4 July 2008, pp. 1–161. ICCAT, Madrid, Spain.

tonnes in 2000. Subsequently the annual catch remained fairly constant but started to decline in 2008, with 19,700 recorded for 2009.

Spawning stock biomass peaked at 305, 000 tonnes in 1958 (Figure 8.4). After a dip, another peak of 297,000 tonnes was observed in 1974. The stock has ever since been in more or less continuous decline. In 2009, stock size was estimated at 100,000 tonnes, the lowest level ever observed.

8.2.1 THE BIOECONOMIC MODEL

The Bjørndal and Brasão bioeconomic model consists of a model of population dynamics and of an economic model. Simulation chooses the total allowable catch quotas (TACs) and the best combination of gear types in order to maximize the net present value from the fisheries.

The model of population dynamics for the East Atlantic bluefin tuna consists of an age-structured, multi-gear, discrete time model of the Beverton–Holt type (see chapter 1). The model is solved for each of 60 time periods (years). An interesting feature of the model is that non-linear system of five (number of gear types) and ten age classes. A new cohort is recruited to the fishery every year. A model as complex as this one is necessary to account for the number of sub-fisheries involved, representing different technologies, and the year-class structure of the stock. As the biological model conceptually is very similar to that for Norwegian spring spawning herring, it will not be considered any further here.

Bjørndal and Brasão (2006) examine stock evolution under natural conditions. In the absence of harvesting, after a couple of decades the stock will approach a level exceeding 1.2 million tonnes. This can be considered the carrying capacity of the stock.

In the economic model, five different gear types, $s = 1, \ldots, 5$, are considered: the long line (LL), the purse seine (PS), the trap, the bait boat (BB), and the remainder. The reason for specifying gear types is that economic efficiency very much depends on the combination of technologies employed in the fishery. Moreover, most of the countries that are active in the fishery specialize in one technology.

The model is set out in equations (8.7)—(8.11):

$$Rev_{t,s} = (1 - \gamma) p_s H_{t,s} \tag{8.7}$$

$$H_{t,s} = q_s E_{t,s} B_t^{\beta_s} \tag{8.8}$$

$$\text{Cost}_{t,s} = b_s E_{t,s} \tag{8.9}$$

$$\Pi_{t,s} = Rev_{t,s} - \text{Cost}_{t,s} \tag{8.10}$$

$$TNPV = \sum_{s=1}^{S} \sum_{t=1}^{T} \left(\frac{1}{1+i} \right)^t \Pi_{t,s} \tag{8.11}$$

$Rev_{t,s}$ represents revenue in year t to gear group s. This is equal to price p_s, which varies by gear type, times catch $H_{t,s}$. This represents gross income. Fishermen are remunerated according to the share system, and $(1-\gamma)$ is the share of gross revenue that accrues to the boat owner (equation 8.7). The harvest (production) function is given by equation (8.8), where E_t is effort in year t and B_t is stock size in year t. The stock output elasticity, β_s, may be different from one, and the catchability coefficient q_s also varies with gear type. Equation (8.9) gives costs per gear type, which is equal to cost per unit effort (b_s) times effort ($E_{t,s}$). Equation (8.10) gives net profits per gear type in year t. Then, equation (8.11) gives total net present value (TNPV) summed over time for all five gear types and where i, as before, is the periodic discount rate.

The link between the model of population dynamics and the economic model is established through equation (8.8), which gives the harvest (production) function.

For long line, bait boat, and purse seine—whose catches do not depend much on the existing stock—a low stock-output elasticity of 0.2 is assumed. For the more traditional gear types—trap and remainder—which are more stock dependent, the value is assumed to be 0.8. This means that harvesting by some of the most important gears is quite unresponsive to changes in stock size.

As already indicated, prices and costs vary with gear type. Trap gives by far the highest price followed by remainder, long line, purse seine, and bait boat. The high price for trap catches is because of the superior quality that can be attributed to the use of this technology. It should, however, be pointed out that catches by trap and remainder are constrained by natural availability of the stock.

8.2.2 OPTIMAL MANAGEMENT

Bjørndal and Brasão (2006) examined the optimal pattern of catches that maximizes the total net present value of the fishery, i.e., equation (8.11), subject to the model of population dynamics and other constraints, as specified below.

Two scenarios are considered:

A. The status quo fleet—a constant relative gear type structure as of 1995, where all current gear types remain in the fishery.
B. A flexible gear structure, consisting of the most efficient gear types, with upper limits on the harvests by trap and remainder.

In the analysis, the maximum annual trap catch is 10,000 tonnes and maximum remainder catch is 5,000 tonnes per year. This is due to the fact that these two gear types can only harvest tuna at certain times of the year and have access only to parts of the stock. The maxima imposed are the highest catch quantities ever observed for these two gears. The discount rate is set at 4 per cent in both scenarios.[10]

Results are presented in Table 8.3. The total net present value (TNPV) results show that in fact the initial gear structure (scenario A) is not optimal. Indeed, optimality implies that two gears should be eliminated, namely, bait boat and purse seine (scenario B); this would increase TNPV from $937

Table 8.3. Comparison of alternative management scenarios, 4% discount rate

	Scenario A: All Gears	Scenario B: Long line, Trap, and Remainder
Total net present value (mill. US$)	937	3,040
Moratorium period (years)	10[a]	[13, 3, 4]
Optimal steady state stock (tonnes)	499,510–800,000[b]	811,130
Optimal steady state harvest[c] (tonnes)	55,000	[35,000, 10,000, 5,000]

a Moratorium from years 1 to 10, 23 to 28 and 41 to 46.

b These are the stock levels during the last (lower stock level) and the first year of the fishing period, respectively.

c Harvest levels are rounded off to the nearest 1,000 tonnes.

million to $3,040 million. Thus, the current gear structure of the fishery is inefficient and, if maintained, would result in a loss in present value compared with the optimal policy structure.

The analysis indicates severe overexploitation of the fishery. The optimal stock level is seen to vary between 500,000 tonnes and 800,000 tonnes, while the actual stock level was only 100,000 tonnes in 2009. Consequently, the case for rebuilding the stock is very strong, as the sustainability of the stock is threatened and the costs of not instituting a recovery programme are substantial. Complete stock collapse is predicted if there is no recovery programme, such as a moratorium on fishing or possibly a more gradual approach to a steady state given by an extremely low TAC.

Scenario A is shown to give rise to pulse fishing (see Clark, 1985). There is a moratorium for the first ten years, which is followed by fishing for 13 years, then a further moratorium of five years, and so on. In other words, the cycle is 13 years of fishing followed by a five-year moratorium. It is assumed that both produced capital and human capital in the fishery are highly malleable.[11]

During the moratorium the stock increases to a level of about 800,000 tonnes. As fishing commences, stock size is gradually reduced to a level of 499,500 tonnes at the point in time when the new moratorium is imposed. During fishing periods, harvest is 55,000 tonnes.

In scenario B, the pattern of catches is characterized by a 13-year moratorium for long line, a three-year moratorium for trap, and a four-year moratorium for remainder. Thereafter, long line attains 35,000 tonnes, trap 10,000 tonnes, and remainder 5,000 tonnes, i.e., a total annual harvest of 50,000 tonnes. The brief moratorium period declared for the trap is explained by the high profitability of this gear type followed by remainder, which has the second highest profitability.

In the case of scenario B, the stock gradually approaches 811,000 tonnes, which can be considered the steady state stock level that maximizes the total net present value of the fishery.

It should be noted that purse seine is one of the gears that is eliminated as part of an optimal gear structure. Purse seine is currently the most important gear type. However, it has lower selectivity than any of the other gear types. Purse seine and bait boat both target young bluefin tuna, with consequent effects on the stock age structure. Thus, from an ecological point of view there are also arguments for eliminating this technology.

It is well known that, in a fishery where price is dependent on quantity, an optimal policy will often involve some harvest even if the stock is low in order to take advantage of the high price (see chapter 3). On the other hand, as stock increases, increasing catches will be constrained by the declining price (Grafton et al., 2000). In the Bjørndal–Brasão analysis, price has been assumed constant. Qualitatively, however, the high profitability of trap plays a role somewhat similar to that of a quantity dependant price. In scenario B, the initial moratorium for trap is very brief, only three years, despite the fact that the initial stock is much depleted. Furthermore, while catches of long line are reduced in later years, those of trap are always maintained at their maximum level due to the high profitability of this gear.

8.2.3 POLICY IMPLICATIONS

The empirical analysis has resulted in a number of novel and interesting results with important consequences for an RFMO:

First, the optimal stock level varies between roughly 500,000 and 800,000 tonnes. This compares with a stock level of 150,000 tonnes in 2000.
Second, to rebuild the stock, draconian measures are called for, namely outright moratoria over fairly lengthy periods.[12]
Third, the cost of inefficient gear structure is very high indeed.
Fourth, generally speaking, non-constant policies are called for as both scenarios call for pulse fishing.

We have seen that the optimal policy for the bluefin tuna fishery, on the one hand, is to shut down some of the existing gear types, namely bait boat and purse seine, and, on the other hand, to declare a temporary harvest moratorium. A moratorium can be expected to lead to the exit from the sector of a number of fishermen. Moreover, as the moratorium periods are different for each gear type, those excluded from the fishery or with a long moratorium may have incentives to harvest with gear types with shorter moratorium periods.

Policy recommendations on the bluefin tuna fishery require that all these issues be taken into account, sooner rather than later. If nothing is done, the stock will be reduced to such low levels that the sustainability of the fishery is threatened. Only draconian measures will guarantee the long-term sustainability of the stock and the fishery.

One difficulty to be encountered in the restoration of the stock arises from the fact that over 50 states participated in the fishery, when the resource was healthy. Currently 25–30 are active. Should the stock improve and, as a consequence, the distribution area be extended, there will be a strong incentive for more states to join the fishery. This incentive to join will be enhanced by the high value of the tuna.[13]

8.3 **Summary and conclusions**

Straddling fish stocks (broadly defined) present us with the most difficult capture fisheries management problem that we have encountered in our examination of the management of world capture fisheries. In chapter 7, we discussed in detail the economics of the management of straddling fish stocks (broadly defined), and the policy implications arising therefrom. In this chapter, we have put the theory to the test, by exploring empirical studies of two such fisheries, the Norwegian spring spawning herring fishery and the East Atlantic bluefin fishery. Both cases involved fishery resources that had been heavily overexploited in the past, and in which a major resource investment programme was seen to be very much in order. There has been a marked difference in the success of the two resource investment programmes.

First, let us note that the empirical studies validate our analysis of non-cooperative management of straddling fish stocks (broadly defined). The Prisoner's Dilemma in the non-cooperative management of both resources is starkly evident. The empirical studies reveal that, in both cases, the combined threat point payoffs are far, far below the potential Grand Coalition payoffs arising from fully effective cooperative resource management. Indeed, in both cases, non-cooperative resource management carries with it the threat of outright destruction of the resource.

With respect to cooperative resource management, the empirical studies of Norwegian spring spawning herring illustrate the importance of side payments, or at least side payment-like arrangements, for ensuring the long-term stability of the cooperative resource management arrangement. The Norwegian spring spawning herring case also demonstrates the danger posed by time inconsistent cooperative resource management arrangements.

The empirical study on the East Atlantic bluefin tuna fishery (see Table 8.4) illustrates the threats to cooperative resource management posed by: (i) large numbers of players; (ii) the prospect of future new members; (iii) lack of monitoring and enforcement of regulations; and (iv) illegal and unregulated fishing. The East Atlantic bluefin tuna case, one can add, clearly validates the partition function game analysis discussed in chapter 7.[14]

Finally, we need to ask, in light of our discussion in chapter 7 and this chapter, why the cooperative management of the Norwegian spring spawning herring fishery, in contrast to that of the East Atlantic bluefin fisheries, was so successful. Recall that the Norwegian spring spawning herring resource is now at levels not seen since the 1950s—a very successful resource investment programme indeed.

To begin, we take note of the fact that, when the resource crashed in the early 1970s, the remnants of the resource were confined to Norwegian waters. Thus, at that critical stage the resource ceased to be shared in nature. The Norwegians alone were responsible for initiating the resource investment programme, a programme that was to extend for over 20 years.[15]

Table 8.4. Bluefin tuna catches for East Atlantic and Mediterranean

Year	East Atlantic	Mediterranean	Total
1970	5,972	4,694	10,666
1971	4,761	6,195	10,956
1972	4,733	5,954	10,687
1973	4,687	6,051	10,738
1974	6,071	13,056	19,127
1975	9,971	11,241	21,212
1976	5,120	17,073	22,193
1977	6,921	11,797	18,718
1978	5,799	8,846	14,645
1979	4,767	7,456	12,223
1980	4,211	10,039	14,250
1981	3,259	10,515	13,774
1982	6,702	15,706	22,408
1983	8,152	13,650	21,802
1984	7,395	17,032	24,427
1985	4,807	17,203	22,010
1986	4,687	14,560	19,247
1987	4,456	13,764	18,220
1988	6,951	17,167	24,118
1989	5,433	15,628	21,061
1990	6,040	17,207	23,247
1991	6,556	19,872	26,428
1992	7,619	24,230	31,849
1993	9,367	24,901	34,268
1994	6,930	39,810	46,740
1995	9,650	37,640	47,290
1996	12,663	38,144	50,807
1997	13,539	33,616	47,155
1998	11,376	28,342	39,718
1999	9,628	22,828	32,456
2000	10,528	23,238	33,766
2001	10,086	24,519	34,605
2002	10,347	23,424	33,771
2003	7,362	23,801	31,163
2004	7,410	23,971	31,381
2005	9,036	26,810	35,846
2006	7,535	23,154	30,689
2007	8,037	26,479	34,516
2008	7,645	16,205	23,850
2009	6,684	13,016	19,700

Source: ICCAT, *Statistical Bulletin*, Vol. 40 (2011).

The resource investment programme's success was first indicated by the fact that, by the mid-1990s, the resource was once more a straddling one. Cooperative resource management became mandatory. Fortunately, the number of players in the emerging cooperative fisheries game was, and is, small, being only five—the EU, the Faeroe Islands, Iceland, Norway, and Russia. We pointed out, let it be added, that, while NEAFC constitutes the formal RFMO, the de facto RFMO (arrangement) consists of the five coastal states.

The five were, and are, faced with no significant unregulated fishing threat, nor are they faced with any perceptible threat of potential new members. One can conjecture that the absence of these threats is largely due to geography combined with the fact that herring distribution on the open seas is very limited. Only a small fraction of total harvest is captured in the Banana Hole.[16] It is questionable whether it would be economically feasible for prospective new member vessels to operate in the herring fishery, unless they were granted access to the EEZs of the five. Moreover, it is important to add the fact that two of the players, the EU and Russia, are very powerful politically. While legal experts may debate the legal status of the resource, when it is in the high seas enclave the Banana Hole, this cooperatively managed straddling fish stock looks remarkably similar to a cooperatively and successfully managed transboundary fish stock. In other words, to all intents and purposes, the five appear to have de facto collective property rights to the resource (Munro, 2011).

As a final comment, a recurrent theme throughout this book has been the need for dynamic (capital theoretic) economic models, in analysing the economics of capture fisheries management. It should now be obvious that it would be virtually impossible to analyse, with any hope of success, the economics of the management of shared fish stocks within the context of static economic models of the fishery.

BIBLIOGRAPHY

As noted above, the sharing agreement for Norwegian spring spawning herring broke down due to the fact that the migration pattern in the 1990s was different from the historical pattern which formed the basis for determining the catch shares. These changes are likely due to temperature change, which again may affect the North Atlantic current. This topic is addressed by Sissener and Bjørndal (2004).

In section 8.1, there was discussion of the stock level below which there would be no stock migration. Hannesson (2006), in a game theoretic model, analyses the strategic advantage Norway might have from being in a position to manage the stock itself by fishing it down to the level above which it starts to migrate out of the Norwegian EEZ.

MacKenzie et al. (2009) analysed the bluefin tuna fishery on the basis of the current management plan for the fishery. The authors predicted that the adult population in 2011 would likely be 75 per cent lower relative to 2005 and showed that quotas in some intervening years would allow the fishery to capture legally all of the adult fish. In other words, if the current management plan is maintained, the stock is at risk of collapse.

NOTES

1. While NEAFC constitutes the formal RFMO in this case, the de facto RFMO consists of the five. The 1995 UNFSA calls for coastal states and DWFSs to cooperate in the management of straddling stocks (broadly defined) 'either directly or through subregional or regional fisheries management organizations or *arrangements* [italics added]' (UN, 1995, Article 8(1)). Article 1(a) defines an 'arrangement' as a cooperative mechanism established in accordance with the 1982 Convention and the 1995 UNFSA by two or more states for the purpose of conserving and managing straddling stocks (broadly defined) in a subregion or region (UN, 1995, Article 1(a)). Legal experts Örebech, Sigurjonsson, and McDorman (1998, p. 124) make it clear that the 'five' clearly fit within the 1995 UNFSA definition of 'arrangement', a definition that does not require a formal organization with its own secretariat and other accoutrements (see also Bjørndal, 2009).
2. As shown in Table 8.1, total catch exceeded TAC in 2008 and 2009, but by small amounts.
3. See Bjørndal (2009) on NEAFC. In 2005, the total landings in the Convention Area, i.e., the EEZs of the coastal states, and in the Regulatory Area, i.e., the high seas, were approximately 1,254,000 tonnes, and 195,000 tonnes, respectively.
4. Why the success? This will be discussed at a later point.
5. See Bjørndal et al. (2004) for details.
6. It must be noted that the state of non-migration has been observed only once in recorded history. The assumption about the stock level of non-migration is therefore somewhat arbitrary. Moreover, should the stock once more be driven to near extinction, there is no guarantee that it will recover to high levels as it did in the 1990s.
7. This is based on Bjørndal and Brasão (2006).
8. ICCAT Contracting parties (web page accessed 5th September 2011) <http://www.iccat.es/en/contracting.htm>.
9. According to the report by MRAG and Fisheries Ecosystems Restoration Research (2008), 'The Global Extent of Illegal Fishing' <www.mrag.co.uk/Documents/ExtentGlobalIllegalFishing.pdf> (accessed 3 November 2011), unreported catches may correspond to 50 per cent of reported catches. In 2006 unreported catches were estimated at 17,000 tonnes. According to Marion et al. (2010), 'Experts estimate that 20,000 tonnes of bluefin tune were caught illegally each year in the ICCAT convention area.' Presumably this refers to years up to 2010.

10. See Bjørndal and Brasão (2006) for other scenarios as well as results for a 10 per cent discount rate.
11. This assumption is not unreasonable for the DWFS fleets in the fishery. It is admittedly very questionable for the coastal state fleets. If the malleability of capital assumption were to be relaxed, the complexity of the model would be much increased.
12. See n. 11.
13. Recall our discussion of the new member problem in chapter 7.
14. Partition function games analysis was introduced to fisheries economics by Pintassilgo (2003). In this article, Pintassilgo develops an empirical model of the East Atlantic bluefin tuna fisheries to demonstrate the relevance of partition function games to the analysis of straddling fish stock (broadly defined) management (Pintassilgo, 2003).
15. In a paper prepared for the OECD programme on the economics of rebuilding fisheries, Sandberg (2010) of the Norwegian Directorate of Fisheries, raises the question of whether the resource investment was worth it, in economic terms, from the Norwegian perspective—did the investment payoff exceed the cost on investment? The theoretical framework for his analysis is provided by the Clark–Munro linear autonomous model that we examined in chapter 3. His conclusion is that the resource investment programme was well worth it from the Norwegian prospective, even though the payoff was decades in appearing.
16. According to Bjørndal (2009), in 2006 high sea catches represented less than 16 per cent of the total. When EU countries fished at full capacity on the high seas in 1996, they landed only a moderate quantity of herring.

REFERENCES

ACFM. 2008. *The Herring Assessment Working Group Report.* Copenhagen: International Council for the Exploitation of the Sea (ICES).

Ackley, G. 1978. *Macroeconomics: Theory and Policy.* New York: Macmillan.

Alexander, L. M. and Hodgson, R. D. 1975. 'The Impact of the 200-Mile Economic Zone on the Law of the Sea', *San Diego Law Review,* 12, 569–99.

Andersen, P. 1983. "On Rent on Fishing Grounds": A Translation of Jens Warming's 1911 Article, with an Introduction', *History of Political Economy,* 15(3), 391–6.

——and Sutinen, J. G. 1984. 'Stochastic Bioeconomics: A Review of Basic Methods and Results', *Marine Resource Economics,* 1, 117–36.

Anderson, J., Curtis, H., Stewart, A.-M., and McShane, H. 2008. *2006 Economic Survey of the UK Fishing Fleet.* Edinburgh: Seafish.

Anderson, L. and Seijo, J. C. 2010. *Bioeconomics of Fisheries Management.* Oxford: Wiley-Blackwell.

Armstrong, C. W. 1994. 'Cooperative Solutions in a Transboundary Fishery: The Russian–Norwegian Co-management of the Arcto-Norwegian Cod Stock', *Marine Resource Economics,* 9(4), 329–51.

——and Flaaten, O. 1991. 'The Optimal Management of a Transboundary Renewable Resource: The Arcto-Norwegian Cod Stock'. In: R. Arnason and T. Bjornadal (eds), *Essays on the Economics of Migratory Fish Stocks.* New York: Springer, pp. 137–52.

——Eide, A., Flaaten, O., and Heen, K. 2011. 'The Norwegian Northeast Arctic Cod Fishery'. In: OECD, *Economics of Rebuilding Fisheries: Case Studies of Fisheries Rebuilding Plans,* Part 2. Paris: OECD, pp. 202–19.

Arnason, R. 1990. 'Minimum Information Management in Fisheries', *Canadian Journal of Economics,* 23, 630–53.

——2007. 'Property Rights Quality and the Economic Efficiency of Fisheries Management Regimes: Some Basic Results'. In: T. Bjørndal, D. V. Gordon, R. Arnason, and U. R. Sumaila (eds), *Advances in Fisheries Economics: Festschrift in Honour of Professor Gordon R. Munro.* Oxford: Wiley-Blackwell, pp. 32–58.

——2010. 'Individual Transferable Quotas: Attributes, Outcomes and Allocation', PowerPoint presentation prepared for the Symposium on Fishing Quotas and Quota Allocation, Fisheries Committee of the Chilean Congress, Valparaiso, 29 July 2010.

——Magnusson, G., and Agnarsson, S. 2000. 'A Game Theoretic Model of the Atlanto-Scandian Herring Fishery', *Marine Resource Economics,* 15(4), 293–320.

Arrow, K. J. 1964. 'The Role of Securities in the Optimal Allocation of Riskbearing', *Review of Economic Studies,* 31, 91–6.

——1968. 'Economic Equilibrium'. In: D. L. Sills (ed.), *International Encyclopedia of the Social Sciences 4.* London and New York: Macmillan and the Free Press, pp. 376–88.

Asche, F. and Bjørndal, T. 2011. *The Economics of Salmon Aquaculture,* 2nd edn. Oxford: Wiley-Blackwell.

Australia. 1989. *New Directions For Commonwealth Fisheries Management in the 1990s: A Government Policy Statement.* Canberra: Government of Australia.

Bacharach, M. 1976. *Economics and the Theory of Games.* London: Macmillan.

Bailey, M., Sumaila, U. R., and Lindroos, M. 2010. 'Application of Game Theory to Fisheries Over Three Decades', *Fisheries Research,* 102: 1–8.

Barrett, S. 2003. *Environment and Statecraft: The Strategy of Environmental Treaty Making.* Oxford: Oxford University Press.

Batstone, C. J. and Sharp, B. 2003. 'Minimum Information Management Systems and ITQ Fisheries Management', *Journal of Environmental Economics and Management*, 45(2), 492–504.

Baumol, W. J. and Oates, W. E. 1998. *The Theory of Environmental Policy*, 2nd edn. Cambridge: Cambridge University Press.

Berck, P. and Perloff, J. 1984. 'An Open Access Fishery with Rational Expectations', *Econometrica*, 52, 489–506.

Bjørndal, T. 1987. 'Production Economics and Optimal Stock Size in a North Atlantic Fishery', *Scandinavian Journal of Economics*, 89, 145–64.

——1988. 'The Optimal Management of North Sea Herring', *Journal of Environmental Economics and Management*, 15, 9–29.

——2002. 'Production Economics and Optimal Stock Size in a North Atlantic Fishery'. In: F. R. Førsund, T. J. Klette, K. Løyland, and K. G. Salvanes (eds), *Microeconometrics of Production: Norwegian Contributions.* Oslo: Cappelen, pp. 141–60.

——2008. 'Rent in the Fishery for Norwegian Spring Spawning Herring'. Study prepared for the FAO–World Bank Project on Rent Estimates in Fisheries.

——2009. 'Overview, Roles, and Performance of the North East Atlantic Fisheries Commission (NEAFC)', *Marine Policy*, 33, 685–97.

——and Asche, F. 2011. *The Economics of Salmon Aquaculture*, 2nd edn. Oxford: Blackwell.

——and Bezabih, M. 2008. *The Economic Benefits of Fisheries Management: Assessing Potential Economic Rents in Sea Fisheries—The Case of Western Channel Sole.* DEFRA report. <http://www.defra.gov.uk/marine/pdf/fisheries/200805econ-fishman-case.pdf >.

————2010. 'Resource Rents and Management Regimes: The Case of the Western Channel Sole Fishery', *Marine Resource Economics*, 25, 325–31.

——and Brasao, A. 2006. 'The East Atlantic Bluefin Tuna Fisheries: Stock Collapse or Recovery?' *Marine Resource Economics*, 21(2), 193–210.

——and Conrad, J. M. 1987a. 'The Dynamics of an Open Access Fishery', *Canadian Journal of Economics*, 20, 74–85.

————1987b. 'Capital Dynamics in the North Sea Herring Fishery', *Marine Resource Economics*, 4, 63–74.

——and Lindroos, M. 2004. 'International Management of North Sea Herring', *Environmental and Resource Economics*, 29, 83–96.

————2012. 'Cooperative and Non-Cooperative Management of the Northeast Atlantic Cod Fishery', *Journal of Bioeconomics*, 14(1), 41–60.

——and Martin, S. 2007. *The Relevance of Bioeconomic Modelling to RFMO Resources: A Survey of the Literature.* Recommended Best Practices for Regional Fisheries Management Organizations: Technical Study No. 3. London: Chatham House.

——and Munro, G. R. 1998. 'The Economics of Fisheries Management: A Survey'. In: T. Tietenberg and H. Folmer (eds), *The International Yearbook of Environmental and Resource Economics 1998/1999.* Cheltenham: Edward Elgar, pp. 153–88.

——Gordon, D. V., Kaitala, V., and Lindroos, V. 2004. 'International Management Strategies for a Straddling Fish Stock: A Bio-Economic Simulation Model of the Norwegian Spring-Spawning Herring Fishery', *Environmental and Resource Economics*, 29(4), 435–57.

————Arnason, R., and Sumaila, U. R. (eds) 2007. *Advances in Fisheries Economics: Festschrift in Honour of Professor Gordon R. Munro.* Oxford: Wiley-Blackwell.

————and Bezabih, M. 2010. 'Rents in the North Sea Herring Fishery in Area IV'. Mimeo. Portmouth: CEMARE, University of Portsmouth.

———2011. 'Measuring Potential Profits in a Bioeconomic Model of the Mixed Demersal Fishery in the North Sea', *Journal of Bioeconomics* (published online 2 October 2011).

Bromley, D. W. 2009. 'Abdicating Responsibility: The Deceits of Fisheries Policy', *Fisheries*, 34(6), 280–90.

Brown, G. 1974. 'An Optimal Program for Managing Common Property Resources with Congestion Externalities', *Journal of Political Economy*, 82, 163–73.

Buergenthal, T. and Murphy, S. D. 2002. *Public International Law*. New York: St. West Group.

Burke, W. T. 1983. '1982 Convention on the Law of the Sea Provisions on Conditions of Access to Fisheries Subject to National Jurisdiction'. In: FAO, *Report of the Expert Consultation on the Conditions of Access to the Fish Resources of the Exclusive Economic Zone, Rome, 11–15 April 1983*, FAO Fisheries Report No. 293. Rome: Food and Agriculture Organization of the United Nations, pp. 23–42.

——1994. *The New International Law of Fisheries: UNCLOS 1982 and Beyond*. Oxford: Clarendon Press.

Caddy, John F. 1997. 'Establishing a Consultative Mechanism or Arrangement for Managing Shared Stocks Within the Jurisdiction of Contiguous States'. In: D. Hancock (ed.), *Taking Stock: Defining and Managing Shared Resources*. Australian Society for Fish Biology and Aquatic Resource Management Association of Australasia Joint Workshop Proceedings, Darwin, NT, 15–16 June 1997. Sydney: Australian Society for Fish Biology, pp. 81–123.

Canada. 1982. *Navigating Troubled Waters: A New Policy for the Atlantic Fisheries*. Report of the Task Force on Atlantic Fisheries. Ottawa: Canadian Government Publishing Centre.

——1997. Department of Fisheries and Oceans, 'Pacific Salmon Treaty: Moving Towards Equity and Conservation', paper prepared by Bud Graham, Director of Fisheries Management, Department of Fisheries and Oceans, Pacific Region, Vancouver.

——2002. Independent Panel on Access Criteria for the Atlantic Coast Commercial Fishery, *Report of the Independent Panel on Access Criteria*, Ottawa.

Cancino, J. P., Uchida, H., and Wilen, J. E. 2007. 'TURFs and ITQs: Collective vs. Individual Decision Making', *Marine Resource Economics*, 22, 391–406.

Casey, K. E., Dewees, C. M., Turris, B. R., and Wilen, J. E. 1995. 'The Effects of Individual Vessel Quotas in the British Columbia Halibut Fishery', *Marine Resource Economics*, 10, 211–30.

Chan, K. S. Y. 1978. 'The Economic Consequence of the 200-mile Seabed Zone', *Revue canadienne d'economique*, 11, 314–18.

Charles, A. T. 1983. 'Optimal Fisheries Investment Under Uncertainty', *Canadian Journal of Fisheries and Aquatic Sciences*, 40, 2080–91.

——1992. 'Fishery Conflicts: A Unified Framework', *Marine Policy*, 16(5), 379–93.

——2001. *Sustainable Fishery Systems*. Fish and Aquatic Resources Series 5. Oxford: Blackwell Science.

——2007. 'Adaptive Co-Management for Resilient Resource Systems: Some Ingredients and the Implications of their Absence'. In: D. Armitage, F. Berkes, and N. Doubleday (eds), *Adaptive Co-Management*. Vancouver: UBC Press.

——and Munro, G. R. 1985. 'Irreversible Investment and Optimal Fisheries Management: A Stochastic Analysis', *Marine Resource Economics*, 1, 247–64.

Christy, F. T., Jr. 1973. 'Fisherman quotas: A Tentative Suggestion for Domestic Management'. Occasional Paper 19. Kingston: Law of the Sea Institute, University of Rhode Island.

——1982. 'Territorial Use Rights in Marine Fisheries: Definitions and Conditions'. FAO Fisheries Technical Paper 227. Rome: Food and Agriculture Organization of the United Nations, 118–35.

Christy, F. T., Jr. 2000. 'Common Property Rights: An Alternative to ITQs'. In: R. Shotten (ed.), *Use of Property Rights in Fisheries Management*, FAO Fisheries Technical Paper 404/1, Rome.

——and Scott, A. D. 1965. *The Common Wealth in Ocean Fisheries.* Baltimore: Johns Hopkins University Press.

Clark, C. W. 1973. 'The Economics of Overexploitation', *Science*, 181, 630–4.

——1976. *Mathematical Bioeconomics: The Optimal Management of Renewable Resources.* New York: Wiley-Interscience.

——1980. 'Restricted Access to a Common Property Resource'. In: P. Liu (ed.), *Dynamic Optimization and Mathematical Economics.* New York: Wiley-Interscience, pp. 117–32.

——1985. *Bioeconomic Modelling and Fisheries Management.* New York: Wiley-Interscience.

——1990. *Mathematical Bioeconomics: The Optimal Management of Renewable Resources.* 2nd edn. New York: John Wiley.

——2006. *The Worldwide Crisis in Fisheries: Economic Models and Human Behavior.* Cambridge: Cambridge University Press.

——2010. *Mathematical Bioeconomics: The Mathematics of Conservation.* 3rd edn. New York: John Wiley.

——and Lamberson, R. 1982. 'An Economic History and Analysis of Pelagic Whaling', *Marine Policy*, 6, 103–20.

——and Munro, G. R. 1978. Renewable Resources and Extinction. *Journal of Environmental Economics and Management*, 5, pp. 198–205.

————1980. 'Fisheries and the Processing Sector: Some Implications for Management Policy', *The Bell Journal of Economics*, 11, 603–16.

————1982. 'The Economies of Fishing and Modern Capital Theory: A Simplified Approach'. In: L. J. Mirman and D. F. Spulber (eds), *Essays in the Economics of Renewable Resources.* Amsterdam: North-Holland, pp. 31–54.

——Clarke, F. H. and Munro, G. 1979. 'The Optimal Management of Renewable Resource Stocks: Problems of Irreversible Investment', *Econometrica*, 47, 25–47.

——Munro, G., and Sumaila, U. R. 2005. 'Subsidies, Buybacks and Sustainable Fisheries', *Journal of Environmental Economics and Management*, 50(1), 47–58.

——————2010a. 'Limits to the Privatization of Fishery Resources', *Land Economics*, 86(2), 209–18.

——————2010b. 'Limits to the Privatization of Fishery Resources: Reply', *Land Economics*, 86(3), 614–18.

Clarke, F. H. and Darrough, M. N. 1983. 'Optimal Employment Contracts in a Principal-Agent Relationship', *Journal of Economic Behavior and Organization*, 4(2–3), 69–90.

——and Munro, G. R. 1987. 'Coastal States, Distant Water Fishing Nations and Extended Jurisdiction: A Principal Agent Analysis', *Natural Resource Modeling*, 2, 81–107.

————1991. 'Coastal States and Distant Water Fishing Nations: Conflicting Views of the Future', *Natural Resource Modeling*, 5(3), 345–69.

Coglan, L. and Pascoe, S. 1999. 'Separating Resource Rents from Intra-marginal Rents in Fisheries', *Agricultural and Resource Economics Review*, 28, 219–28.

Commission on Pacific Fisheries Policy. 1982. *Turning the Tide: A New Policy for Canada's Pacific Fisheries: Final Report.* Canada: Commission on Pacific Fisheries Policy.

Connor, R. and Shallard, B. 2010. 'Evolving Governance in New Zealand Fisheries'. In: R. Q. Grafton, R. Hilborn, D. Squires, M. Tait, and M. J. Williams (eds), *Handbook of Marine Fisheries Conservation and Management.* New York: Oxford University Press, pp. 347–59.

Conrad, J. M. and Clark, C. W. 1987. *Natural Resource Economics: Notes and Problems.* Cambridge: Cambridge University Press.

Copes, P. 1972. *The Resettlement of Fishing Communities in Newfoundland, Ottawa.* Canada: Canadian Council of Rural Development.

——1986. 'A Critical Review of Individual Quota as a Device in Fisheries Management', *Land Economics*, 63(3), 278–91.

Costello, C., Gaines, S. D., and Lynham, J. 2008. 'Can Catch Shares Prevent Fisheries Collapse?' *Science*, 321, 1678–81.

Crutchfield, J. 1975. 'An Economic View of Maximum Sustainable Yield'. In: P. Roedel (ed.), *Optimum Sustainable Yield as a Concept in Fisheries Management.* American Fisheries Society Special Publication No. 9. Washington, DC: American Fisheries Society, pp. 13–19.

——1979. 'Economic and Social Implications of the Main Policy Alternatives for Controlling Fishing Effort', *Journal of the Fisheries Research Board of Canada*, 36, 742–52.

——and Pontecorvo, G. 1969. *The Pacific Salmon Fisheries: A Study of Irrational Conservation.* Baltimore: Johns Hopkins University Press.

——and Zellner, A. 1962. *Economic Aspects of the Pacific Halibut Fishery.* Washington, DC: Department of the Interior.

————(eds) 2003. *The Economics of Marine Resources and Conservation Policy: The Pacific Halibut Case Study with Commentary.* Chicago: University of Chicago Press.

Cunningham, S., Nguer, A., and Thiam, I. 1995. 'Monopsonistic Fisheries Management: The Case of Mauritania'. In: David S. Liao (ed.), *International Cooperation for Fisheries and Aquaculture Development*, Vol. 1. Taiwan: National Taiwan Ocean University, pp. 305–17.

——Iyaye, S., and Zeine D. 2005. 'The Experience of the Mauritanian Fish Trading Company (SMCP) in the Management of the Fisheries Sector in Mauritania'. In: S. Cunningham and T. Bostock (eds), *Successful Fisheries Management: Issues, Case Studies and Perspectives.* Delft: Eburon Publishers, pp. 75–89.

Curtis, R. and Squires, D. 2007. *Fisheries Buybacks.* Oxford: Blackwell.

Dorfman, R. and Dorfman, N. 1972. *Economics of the Environment.* New York: W. W. Norton (3rd edn. 1993).

Doulman, D. J. 1987. *Options for U.S. Fisheries Investment in the Pacific Islands Region.* Issue 8 of Research Report Series, Pacific Islands Development Program. Honolulu: East-West Center.

Doyle, M., Singh, R., and Weninger, Q. 2007. 'Fisheries Management with Stock Uncertainty and Costly Capital Adjustment'. In: T. Bjørndal, D. V. Gordon, R. Arnason, and U. R. Sumaila (eds), *Advances in Fisheries Economics: Festschrift in Honour of Professor Gordon R. Munro.* Oxford: Wiley-Blackwell, pp. 137–55.

Dupont, D. and Gordon, D. V. 2007. 'Shadow Prices for Fishing Quotas: Fishing with Econometrics'. In: T. Bjørndal, D. V. Gordon, R. Arnason, and U. R. Sumaila (eds), *Advances in Fisheries Economics: Festschrift in Honour of Professor Gordon R. Munro.* Oxford: Wiley-Blackwell, pp. 87–106.

Eckert, R. D. 1979. *The Enclosure of Ocean Resources: Economics and the Law of the Sea.* Stanford: Hoover Institute Press.

Eide, A. and Flaaten, O. 1998. 'Bioeconomic Multispecies Models of the Barents Sea Fisheries'. In: T. Rødseth (ed.), *Models for Multispecies Management.* Heidelberg: Physica-Verlag, pp. 141–72.

Flaaten, O. 1988. *The Economics of Multispecies Harvesting: Theory and Applications.* Heidelberg: Springer-Verlag.

——1991. 'Bioeconomics of Sustainable Harvest of Competing Species', *Journal of Environmental Economic and Management*, 20, 163–80.

——2011. *Fisheries Economics and Management.* Tromsø: Norwegian College of Fishery Science, University of Tromsø.

Food and Agriculture Organization of the United Nations. 1992. *Marine Fisheries and the Law of the Sea: A Decade of Change*. FAO Fisheries Circular No. 853. Rome: FAO.
——1994. *World Review of Highly Migratory Species and Straddling Stocks*. FAO Fisheries Technical Paper 337. Rome: FAO.
——1995. 'Code of Conduct for Responsible Fisheries'. Rome: FAO.
——1998. *Managing Fishing Capacity: Selected Papers on Underlying Concepts*. FAO Fisheries Technical Paper 386. Rome: FAO.
——2001a. *International Plan of Action to Prevent, Deter and Eliminate Illegal, Unreported and Unregulated Fishing*. Rome: FAO.
——2001b. *Report of the Reykjavik Conference on Reponsible Fisheries in the Marine Ecosystem, Reykjavik, Iceland, 1–4 October 2001*. FAO Fisheries Report No. 658. Rome: FAO.
——2002. *Report of the Norway-FAO Expert Consultation on the Management of Shared Fish Stocks Bergen, Norway, 7–10 October 2002*. FAO Fisheries Report No. 695. Rome: FAO.
——2005. *Review of the State of World Marine Fishery Resources*. Fisheries Technical Paper 457. Rome: FAO.
——2010. *The State of World Fisheries and Aquaculture 2010*. Rome: FAO.
——2011. 'The Ecosystem Approach to Fisheries Management'. Available at <http://www.fao.org/fishery/topic/13261/en>.
Fraser, A. 1979. 'Limited Entry: Experience of the British Columbia Salmon Fishery', *Journal of the Fisheries Research Board of Canada*, 36, 754–63.
Gaudet, G. 2007. 'Natural Resource Economics Under the Rule of Hotelling', *Canadian Journal of Economics*, 40(4), 1033–59.
Gillett, R. and Cartwright, I. 2010. *The Future of Pacific Island Fisheries*. Noumea, New Caledonia: Secretariat of the Pacific Community and Honiara: The Forum Fisheries Agency.
Gooday, P., Kompas, T., Che, N., and Curtotti, R. 2010. 'Harvest Strategy Policy and Stock Rebuilding for Commonwealth Fisheries in Australia'. In: *OECD, The Economics of Rebuilding Fisheriess: Workshop Proceedings*. Paris: Organisation for Economic Co-operation and Development (OECD), pp. 113–39.
Gordon, H. S. 1953. 'An Economic Approach to the Optimum Utilization of Fishery Resources', *Journal of the Fisheries Research Board of Canada*, 10(7), 442–57.
——1954. 'The Economic Theory of a Common Property Resource: The Fishery', *Journal of Political Economy*, 62, 124–42.
——1956. 'Obstacles to Agreement on Control in the Fishing Industry'. In: R. Turvey and J. Wiseman (eds), *The Economics of Fisheries*. Rome: FAO, pp. 65–72.
Grafton, R. Q. 1995. 'Rent Capture in a Rights-Based Fishery', *Journal of Environmental Economics and Management*, 28(1), 48–67.
——and Nelson, H. W. 2007. 'The Effects of Buy-back Programs in the British Columbia Salmon Fishery'. In: R. Curtis and D. Squires (eds), *Fisheries Buybacks*. Oxford: Blackwell Publishing, pp. 191–202.
——Kompas, T., and Hilborn, R. 2007. 'Economics of Overexploitation Revisited', *Science*, 318, 1601.
————2010. 'Limits to the Privatization of Fisheries Resources: Comment', *Land Economics*, 86, 609–13.
——Nelson, H. W., and Turris, B. R. 2007. 'Resolving the Class II Common Property Problem: The Case of the B.C. Groundfish Trawl Fishery'. In: T. Bjørndal, D. V. Gordon, R. Arnason, and U. R. Sumaila (eds), *Advances in Fisheries Economics: Festschrift in Honour of Professor Gordon R. Munro*. Oxford: Wiley-Blackwell, pp 59–73.

——Squires, D., and Fox, K. J. 2000. 'Private Property and Economic Efficiency: A Study of a Common-Pool Resource', *Journal of Law and Economics*, 43(2), 679–712.

Gray, K. 1991. 'Property in Thin Air', *Cambridge Law Journal*, 50, 252–307.

Gréboval, D. and Munro, G. 1999. 'Overcapitalization and Excess Capacity in World Fisheries: Underlying Economics and Methods of Control'. In: D. Gréboval (ed.), *Managing Fishing Capacity: Selected Papers on Underlying Concepts and Issues.* FAO Fisheries Technical Paper No. 386. Rome: FAO, pp. 1–48.

Hanna, S. 2010. 'Managing the Transition: Distributional Issues of Fish Stock Rebuilding', in OECD, *The Economics of Rebuilding Fisheries: Workshop Proceedings.* Paris: Organisation for Economic Co-operation and Development (OECD), pp. 141–66.

Hannesson, R. 1997. 'Fishing as a Supergame', *Journal of Environmental Economics and Management*, 32, 309–22.

——2006. 'Sharing the Herring: Fish Migrations, Strategic Advantage, and Climate Change'. In R. Hannesson, M. Barange, and S. Herrick (eds), *Climate Change and the Economics of the World's Fisheries: Examples of Small Pelagic Stocks.* Cheltenham: Edward Elgar, pp. 66–99.

Hicks, J. R. 1946. *Value and Capital*, 2nd edn. Oxford: Oxford University Press.

Hilborn, R. and Sibert, J. 1988. 'Is International Management of Tuna Necessary?' *Marine Policy*, 12, 31–39.

——Maguire, J. J., Parma, A. M., and Rosenberg, A. A. 2001. 'The Precautionary Approach and Risk Management: Can They Increase the Probability of Success in Fisheries Management?' *Canadian Journal of Fisheries and Aquatic Sciences*, 58, 99–107.

Holland, D. 2010. 'Self Selecting Sectors for the New England Fisheries', PowerPoint presentation prepared for the Gulf of Maine Research Institute, Portland, Maine.

——Gudmundsson, E., and Gates, J. 1999. 'Do Fishing Vessel Buyback Programs Work? A Survey of the Evidence', *Marine Policy*, 23(1), 47–69.

Horwood, J. 1999. 'The Precautionary Approach', *Fishing News*, 30 April.

Huppert, D. 1995. *Why the Pacific Salmon Treaty Failed to End the Salmon Wars.* Seattle: School of Marine Affairs, University of Washington, SMA 95-1.

——2007. 'Auctions of IFQs as a Means to Share the Rent'. In: T. Bjørndal, D. V. Gordon, R. Arnason, and U. R. Sumaila (eds), *Advances in Fisheries Economics: Festschrift in Honour of Professor Gordon R. Munro.* Oxford: Wiley-Blackwell, pp. 74–86.

Jensen, M. C, 1986. 'The Takeover Controversy: Analysis and Evidence', *Midland Corporate Finance Journal*, 4, 6–32.

Jörgensen, H. and Jensen, C. 1999. 'Overcapacity, Subsidies and Local Stability'. In: A. Hatcher and K. Robinson (eds), *Overcapacity, Overcapitalization and Subsidies in European Fisheries.* University of Portsmouth: Centre for the Economics and Management of Aquatic Resources, pp. 239–52.

Kaitala, V. T. 1985. *Game Theory Models of Dynamic Bargaining and Contracting in Fisheries Management.* Helsinki: Institute of Mathematics, Helsinki University of Technology.

——and Lindroos, M. 2007. 'Game Theoretic Application to Fisheries'. In: A. Weintraub, C. Romero, T. Bjørndal, and R. Epstein (eds), *Handbook of Operations Research in Natural Resources.* New York: Springer, pp. 201–16.

——and Munro, G. R. 1997. 'The Conservation and Management of High Seas Fishery Resources', *Natural Resource Modeling*, 10, 87–108.

——and Pohjola, M. 1988. 'Optimal Recovery of a Shared Resource Stock: A Differential Game Model with Efficient Memory Equilibria', *Natural Resource Modeling*, 3, 91–117.

Keynes, J. M. 1936. *The General Theory of Employment, Interest and Money.* London: Macmillan, reprinted 2007, Palgrave Macmillan.

Khwaja, S. and Cox, A. 2010. 'Rebuilding Fisheries: An Overview of Issues and Policy Approaches in the OECD'. In: OECD, *The Economics of Rebuilding Fisheries: Workshop Proceedings*. Paris: Organisation for Economic Co-operation and Development (OECD), pp. 5–26.

Kobayashi, S. 2007. 'A Dynamic Game on Renewable Natural Resource Exploitation and Markov Perfect Equilibrium'. In: T. Bjørndal, D. V. Gordon, R. Arnason, and U. R. Sumaila (eds), *Advances in Fisheries Economics: Festschrift in Honour of Professor Gordon R. Munro*. Oxford: Wiley-Blackwell, pp. 222–36.

Kompas, T., Grafton, R. Q., Che, N., and Gooday, P. 2009. *Development of Methods and Information to Support the Assessment of Economic Performance in Commonwealth Fisheries*. Canberra: ABARE.

Kronbak, L. G. and Lindroos, M. 2006. 'An Enforcement-Coalition Model: Fishermen and Authorities Forming Coalitions', *Environmental and Resource Economics*, 35, 169–94.

Kurien, J. and Willmann, R. 2009. 'Special Considerations for Small-Scale Fisheries Management in Developing Countries'. In: K. L. Cochrane and S. M. Garcia (eds), *A Fishery Manager's Guidebook*. Rome: FAO and Oxford: Blackwell Publishing, pp. 404–24.

Kydland, F. and Prescott, E. 1977. 'Rules Rather than Discretion: The Inconsistency of Optimal Plans', *Journal of Political Economy*, 85, 473–90.

Lane, D. E. and Stephenson, R. L. 1998. 'A Framework for Risk Analysis in Fisheries Decision-Making', *ICES Journal of Marine Science*, 55, 1–13.

Lauck T., Clark C. W., Mangel M., and Munro G. R. 1998. 'Implementing the Precautionary Principle in Fisheries Management Through Marine Reserves', *Journal* of *Applied Ecology*, 8 (Suppl.), S72–S78.

Lewis, T. R. 1981. 'Exploitation of a Renewable Resource Under Uncertainty', *Canadian Journal of Economics*, 14(3), 422–39.

Li, E. A. L. 1999. 'Cooperative High-Seas Straddling Stock Agreement as a Characteristic Function Game', *Marine Resource Economics*, 13, 247–58.

Libecap, G. D. 2007. 'Assigning Property Rights in the Common Pool: Implications of the Prevalence of First-Possession Rules for ITQs in Fisheries 2007', *Marine Resource Economics*, 22(1), 407–24.

——2010. 'Rights-Based Management of Tuna Fisheries: Lessons from the Assignment of Property Rights on the Western US Frontier'. In: R. Allen, J. A. Joseph, and D. Squires (eds), *Conservation and Management of Transnational Tuna Fisheries*. Oxford: Wiley-Blackwell.

Lindroos, M., Kaitala, V., and Kronbak, L. G. 2007. 'Coalitions in Fisheries Games'. In: T. Bjørndal, D. V. Gordon, R. Arnason, and U. R. Sumaila (eds), *Advances in Fisheries Economics: Festschrift in Honour of Professor Gordon R. Munro*. Oxford: Wiley-Blackwell, pp. 184–95.

Little, I. M. D. and Mirrlees, J. A. 1968. *Manual of Industrial Project Analysis in Developing Countries*. Paris: Development Centre of the Organisation for Economic Co-Operation and Development.

Lodge, M. W., Anderson, D., Løbach, T., Munro, G., Sainsbury, K. and Willock, A. 2007. *Recommended Best Practices for Regional Fisheries Management Organizations: Report of an Independent Panel to Develop a Model for Improved Governance by Regional Fisheries Management Organizations*. London: Chatham House.

Logan, R. M. 1974. *Canada, the United States and the Third Law of the Sea Conference*. Montreal: The Canadian-American Committee, C. D. Howe Research Institute and The National Planning Association.

McConnell, K. E. and Norton, V. J. 1978. 'An Evaluation of Limited Entry and Alternative Fishery Management Schemes'. In: R. B. Rettig and J. J. C. Gintner (eds), *Limited Entry as a Fishery Management Tool.* Seattle: Washington Sea Grant, University of Washington Press.

McKelvey, R. 1985. 'Decentralized Regulation of a Common Property Renewable Resource Industry with Irreversible Investment', *Journal of Environmental Economics and Management*, 12, 287–307.

——1987. 'Fur Seal and Blue Whales: The Bioeconomics of Extinction'. In: Y. Cohen (ed.), *Applications of Control Theory in Ecology.* Heidelberg: Springer-Verlag, pp. 52–82.

——Golubtsov, R. V., Cripe, G., and Miller, K. A. 2007. 'The Incomplete Information Stochastic Split-Stream Model: An Overview'. In: T. Bjørndal, D. V. Gordon, R. Arnason, and U. R. Sumaila (eds), *Advances in Fisheries Economics: Festschrift in Honour of Professor Gordon R. Munro.* Oxford: Wiley-Blackwell, pp. 159–83.

MacKenzie, B. R., Mosegaard, H., and Rosenberg, A. A. 2009. 'Impending Collapse of Bluefin Tuna in the Northeast Atlantic and Mediterranean', *Conservation Letters*, 2, 26–35.

McRae, D. M. and Munro, G. R. 1989a. *Canadian Oceans Policy: National Strategies and the New Law of the Sea.* Vancouver: University of British Columbia Press.

————1989b. 'Coastal State "Rights" Within the 200-mile Exclusive Economic Zone', in P. A. Neher, R. Arnason, and N. Mollett (eds), *Rights Based Fishing.* Dordrecht, Kluwer Academic Publishers, pp. 97–112.

Maloney, D. G. and Pearse, P. H. 1979. 'Quantitative Rights as an Instrument for Regulating Commercial Fisheries', *Journal of the Fisheries Research Board of Canada*, 36, 859–66.

Manning, P. 2005. 'The Namibian Hake Fishery'. In: S. Cunningham and T.Bostock (eds), *Successful Fisheries Management: Issues, Case Studies and Perspectives.* Delft, Netherlands: Eburon Academic Publishers, pp. 169–96.

Marine Conservation Society. 2007. 'Fish Online: Sole'. <http://www.fishonline.org/search/advanced/?step=2&fish_id=31>.

Marion, G., Furtado, T., Proaño, L., Corridoni, L., Al Musalli, M. and Blanca, M. 2010. 'Overfishing and the Case of the Atlantic Bluefin Tuna', paper presented at the 3rd UPC International Seminar on Sustainable Technology Development, 11th–18th June 2010, Universitat Politecnica de Catalunya. Available at <http://www.upc.edu/sostenible2015/menu3/Seminaris/Seminari_STD_10/docs/presentacions/grup%206.pdf> (accessed 3 November 2011).

Marshall, A. 1927. *Principles of Economics*, 8th edn. London: Macmillan.

Matthiasson, T. and Agnarsson, H. 2010. 'Property Rights in Icelandic Fisheries'. In: R. Q. Grafton, D. Squires, M. Tait, and M. Williams (eds), *Handbook of Marine Fisheries Conservation and Management.* New York: Oxford University Press, pp. 299–308.

Mesterton-Gibbons, M. 1993. 'Game-Theoretic Resource Modeling', *Natural Resource Modeling*, 7, 93–147.

Milazzo, M. 1998. *Subsidies in World Fisheries: A Reexamination.* World Bank Technical Paper No. 406, Fisheries Series. Washington DC: World Bank.

Miles, E. and Burke, W. 1989. 'Pressures on the United Nations Convention on the Law of Sea of 1982 Arising from the New Fisheries Conflicts: The Problem of Straddling Stocks', *Ocean Development and International Law*, 20, 343–57.

Miller, K. A. 2007. 'Fish Stew: Uncertainty, Conflicting Interests and Climate Regime Shifts'. In: T. Bjørndal, D. V. Gordon, R. Arnason, and U. R. Sumaila (eds), *Advances in Fisheries Economics: Festschrift in Honour of Professor Gordon R. Munro.* Oxford: Wiley-Blackwell, pp. 207–21.

Miller, K. A. and Munro, G. R. 2004. 'Climate and Cooperation: A New Perspective on the Management of Shared Fish Stocks', *Marine Resource Economics*, 19, 367–93.

————McDorman, T., McKelvey, R., and Tyedmers, P. 2001. *The 1999 Pacific Salmon Agreement: A Sustainable Solution?* Canadian-American Public Policy, Occasional Papers No. 47. Orono: Canadian-American Centre, University of Maine.

Moma, A. 2007. *The Potential of Artisanal Hilsa Fishery: Economically Efficient Fisheries Policy.* Department of Fisheries, Bangladesh and University of Iceland/United Nations University.

Munro, G. R. 1979. 'The Optimal Management of Transboundary Renewable Resources', *Canadian Journal of Economics*, 3, 271–96.

——1980. *A Promise of Abundance: Extended Fisheries Jurisdiction and the Newfoundland Economy.* Ottawa: Economic Council of Canada.

——1982. 'Fisheries, Extended Jurisdiction and the Economics of Common Property Resources', *Canadian Journal of Economics*, 15, 405–25.

——1983. 'Foreign Access to EEZs and the Derivation of Coastal State Benefits: Methods and Techniques'. In: Food and Agricultural Organization of the United Nations (ed.), *Report of the Expert Consultation on Conditions of Access to Fish Resources of the Exclusive Economic Zone.* Rome: FAO, pp. 143–56.

——1987. 'The Management of Shared Fishery Resources under Extended Jurisdiction', *Marine Resource Economics*, 3, 271–96.

——1990. 'The Optimal Management of Transboundary Fisheries: Game Theoretic Considerations', *Natural Resource Modeling*, 4, 403–26.

——1992. 'Evolution of Canadian Fisheries Management under the New Law of the Sea: International Dimensions'. In: C. Cutter and M. Zacher (eds), *Canadian Foreign Policy and International Economic Regimes.* British Columbia: University of British Columbia Press, pp. 289–310.

——1994. 'Coastal States and Distant Water Fleets Under Extended Jurisdiction: The Search for Optimal Incentive Schemes', *Annals of the International Society of Dynamic Games*, 1, 301–17.

——2000. 'On the Management of Shared Fishery Resources'. In: A. Hatcher and D. Tingley (eds), *International Relations and the Common Fisheries Policy.* University of Portsmouth: Centre for the Economics and Management of Aquatic Resources, pp. 149–62.

——2007. 'Internationally Shared Fish Stocks, the High Seas, and Property Rights in Fisheries', *Marine Resource Economics*, 22(4), 425–43.

——2009. 'Game Theory and the Development of Resource Management Policy: The Case of International Fisheries', *Environment and Development Economics*, 14, 7–27.

——2010a. *From Drain to Gain in Capture Fisheries Rents: A Synthesis Study.* FAO Fisheries and Aquaculture Technical Paper 538, Rome: FAO.

——2010b. 'The Economics of Overcapacity and the Management of Capture Fishery Resources: A Review', *International Review of Environmental and Resource Economics*, 4(2), 93–122.

——2011. 'On the Management of Living Marine Resources', paper prepared for the Danish Council on Environmental Economics Conference 2011, Skodsborg, 5–6 September, <http://www.dors.dk/sw8608.asp>.

——and Clark, C. W. 1979. *The Economics of Fishing: A Dynamic Approach: Supplement to The Economics of Fishing and the Developing World: A Malaysian Case Study by G. R. Munro and Chee Kim Loy.* Pulau Pinang, Malaysia: Penerbit Universiti Sains.

————2003. 'Fishing Capacity and Resource Management Objectives'. In: S. Pascoe and D. Gréboval (eds), *Measuring Capacity in Fisheries.* FAO Fisheries Technical Paper No. 445. Rome: FAO, pp.13–34.

——and Loy, C. K. 1978. *The Economics of Fishing and the Developing World: A Malaysian Case Study.* Pulau Pinang, Malaysia: Penerbit Universiti Sains.

——and Scott, A. D. 1985. 'The Economics of Fisheries Management'. In: A. V. Kneese and J. L. Sweeney (eds), *Handbook of Natural Resources and Energy Economics*, vol. II. Amsterdam: North Holland, pp. 623–76.

——and Sumaila, U. R. 2002. 'The Impact of Subsidies Upon Fisheries Management and Sustainability: The Case of the North Atlantic', *Fish and Fisheries*, 3, 233–90.

————2011. 'Lutter Contre la Pêche Illégale'. In: P. Jacquet, R. Pachuari, and L. Tubiana (eds), *Regards Sur la Terre 2011: Océans La Nouvelle Frontière.* Paris: Armand Colin, pp. 265–73.

——McDorman, T., and R. McKelvey. 1998. *Transboundary Fishery Resources and the Canadian–United States Pacific Salmon Treaty.* Canadian-American Public Policy, Occasional Papers No. 33. Orono: Canadian-American Centre, University of Maine.

——Turris, B., Clark, C., Sumaila, U. R., and Bailey M. 2009. *Impacts of Harvesting Rights in Canadian Pacific Fisheries.* Statistical and Economic Analysis Series, Economic Analysis No. 1–3. Ontario: Economic Analysis and Statistics Branch.

——Van Houtte, A., and Willmann, R. 2004. *The Conservation and Management of Shared Fish Stocks: Legal and Economic Aspects.* Food and Agriculture Organization, Fisheries Technical Paper No. 465. Rome: FAO.

Nash, J. 1951. 'Non-Cooperative Games', *Annals of Mathematics*, 54(2), 286–95.

——1953. 'Two-Person Cooperative Games', *Econometrica*, 21, 128–40.

National Research Council. 1999. *Sustaining Marine Fisheries.* Washington, DC: National Academy Press.

Neher, P. A. 1990. *Natural Resource Economics: Conservation and Exploitation.* Cambridge: Cambridge University Press.

Northwest Atlantic Fisheries Organization. 2011. <http://www.nafo.int>.

Nøstbakken, L. 2006. 'Regime Switching in a Fishery with Stochastic Stock and Price', *Journal of Environmental Economics and Management*, 51, 231–41.

——2008. 'Stochastic Modelling of the North Sea Herring Fishery under Alternative Management Regimes', *Marine Resource Economics*, 22, 65–86.

——and Bjørndal, T. 2003. 'Supply Functions for North Sea Herring', *Marine Resource Economics*, 18, 345–61.

——and Conrad, J. M. 2007. 'Uncertainty in Bioeconomic Modelling'. In: A. Weintraub, R. Epstein, T. Bjørndal, and C. Romero (eds), *Handbook on Operations Research in Natural Resources.* Dordrecht: Kluwer, pp. 217–35.

OECD. 2000. *Transition to Responsible Fisheries: Economic and Policy Implications.* Paris: OECD.

——2009. *Strengthening Regional Fisheries Management Organizations.* Paris: OECD.

——2010. *The Economics of Rebuilding Fisheries: Workshop Proceedings.* Paris: OECD.

Örebech, P., Sigurjonsson, K., and McDorman, T. L. 1998. 'The 1995 United Nations Straddling and Highly Migratory Fish Stocks Agreement: Management, Enforcement and Dispute Settlement', *International Journal of Marine and Coastal Law*, 15, 361–78.

Orrego Vicuña, F. 1999. *The Changing International Law of International Fisheries.* Cambridge: Cambridge University Press.

Oxford Dictionaries Online. 2011. <http://oxforddictionaries.com>.

Oxford University Press. 2011. *Oxford Dictionaries.* Oxford: Oxford University Press. <http://oxforddictionaries.com>.

Pearce, D. W. and Turner, R. K. 1990. *Economics of Natural Resources and the Environment.* Exeter: BPCC Wheatsons Ltd.

Pigou, A. C. 1912. *Wealth and Welfare.* London: Macmillan.

Pintassilgo, P. 2003. 'A Coalition Game Approach to the Management of High Seas Fisheries in the Presence of Externalities', *Natural Resource Modeling*, 16(2), 175–97.

——and Costa Duarte, C. 2000. 'The New-Member Problem in the Cooperative Management of High Seas Fisheries', *Marine Resource Economics*, 15(4), 361–78.

——and Lindroos, M. 2008. 'Coalition Formation in High Seas Fisheries: A Partition Function Approach', *International Game Theory Review*, 10, 303–17.

——Finus, M., Lindroos, M., and Munro, G. 2010. 'Stability and Success of Regional Fisheries Management Organizations', *Environmental and Resource Economics*, 46(3) 377–402.

Prescott, E. C. 2004. 'The Transformation of Macroeconomic Research and Policy', <http://www.nobelprize.org/nobel_prizes/economics/laureates/2004/prescott-lecture.pdf>.

Reed, W. J. 1979. 'Optimal Escapement Levels in Stochastic and Deterministic Harvesting Models', *Journal of Environmental and Economic Management*, 6, 350–63.

Reinheimer, H. 1913. *Evolution by Co-operation: A Study in Bioeconomics*. London: Kegan, Paul, Trench, and Trübner.

Robinson, J. 1965. *The Economics of Imperfect Competition*. London: Macmillan.

Roth, A. 1979. *Axiomatic Models of Bargaining*. Lecture Notes in Mathematics and Mathematical Systems 170. Heidelberg: Springer Verlag.

Roy, N. 1996. 'What Went Wrong and What Can we Learn From it?' In: D. V. Gordon and G. R. Munro (eds), *Fisheries and Uncertainty: A Precautionary Approach to Resource Management*. Calgary: University of Calgary Press, pp. 15–26.

Safina, C., Rosenberg, A. A., Meyers, R. A., Quinn II, T. J., and Collie, J. S. 2005. 'U.S. Ocean Recovery: Staying the Course', *Science*, 309, 707–8.

Salagrama, V. 2005. *Investigating the Linkages between Fisheries, Poverty and Growth: A Case Study for India*. London: Department for International Development (DFID).

San Martín, G., Parma, A. M., and Orensanz, J. M. L. 2010. 'The Chilean Experience with Territorial Use Rights in Fisheries'. In: R. H. R. Q. Grafton, D. Squires, M. Tait, and M. Williams (eds), *Handbook of Marine Fisheries Conservation and Management*. New York: Oxford University Press, pp. 324–37.

Sanchirico, J. N. and Wilen, J. E. 1999. 'Bioeconomics of Spatial Exploitation in a Patchy Environment', *Journal of Environmental Economics and Management*, 37, 129–50.

————2005. 'Optimal Spatial Management of Renewable Resources: Matching Policy Scope to Ecosystem Scale', *Journal of Environmental Economics and Management*, 50(1), 23–46.

Sandberg, P. 2010. 'Rebuilding the Stock of Norwegian Spring Spawning Herring: Lessons Learned'. In: OECD, *The Economics of Rebuilding Fisheries: Workshop Proceedings*. Paris: Organisation for Economic Co-operation and Development, pp. 219–33.

Sappington, D. E. M. 1991. 'Incentives in Principal-Agent Relationships', *Journal of Economic Perspectives*, 5(2), 45–66.

Schaefer, M. B. and Beverton, R. H. J. 1963. 'Fishing Dynamics: Their Analysis and Interpretation'. In: M. N. Hill (ed.), *The Sea*, vol. II. New York: Interscience, pp. 464–83.

Schrank, W. E. and Keithly, W. R. 1999. 'Thalassorama: The Concept of Subsidies', *Marine Resource Economics*, 14, 151–64.

Scott, A. D. 1955. 'The Fishery: The Objectives of Sole Ownership', *Journal of Political Economy*, 63, 116–24.

——1955 and 1983. *Natural Resources: The Economics of Conservation*. Toronto: University of Toronto Press (1955 edn.); Ottawa: Carleton University Press (1983 edn.).

——2007. 'Phases in the Evolution of Property in Sea Fisheries'. In: T. Bjørndal, D. V. Gordon, R. Arnason, and U. R. Sumaila (eds), *Advances in Fisheries Economics: Festschrift in Honour of Professor Gordon R. Munro*. Oxford: Wiley-Blackwell, pp. 17–31.

Serdy, A. 2007. 'Trading of Fishery Commission Quota Under International Law', *Ocean Yearbook*, 21, 265–88.

Shapley, L. S. 1953. 'A Value for n-Person Games'. In: A.W. Tucker and R. D. Luce (eds), *Contributions to the Theory of Games II*. Princeton: Princeton University Press, pp. 307–17.

Sharp, B. M. H. and Batstone, C. 2007. 'Factor Use and Productivity Change in a Rights Based Fishery'. In: T. Bjørndal, D. V. Gordon, R. Arnason, and U. R. Sumaila (eds), *Advances in Fisheries Economics: Festschrift in Honour of Professor Gordon R. Munro*. Oxford: Wiley-Blackwell, pp. 257–69.

Sinclair, S. 1960. *Licence Limitation—British Columbia: A Method of Economic Fisheries Management*. Ottawa: Department of Fisheries.

——1978. *A Licensing and Fee System for the Coastal Fisheries of British Columbia*. Ottawa: Department of Fisheries and Oceans.

Sissener, E. and Bjørndal, T. 2004. 'Climate Change and the Migratory Pattern for Norwegian Spring Spawning Herring', *Marine Policy*, 29, 299–309.

Solow, R. M, 1974. 'The Economics of Resources or the Resources of Economics', *American Economic Review*, 64(2), 1–14.

South Pacific Forum Fisheries Agency. 2011. <http://www.ffa.int/>.

Squires, D., Grafton, Q., Ferdous, M., and Haji, I. 1998. *Where the Land Meets the Sea: Integrated Sustainable Fisheries Development and Artisanal Fishing*. Discussion paper 98-26. Department of Economics, University of California, San Diego.

Stokke, O. S. 2003. *Trade Measures and Climate Compliance: Institutional Interplay between WTO and the Marrakesh Accords*. Lysaker: Fridtjof Nansen Institute.

Sumaila, U. R. 1995. 'Irreversible Capital Investment in a Two-Stage Bimatrix Game Model', *Marine Resource Economics*, 10(3), 163–83.

——Teh, L., Watson, R., Tyedmers, P., and Pauly, D. 2008. 'Fuel Price Increase, Subsidies, Overcapacity and Resource Sustainability', *ICES Journal of Marine Science*, 65(6): 832–40.

Sutinen, J. and Andersen, P. 1985. 'The Economics of Fisheries Law Enforcement', *Land Economics*, 61, 387–97.

Thomson, W. 1989. 'Competitive Models of Bargaining'. Working Paper No. 177. University of Rochester Centre for Economic Research.

Townsend, R., Shotten, R. and Uchida, H. (eds) 2008. *Case Studies in Fisheries Self-governance*. Food and Agriculture Organization, Fisheries Technical Paper No. 504. Rome: FAO.

Treaty between the Government of Canada and the Government of the United States of America Concerning Pacific Salmon, March 1985.

Tsoa, E. 1996. 'The Collapse of the Northern Cod Fishery: Predator–Prey and Other Considerations'. In: D. V. Gordon and G. R. Munro (eds), *Fisheries and Uncertainty: A Precautionary Approach to Resource Management*. Calgary: University of Calgary Press, pp. 45–59.

Turris, B. 2000. 'A Comparison of British Columbia's ITQ Fisheries for Groundfish and Sablefish: Similar Results from Programmes with Differing Objectives, Designs and Processes'. In: R. Shotton (ed.), *Use of Property Rights in Fisheries Management*. FAO Fisheries Technical Paper 404/1. Rome: FAO, pp. 254–61.

United Nations. 1982. *United Nations Convention on the Law of the Sea*. UN Doc. A/Conf.62/122.

——1993. United Nations Conference on Straddling Fish Stocks and Highly Migratory Fish Stocks. 'Draft Convention on the Conservation and Management of Straddling Fish Stocks on the High Seas and Highly Migratory Fish Stocks on the High Seas'. UN Doc. A/Conf./164/L11/Ref.1.

——1995. United Nations Conference on Straddling Fish Stocks and Highly Migratory Fish Stocks. 'Agreement for the Implementation of the Provisions of the United Nations

Convention on the Law of the Sea of 10 December 1982 Relating to the Conservation and Management of Straddling Fish Stocks and Highly Migratory Fish Stocks' (United Nations Fish Stocks Agreement (UNFSA)). UN Doc. A/Conf./164/37.

Vestergaard, N. 2010. 'Principal-Agent Problems in Fisheries'. In: R. Q. Grafton, R. Hilborn, D. Squires, M. Tait, and M. Williams (eds), *Handbook of Marine Fisheries Conservation and Management.* Oxford: Oxford University Press, pp. 563–71.

Vousden, N. 1973. 'Basic Theoretical Issues of Resource Depletion', *Journal of Economic Theory,* 6(2), 126–43.

Walters, C. J. and Hilborn, R. 1978. 'Ecological Optimisation and Adaptive Management', *Annual Review of Ecology and Systematics,* 9, 157–88.

——and Martell, S. J. D. 2004. *Fisheries Ecology and Management.* Princeton: Princeton University Press.

Warming, J. 1911. 'Om Grundrente af Fiskegrunde', *Nationaløkonomisk Tidsskrift,* 49, 499–505.

Warui, S. W. 2008. *Rents and Rents Drain in the Lake Victoria Nile Perch Fishery.* Kenya, Ministry of Livestock and Fisheries Development; and University of Iceland/United Nations University.

Weitzman, M. L. 2002. 'Landing Fees vs. Harvest Quotas with Uncertain Fish Stocks', *Journal of Environmental Economics and Management,* 43, 325–38

Western Central Pacific Fisheries Commission (WCPFC). 2011. <http//www.wcpfc.int >.

Whitmarsh, D. 2011. *Economic Management of Marine Living Resources: A Practical Introduction.* London: Earthscan.

Wilen, J. E. 1985. 'Towards a Theory of the Regulated Fishery', *Marine Resource Economics,* 1, 69–388.

World Bank. 2005. *Where is the Wealth of Nations? Measuring Capital for the 21st Century.* Washington, DC: World Bank. <http://siteresources.worldbank.org/INTEEI/9780199576753 258964/20748034/All/pdf >.

——2009. *The Sunken Billions: The Economic Justification for Fisheries Reform.* Washington: World Bank.

Young, B. de, Peterman, R. M., Dobell, A. R., Pinkerton, E., Breton, Y., Charles, A. T., Fogarty, M. J., Munro, G. R., and Taggart, C. T. 1999. *Canadian Marine Fisheries in a Changing and Uncertain World.* Report prepared for the Canadian Global Change Program of the Royal Society of Canada, Canadian Special Publication of Fisheries and Aquatic Sciences 129. Ottawa: Royal Society of Canada.

INDEX

The manufacturer's authorised representative in the EU for product safety is Oxford University Press España S.A. of el Parque Empresarial San Fernando de Henares, Avenida de Castilla, 2 – 28830 Madrid (www.oup.es/en or product.safety@oup.com). OUP España S.A. also acts as importer into Spain of products made by the manufacturer.

www.ingramcontent.com/pod-product-compliance
Ingram Content Group UK Ltd.
Pitfield, Milton Keynes, MK11 3LW, UK
UKHW021135160926
475595UK00020B/69
* 9 7 8 0 1 9 9 5 7 6 7 5 3 *